PRAISE FOR
THE SCOND DECLARATION

"The impact of this book on mankind and the future world will outclass all the past literature on futurology."

—Zhong Peizhang
Former Director General of Information for China
Department of the CPC Central Committee

"A wise girl thinks wisely, utters her wisdom, shows her wise concern for mankind and displays her great talent."

—Zhang Liwen,
Principal Tutor of Ph.D candidates, Renmin University
President of Confucius Research Institute

"That a little girl has turned out such a great work in China's academic world, lacking fruits in futurology, is really "a shocking matter"!

The old Toffler's future prospect of 10–50 years has predicted many things, too. Little girl Wang Xiaoping is so bodacious that she brought up her own theory on death and the ultimate objectives of human development, a forbidden zone in futurology."

—Xu Guotai
Expert in innovative thinking, "King of Thinking" in China

"*The Second Declaration* is a unique futurological book full of forward-looking, avant-garde, groundbreaking and challenging new ideas, new viewpoints and new theories. It is an epitome of new ideas. It has combined the three elements organically, namely, "valuing people's development, people's overall development and people's sustainable development," building up a brand-new human-centered development theory and human-centered outlook of the future."

—Qin Linzheng
Executive Vice-Chairman of The China Society of Future Research and
Director of The Institute of Future Research

"Wang Xiaoping is courageous enough to break away from mundane traditional ideas and walk independently, and also bold enough to open her mind to society. *The Second Declaration* is yet another great book of hers on futurology that follows the publication of her *Ability Panic* which propagates the Dacheng wisdom, and from this we can trace her mental journey in growing into an unusually wise girl. Her viewpoint has transcended Toffler's foresight on the revolution of new science, and also coincided with Professor Lazlo's worry about the "Great Change".

—Fu Fuchen
Eminent Philosopher and Taoist
Tutor of Ph.D candidates at The Chinese Academy of Social Sciences

"The most courageous and successful aspect of *The Second Declaration* is the daring spirit to challenge traditional ideas and air the voice to realize the dream of living forever. Unexpectedly, the person who uttered such a voice turns out to be Miss Wang, a 21-year-old girl. I cannot but admire her great courage and talent. I hope the publication of the book will arouse people's great attention to life and to anti-aging research. I suggest people of all walks of life should read this book."

—Jian Songbai
Executive Director of The China Anti-aging Academic Committee,
and Director of Longevity Research Institute

"After reading the book, I was amazed a 21-year-old girl could have such encyclopedic knowledge!

The new technological revolution represented by Wang's life technology and information technology will change the life of mankind, change people's working and living styles, resulting in great transformation in industrial structure and economic structure, also leading to great changes in people's social concepts, morals, legal systems and so on. Many things that seem to be useful today will become redundant. Many things that seem to be reasonable today will become absurd. Many things that seem to be moral today will become inhuman. Traces of all of the above can be found Wang Xiaoping's book."

—Zhong Mingrong
Award Winning Economist, Head of Beijing Eyeshot Consultation Center

"Wang Xiaoping's writing is so extensive, while her thinking is so romantic. Her brilliant views and witty words may set people thinking. Maybe, in the eyes of a scholar, her writing is not so sound, but I appreciate her romantic style and original ideas. Her viewpoints on human-centered productive forces and life economy are so accurate and novel, they may enlighten our scholars and enterprisers. Her philosophy of respect of life and advocacy of peace are also agreeable with our mind. I hope Xiaoping will keep the style of "orthodox books with original character", centered on one or two subjects, and make unremitting efforts, so as to achieve greater success."

—Hu Xingdou
Economist, Professor of Economics at
Beijing University of Science and Technology

"I must recommend Wang Xiaoping's book to the students of colleges, middle schools and primary schools of the whole country, because Wang Xiaoping is a model in "research-typed learning", and a paradigm of great accomplishments. Now the educational world is in a boom campaign for "research-typed learning". The appearance of Wang Xiaoping is a miracle indeed."

—Lei Zhenxiao
Eminent Scholar in Personnel Science

Wang Xiaoping's collection of books, are a monograph on learning methods and ways to success, written according to Wang Xiaoping's own experience. This little girl from Chongqing is so ambitious: this time, she wants to discuss the future of mankind.

She even wrote an open letter to Annam, secretary-general of the United Nations, to utter her opinion that the UN should and can play a more active role.

Maybe someone will poke fun at her for her naivety: how can a small girl change the world by her own force? Indeed, a small girl cannot change the world by herself. But seeing a girl in her 20's who vividly express her concern and expectation for mankind;s future, ...will those laughing at her not feel ashamed?

The argument in *The Second Declaration* may be controversial, but the author's sincerity is doubtless. I especially hope the young people in Hong Kong will do more thinking and discussion on this set of books, and learn from the author's broad vision and mind.

—He Zhicheng
Editor-In-Chief, Economy Week Publishing Group

The Second Declaration

Wang Xiaoping

Translated by Liu Guangdi

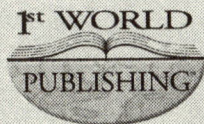

1st WORLD PUBLISHING

The Second Declaration

WANG XIAOPING

© Wang Xiaoping 2007

Published by 1stWorld Publishing
1100 North 4th St. Suite 131, Fairfield, Iowa 52556
tel: 641-209-5000 • fax: 866-440-5234
web: www.1stworldpublishing.com

First Edition

LCCN: 2007933798
SoftCover ISBN: 978-1-4218-9810-0
HardCover ISBN: 978-1-4218-9811-7
eBook ISBN: 978-1-4218-9812-4

CONTENTS

Do you wish to have the beautiful hair of Julia Roberts, the nose of Cleopatra, the eyes of Zhao Wei, the lips of Demi Moore, the stature of Marilyn Monroe, and the wisdom of Athene? You can realize this at your will. Even if you are as ugly as Quasimodo, you don't need to feel sad. By a little adjustment of the gene, you may shape yourself into Tom Cruz...

Today, mankind will use high technology to guide its own evolution, making future humans grow perfect.

Five hundred years ago, in the Ming Dynasty, there was a man called "Wan Hu", who bound himself onto 47 crude rockets, wishing to fly to the sky. He was bombed into pieces in a great explosion. For those who mocked at Wan Hu for his sad result of "being bombed into pieces", could not have imagined that we will be able to fly to the sky and the outer space.

What is more important than the freedom in getting living space is the freedom in getting survival time. Then, how far ahead of us is the prospect of winning the freedom in survival time?

A series of technological revolutions have resulted in all sorts of high and new technologies that might make the First Emperor of Qin Dynasty and Emperor Wu of Han Dynasty admire or envy us beyond description. However,the biggest prison in the world is the prison of the brain. People are prisoners of their ideology. If there is only technological revolution without ideological revolution, people's sufferings might increase automatically.

Why did Hai Rui (a famous righteous official in the Qing Dynasty) do so ruthless and foolish thing to his own child although he was regarded as both wise and virtuous? Why was the self-hurting "three-inch golden lotus" (woman's bound foot) regarded as a symbol of civilization, education and beauty? Why did even the all-mighty kings or queens had to pay money to the gatekeeper?

A cabrilla and many tiny fish were living in a fishbowl. A glass separated the cabrilla from the tiny fish, whenever the cabrilla tried to catch the tiny fish, he knocked against the glass and failed every time. Later, the glass was removed, and the tiny fish were swimming about him, but the cabrilla never tried to touch them again. Thus, this cabrilla actually died in the fishbowl with rich food. In real life, many people take the thinking mode and behavior pattern of the cabrilla, which leads to lots and lots of tragedies. In the 21st century,the most important thing is not money, not resources,and even not technology,but the common ideal that can join up all the people in the world.

"It was the best of times, it was the worst of times, it was the age of wisdom, it was the age of foolishness, it was the epoch of belief, it was the epoch of incredulity, it was the season of Light, it was the season of Darkness, it was the spring of hope, it was the winter of despair, we had everything before us, we had nothing before us, we were all going direct to Heaven, we were all going direct the other way. . . . . . "
When the person is upright, the world is upright. When the mind is upright,the person is upright automatically. To have an upright mind, we must have an upright culture.

Right on the day when God created man, HE regretted! Mankind is the first creature that tries to change its own evolutionary process!
Can mankind prolong its life time? Can it change its track of fate? How can mankind realize its "lofty ambition" to escalate from new mankind, to high mankind and super mankind? The answers to all those questions will affect social development, international relations, world peace,and determine the fate of everyone, including you and me.
A stormy revolution of mankind itself will begin from our hands...

"Youth is gone in an instant, and beauty is lost in a moment!" Xiao Feng chimed in: "I heard this was the sentence most often spoken by women with a deep sigh in those days."

"What a pity! So to speak, countless beautiful girls were disfeatured by the sharp knife of Time! But in those days there was no force to save them!" Gaoqiao Xue, Wen Min, Qiao Er and Lemon got into meditation. They sighed at the fate of the past women, and also felt so happy about the present people who can not only enjoy a long life but also keep young forever! Among them, the oldest is the 667-year-old Qiao Er, and among those admiring Qiao Er are persons ranking from 19 to 37, 57, 97, 177,277, 477, 877... in age.

VIII. Questioning Life 91

A rich man becomes the Chairman of the Board of Directors, he has assets worth millions of dollars, he has risen from abject poverty, because of "desperate struggle" for more than one decade. But now he is lying on the sickbed, on the verge of leaving this world. When some young adorers visited him, he said: "I would rather be a pauper now. If I could exchange, I would use my family property worth millions of dollars in exchange for life..."

"Both Confucius and Robber Zhi have become motes". Confucius was the greatest mahatma but Robber Zhi was the most notorious robber. Yet, whether you are a saint or a robber, you have no way to escape death, and after death, both will turn to dust and ashes.

IX. Revaluation of Longevity. 97

When someone asked Buffett, the famous U.S. investor, "Once you become the top rich man in U.S., what other goals will you have?", he answered without hesitation: "Become the person with the longest life in U.S.!"

Just as the ancient saying goes: "Of myriad things in the universe, longevity is the most valuable." A long life determines the degree of a person's right to existence, his right to enjoyment and his right to development. It determines the degree of satisfaction of the three major needs, namely survival, enjoyment and development. It directly concerns his achievement in academic career, occupation career and everything in life.

X. Reflection on Eternal Living

More than 100 years ago, the enlightenment scholar in the French Revolution, Jean Condorcet, wrote his last work, hiding himself in an attic near Luxemburg of Paris, regardless of the threat of death sentence. Condorcet thought, history and science are just struggles against death. His book was written for the conquest of death. So the last words he wrote were: "Science is meant to conquer death, then, there will be no person dying in future." Seeking eternal living just means striving to overcome ageing and death. Death is the greatest enemy of everyone, and how to defeat this greatest enemy is the biggest problem confronted by mankind.

Sun Wukong, the famous hero in the novel Pilgrimage to the West, rushed into the Dragon Palace to get the Gold-Hooped Pole as his weapon, got into the nether world to tear up the Book of Lives and Deaths, and started a great unrest in Heaven. When he visited his master and sought abilities, what was the only ability he chose? Can you guess?
In comparison with Sun Wukong, we are far more foolish. Even till now, we do not know what should we pursue the most, and what ability we should develop the most. Mankind is crazy about development of internecine abilities, competing wildly with each other in the ability to destroy the same species, but muddleheaded about the ability most worth seeking, casting it aside like an old shoe.

Mr. Lu Xun said: "Experts often talk nonsense."
Today, "old-typed foolish citizens are still many, but new-typed 'foolish citizens' have also reached a certain scale." Old-typed foolish citizens are those illiterate people, but new-typed foolish citizens are those who follow authorities like sheep and believe experts' nonsense blindly.

Although scientists have not been able to do large-scale study on the "longevity gene" in an organized and planned way, the bit-by-bit study has led to many great discoveries, such as the gene of "I have not died", the "Methuselah gene" and so on. There are still other great breakthroughs in the exploration into the secrets of life, such as the stem cell, telomerase, death hormone and so on. In the eyes of today's life sciences, how long on earth can a person live? Can we "live forever" and "rejuvenate"?

In the 18th century, smallpox was an incurable disease, running wild in the whole world, putting people into a helpless position, but has it not died out on our earth? In the 20th century, pulmonary disease was a terrible "incurable disease", but can we still call it "an incurable disease" now?
If Kang Youwei (a famous scholar in the Qing Dynasty) lived in the present time, he would not have become listless and died because of transplantation of an ape's spermary...
Conquest of incurable diseases ... organ transplantation... gene therapy... preventive medicine... What will all these bring to us?

When I was wandering in a park, watching the fountain rushing up like a jade pole, an idea suddenly occurred to me: So long as the conditions are changed, the natural law of "water flows downward" should have been changed to the opposite artificially! "As there is birth, there must be death" is the law we have seen in the present conditions. So long as we create new conditions and change the conditions, this "law" can be changed, too!
In this world,so long as we have the "three sufficients", namely, sufficient intelligence, sufficient investment and sufficient time,any impossible wish may be turned into vivid reality. Eternal life is no exception, either.

The great strategist in ancient China, Sunzi, said: "A wise man's consideration must involve benefit or harm. When it concerns benefit, it will lead to credible practice; when it focuses on harm, it will result in solution to problems." In the face of many problems caused by eternal living, such as population explosion, an ageing society, unemployment, grain shortage, energy source, other resources and so on, what is to be done? How can we give solutions to harm?

Mankind does not need a superficial optimism, but needs a profound optimism; we do not need a foolish way of thinking characterized by "Either this or that", but need a way of thinking that considers both benefit and harm, so as to solve any problem.

Part III — Megatrends of Future Human Development 164

Shao Xiaobo opens his eyes with difficulty. A pretty girl with dark brown hair smiled to him, "Welcome you to the world of 2050! You have 5 minutes to stay. What wish do you have? What do you want to know about the world of 2050?"

Shao Xiaobo opened his mouth in surprise. Luckily, he is an earnest reader of Wesley's novels from childhood, so he says: "Is what you said true?"

The pretty girl smiled without remark. "Your time is limited. Please visit the supermarket nearby. Today is the date for free dispatch of 'body optimizing' products. All the products here are labeled with any of the three markers, "birth improvement by technology", "education improvement by technology and "body improvement by technology", so as to attract people to buy. Oh, you may not understand what a 'body improvement' product is? Look, now Mr. Seim is handing out 'I Am Tender and Kind' hamburgers and 'You Are Vinci' biscuits. As soon you eat the bread, you will have the painting talent of Da Vinci. Also, we have an A-typed fruit juice that can change your eyes into phoenix-typed eyes, and we have a C-typed almond milk that will change your eyes into the shape of a crescent..."

XVII. New Development Theory in the New Century..... 169

If Marx woke up from the tomb, we might give him 24 hours to see the present world: first, let him spend 3 hours rambling on the Internet, then, let him see the miraculous nanometer technology, then, let him listen to the news about deciphering of mankind's gene map, then, let him take a look at a page of *Yangcheng Evening* News that contains more information than the information amount of a 17th-century farmer's whole life. He would surely believe...

"Development is the hard truth", and this famous saying by Deng Xiaoping has already become the tenet in the minds of all the Chinese. In the face of development,what we should think first of all is:What kind of development do we need most? What kind of development should we seek most?

The new economy time needs a new development philosophy.

XVIII. Highest Happiness and Happiness Formula. . . . .

From the epicurean "Hedonism" philosophy to the "Seven Rules" of happiness calculated by Jeremy Benthan, we can see Western philosophers' great enthusiasm for happiness. Our China's serious Confucianism also advocates happiness, as the whole book of Analects of Confucius is alive with the mood of "happiness".

Daylong ecstasy×eternal living without ageing = maximization of happiness·making happiness limitless = highest happiness

XIX. Mankind's Two Deities. . . . .

In the past, Adam Smith called the market "an invisible hand"; Today, we call technology "an invisible head". In the 17th century, Bacon said: "Science has made us closer to God";In the 21st century, we say: Science is just the deity!"

In a time characterized by piling-up of science and technology findings, the large-scale use or small-scale use, advantageous use or disadvantageous use, right use or evil use of science and technology depend completely upon whether we have an ideal society.

XX. First Way to Become Immortal: Birth Improvement through Technology . . . .

What does the invention of "the clever mouse" by scientist Qian Zhuo signify? Does the appearance of a dissolute "rogue" result from acquired moral profligacy or just from an innate character? What is the root of the "frequent crimes" of the Kennedy family? "One ounce of inheritance is more important than one ton of education". In the near future,an eugenic engineering might be practiced to all mankind, and only by that can we truly realize the great ideal proclaimed as early as two centuries ago, "All men are equal by birth". Everyone will get the true equality, that is, get the best natural quality on the equal term, possess all the best basic conditions for self-development on the equal term.

Three ways of health improvement through technology: Gene reconstruction, man-machine integration, and multi-dimension upgrading.
By man-machine integration, the deaf can hear, the blind can see, normal people can have the functions of "clairvoyance", "hearing things thousands of miles away" and so on. These maggots will no longer be maggots, but will enter the mainstream science, becoming reality.
Someone asserts: All geniuses come out of eating. In future,we may use high biological technology to create the most effective, most supernatural and most wonderful "dietary pattern for geniuses", "dietary pattern for longevity", "dietary pattern for youth", "dietary pattern for beauty"... suiting the need of each individual.

One of the important causes of the great difficulty in carrying out the "quality-based education" is that we cannot shake off the bondage of heavy task of "knowledge education". In the face of the great pressure from knowledge explosion, what we can do is only to feed knowledge madly and memorize it madly. With the aid of high and new technology,we can record all the knowledge of the world in a mini chip (nanometer technology has made it possible to store the books as many as those in dozens of Library of Congress of USA, in a space thinner than a hair) and implant it into the human brain, thus, everyone can get a huge knowledge amount at any moment. After achieving "omniscience" easily, we may focus more energy on ability training, so, the dream of "becoming omniscient and omnipotent" will come true.

In ancient Egypt,people worked out the decay preventing mummy, just in order to raise someone from the dead someday. Some present-day Americans have set up corporations of human body refrigeration or corporations of life prolonging.

Unlike the ancient Egyptians' idea to place the hope of revival on gods, the body-preserving technology is to place the hope of "reviving from the dead" on the omnipotent science and technology,so it stakes the hope on the fast developing high technology. Just as Chairman of All-America Refrigeration Technology Association has declared, "The freezing-preservation technology has solved the most difficult problem of mankind, death!"

XXIV. Second Core Technology for Longevity in a Tortuous Way:Mind-preserving Technology

In terms of modern technology, the soul is information,namely, bit. Information belongs to the world, and may not die away along with the flesh. Information can be retrieved,can be caught, can be kept, can be diverted and can be duplicated. If we process the soul in the way of information, we shall be able to catch, divert, duplicate and save it completely.

We have already had the "soul catcher" and the "soul keeper",and will turn out the matching system, the "soul reviver".

XXV. Trial in This Life and Life Encouragement

When we make a comprehensive view of the three major religions in the world, Christianity, Islam and Buddhism,we shall find, in any of them, there is a mechanism of punishing the evil and upholding the good through their own way of "Last Judgment". Can we move the religious "Last Judgment" to reality and establish a present-day bar of trial by way of the high and new social technology system?

Mr. Nan Huaijin (a famous scholar) thinks, all the cultural ideology of the five thousand years of China is based on karma (punitive justice). This "karma" is merely a religious belief, but now we can realize the full and real "punitive justice" in society by way of the high and new social technology.

Part IV — Megatrends of Development of Future Economy

The first place in the front page of *Wall Street Journal* of the U.S.A. on March 12, 2022:

A Nobelist Has Risen From the Dead, an Unprecedented Story and a Start of a New Era!

China Eternal Life Group Company has surpassed the formers

although it is a later comer,and its share price is soaring up, setting a new world record!

Great good news from China Eternal Life Group Company: The said group has successfully revived a human body from refrigeration, and the Nobelist, Professor WWW has become the first person revived in history!

As early as 2009, China Eternal Life Group Company began to search for and offer a fat salary to hire top experts in body preservation all over the world, and started the Eternal Life Revival Project. The gross project investment is 888.8 billion U.S. dollars. After unremitting efforts of 13 years, they have succeeded at last, and people all over the world are overjoyed.

Ever since China Eternal Life Group Company got listed, it has been very popular among the stock investors, especially after the release of this news on the revival, its stock index has been rising again and again, jumping to the most eye-catching position in NASDAQ market. According to someone concerned, China Eternal Life Group Company will buy out Microsoft, Intel, GE...

XXVI. High Technology and High Humanity

What is the essential meaning of the New Economy that has raised such a great unrest in the whole world? After the New Economy, what economy will follow?

What are the most essential five major demands of humanity? Why will the future economy be a union between high technology and high humanity? How to combine high technology with high humanity perfectly?

XXVII. The 1000-year Longevity Industry and Life Economy

In his monumental work Economics, Paul Samuelson, the great master in global economics, says: "You can turn a parrot into an economist,but the precondition must be: Let it understand 'supply' and 'demand'." The demand for life can transcend space and time, transcend national boundaries,and transcend classes, being the strongest and most extensive demand of mankind, that is, it is the greatest common demand of all demands, thus, the 1000-year longevity industry has a boundless market.

The rate of America's health industry in the GDP has risen from 12% to 18%,equivalent to the fact that all the GDP of China is used as health fee in the USA...

XXVIII. God-sleeping Industry and Eternal Life Economy

God-sleeping means taking a long sleep like a god,waiting for resurgence or revival later. Today, the fast progress in nanometer technology, stem cell technology and other high technologies, has offered the chance for us to "revive" some day...
Those who are zealous for revival through refrigeration are just the elite of the time, that is, excellent experts in high technology. People in the Silicon Valley are very confident of the advances in science and technology. They believe, so long as you put plentiful money and technology,you will be able to solve any problem. Why can we not "revive" ourselves?

XXIX. Ecstacy Industry and Pleasure Economy

The rise of pleasure economy and pleasure industry just shows: Our time is undergoing an unprecedented revolution: in the past, only products in kind could increase wealth,but now, "pleasure" can also boost economic development.

XXX. Eugenic Industry and Human-centered Economy

A man wants to have a child. When his fellows invite him to play in the bar, he shakes his head again and again: "No, you go yourself. I have to do something for my future son." Now, the chief reason for letting women return to the home is also for the sake of the child, and the prospect of the child may pull the most feministic Nara back home...
The eugenic industry can not only lead to the appearance of numerous genius prodigies, but also enable them to surpass past genius prodigies in every aspect.

XXXI. Outer Space Industry and Universe Economy

The eminent futurist Alvin Toffler became world famous because of his prediction of information economy and Internet economy as the "Third Wave". Now he regards outer space industry and outer space economy as the "Fourth Wave".
Space tourism ... space farming... space industry... Ask for energy from outer space... Seek homesteads from outer space... Establish

diplomatic relations with high intelligent beings... so as to solve all sorts of problems that cannot be solved on the earth...

Fund pooling is a good way to profit, but intelligence pooling is an even better way to profit. Only by high intelligent persons can we expand financing channels better, increase the total amount of financing and raise the financing efficiency.

However,whether it is fund pooling or intelligence pooling, it must be based on "heart" pooling. In the world, there is nothing more powerful than the heart of a person. Trying to win people's hearts, seize their hearts and change their hearts is the key to everything you do.

In future economy,we should use the "three-in-one" mode of investment, that is, investment in the heart, in intelligence and in funds, do the three together, causing them to circulate in a benign cycle and generate each other.

In 1820,China's economy was still second to none throughout the world, with its GNP accounting for one third of that of the whole world. But why did our No. 1 position slide into the margin before long?

Refusing transformation and refusing industrial revolution,we paid a high price for that, resulting in the one century's disgrace and two centuries' backwardness. We must learn never to make the same mistake again...

Today,in a Western developed country, one big automobile factory's annual output can almost meet the need of all countries in the world for one year. What does this mean?

Material productive forces:Automation and zero-price production... knowledge production force:Intelligentization and brain research... human-centered productive forces:Deity-oriented transformation and genius prodigies coming forth in great numbers...

After the end of the cold war, what mankind have got is not a peaceful world, but we have entered an even more dangerous era. While enjoying the happy days brought about by high technology, we are also faced with all sorts of fatal calamities created by mankind itself. Some scientists assert, the terroristic organization only needs 20 gram virus,to "get rid of" all the 6 billion people of the world...
The greatest terror and the greatest interest will certainly force mankind to make the greatest choice.

TRANSCENDING TOFFLER

夢星

What is the true source of the crisis of mankind? It is the lack of thinking about our own future. The 21st century calls for real thinkers caring about mankind's future. I wish to be such a thinker.

During the last two years, when people asked me what kind of books I was writing, I always answered: Something about futurology.

What amazed me was that people were so ignorant of futurology, and some people did not even know what futurology was. Whenever I faced such a situation, my heart felt pinched, and that made me more determined to write this book well.

In the West just about everyone—whether statesman, entrepreneur, or just an ordinary person, (or just the man on the street?)—cares about the future and longs to know the future. Books on futurology sell extremely well: *The Third Wave* sold 20,000,000 copies, *Megatrends* reached 14,000,000 copies. In the West, futurology is a very popular branch of study, but in China it is a subject few people know! We are a nation that revels in history, but despises the future. We are constantly remaking historical plays, always playing back the romances of Emperor Qianlong or the great deeds of Emperor Kangxi in the golden hour of TV programs. For a time, all the TV channels were filled with the image of the swaying plait and the cry of "Zha"-"Zha" ("Yes" in the Manchu language). In contrast, Hollywood is intent on producing movies on the subject of the future, such as "Star Wars", "Trip to Mars", and "Terminator". As we say in China, *by spying a spot, we can know the whole leopard*. It must be admitted that in our culture, history prevails over the future. We lack the concept of the future, having been accustomed to the way of thinking typical of the agricultural civilization: harking back to the old days. However, if a consideration of the future is lacking in a person, an enterprise, or a nation's thought, it is inevitable that progress will be hindered.

Whoever represents the future will be the one to come out ahead. For instance, imagine you have several production lines of 9-inch black-and-white TV sets, and they are brand new. Certainly they are assets, but will there be many investors ready to buy your stock? Do you have market value for future sales? No! Only by facing the future and embodying the future can you have value, hope and glory. In the 1990s when the Internet age was at hand, Bill Gates was the great magnate in the computer world. Yet he failed to adequately predict the stunning long-term prospect and in developing the speed of computers. He almost missed the huge opportunity in the race for product development.

I am sure many Chinese are familiar with the phrase, *"The past determines the present,"* but few know *"The future determines the present"*! Knowing that the past determines the present is good, but knowing that the future determines the present is even more important; we should value both; looking back and looking forward. We should pay great attention to both history and futurology. Future trends and patterns may determine our current general objectives and general direction, and thus determine what we should do now. Chinese people are in dire need of a revolution in their ways of thinking, and in dire need of an enlightening approach to the future and futurology.

A friend laughed, and said to me: By writing this book, do you wish to be the "Toffler" of China?

Toffler is none other than the author of the futurological book *The Third Wave* that sold 20,000,000 copies worldwide. He is the American writer who brought out the concepts of postindustrial society and information overload and is regarded as the greatest futurist throughout history in the West.

I don't want to be the Toffler of China I wish to surpass Toffler!

Toffler chiefly studied future society (bringing up the concepts of postindustrial society, information overload. and so on) and future economy (putting forward the concept of a "knowledge economy", and so on) but he has done little in the areas of the future of the human beings. Yet, the most important thing is the study of the future of mankind itself. Everything centers around the human element. Both social development and economic development are for the development of mankind, and both future economy and future society should follow and serve future mankind. Hence, I lock my focus of research on the study of the future of human beings. Toffler could not get out of the circle of future economy and future society, whereas what I am thinking about is man's future and mankind's future, and making use of mankind's future development trend as the pivot to study future economic and social development. My aim is to stand at the commanding center of mankind's future development, and make breakthroughs surpassing all previous futurological theories and books.

The greatest value of Toffler's *Third Wave* lies in his point of view that future economy will grow in information economy. *The Second Declaration* not only brings out the five major trends in future economy namely:—life economy, eternal life economy, pleasure economy, human-centered economy and universe economy, but also suggests the corresponding economic theories, such as the new industry theory that has transcended the traditional industry theory and the new productive force theory that has transcended the traditional productive force theory. All of these were never touched by Toffler in his study on the future economy.

Moreover, Toffler only describes the future, whereas what mankind needs more is—to plan the future. Toffler only predicts the future, whereas, what mankind needs more is—to create the future! *The Second Declaration* is meant for macro planning of the most beautiful future for mankind. It not only has descriptions and forecasts but also contains designs and plans for the future.

Most importantly, I think sincerely about the future, from the standpoint of both China and the whole mankind.

The future is approaching us, we "cannot close our eyes", we must have the forward-looking way of thinking characterized by *"The future determines the present"*. Only by accurate prediction of the future can we make the correct strategies, and seize the commanding point of future development at the earliest time? Only in this way can we have frog-leaping development in the real sense, do real transcending and lead ahead in the future, and become the great winners in the future?

INTRODUCTION

夢星

"If possible, go ahead of the time; If not possible, never lag behind it."

Our time, is the greatest revolutionary era throughout history. Unprecedented great revolutions are taking place, and the speed of revolution is getting faster, making people feel that a lot of changes have come all of a sudden. The great changes in East Europe and former USSR were very sudden. The storm of globalization is very sudden. All kinds of fundamental quantum leaps in technology, such as—multi-media technology, Internet technology, nanometer technology, gene technology, have come up all of a sudden. Especially some scientists have announced: *"Judging from the progress of gene deciphering, before the year 2030, mankind will be able to master the gene that controls ageing so as to prolong man's life and realize the wish to live a thousand years or even an eternal life".* Perhaps, this is the greatest surprise for most of the people.

After observing all sorts of great changes in the world, we truly feel that our time has entered an era which is 100 times, 1000 times or 10000 times faster, in which ten years of our time exceeds 100 thousand years of the past. An era of marvelous leaps in the world. An era for high technology to show its invincible might. An era for great explosion of human power.

Everything must be put into such a time background to think and judge.

"Truth is the daughter of time and the daughter of the era".

MYTH AND REALITY

Science is meant to conquer death, then, there will be no person dying in the future.

—Condorcet

In a fairy tale the beautiful Chang'e (goddess in the moon) was the wife of the hero Yi, who shot down nine suns because she had eaten the drug for eternal life and flew into the Palace on the Moon.

Today, the fairy tale of "Chang'e Flying to the Moon" has already become a vivid reality; mankind has not only landed on the moon, but is also marching toward Mars.

There is just such a cartoon in a magazine in the USA: A downpour is going on outside, and the priest in the room cries madly, "My God! It is not a rain from God! It is a rain from us!"

It turns out to be an artificial rainfall to relieve the drought. God's summoning of the wind and rain has been realized by mankind now.

A lot of beautiful myths, in China or elsewhere, in modern or ancient times, have already become or are becoming true. In *Pilgrimage to the West*, Sun Wukong plucked a vellus, gave a puff to it and it changed into uncountable small Sun Wukongs at once. Is clone technology not equally miraculous? Sun Wukong changed himself into an insect to get into Iron-Fan Princess' belly; the nanometer robot will be smaller than an insect, so small that it can travel freely in the blood capillary of a person...

"Schigg, transport me back into the ship of Adventure",—that is a dialogue often appearing in the American science fiction telefilm "Adventures in the Starry Sky". The persons in the film can be transported from one place to another in the flick of an eye. Such a scenario in

science fictions has been regarded by scientists as a would-be reality. You see, after three years' efforts, an Australian research group of overseas Chinese scientist headed by Lin Bingxi has succeeded, for the first time, in diverting a laser beam carrying encrypted radio signals to a place 1 meter away, which marked the first step of remote transportation for mankind. According to the fiction conception of "Adventures in the Starry Sky", "transport" means a process of converting an object into energy at point A, sending it to point B, and then reconverting the energy to the object. "Transporting" can not only enable an object to penetrate a barrier, but also carry a living organism and restructure the organism. A human body is composed of trillions of atoms, thus, it can also be transported, and can even be conveyed from one star to another in a moment.

A decade ago, no one believed remote and instant transportation of a person could be turned into a reality, but now we have taken the crucial first step in this direction.

From ancient times, many people have dreamed for a drug that would enable us to keep awake, so as to have more hours for work or play in a day. How good it would be! We said such an idea is impractical, but now, those who hope to avoid sleep have the possibility of turning it into reality. A new drug called MODAFINIL will shock the whole world soon, and its impact will be far greater than that of *"Weige"* (brand of an afrodyn). It has such an effect that if a healthy person takes one tablet, he can work continually for 40 hours energetically, without getting sleepy. Then, after a sleep of 8 hours, he can take another tablet and can work continually for another 40 hours. This medicine got the attention of Department of High-level National Defense Research Projects of the US Defence Department. The department has always been trying to bring out fighters with controllable metabolism, aiming to cultivate soldiers who can fight 24 hours everyday and continuously for 7 days. The US Office of Defence-related Science and Technology says on its web site that, these soldiers who have no need for sleep, but can still keep a high recognition capability and a strong physical performance, will change the war by the root. The medicine has been available for a short time, and it has not been sold widely in the market yet, but its sales amount of one year alone has reached $ 0.15 billion.

All these facts show us the truth: that science is a miracle, science has boundless divine power, and can turn the most absurd myth into vivid reality.

Science and technology have turned many myths into reality, such as; clairvoyance, clairaudient, going up to the heavens, going down into the earth, man's landing on the moon, test-tube babies, clone technology and so on. All these were regarded as totally impossible in the past. Science enables mankind to play a greater role, close to that of God. With the help of science, all sorts of impossibility will be defeated and removed, and all kinds of great wonders will be revealed out.

In China's myths, the principal protagonist is god. The most essential characteristic of the god in a myth is longevity. *"God lives forever."* (*Shuo Wen Jie Zi*, an ancient Chinese dictionary). *"Those who age but do not die are called god"* (*Shi Ming*). Ge Hong, a famous ancient medical scientist, also stressed in his book *Bao Pu Zi*, *"God sometimes goes to Heaven, sometimes lives on the earth, and never dies"*. Besides eternal life, unlimited freedom and unlimited happiness are also great characteristics of god. In fairy tales, god is worry-free, lives in wonderland full of treasures, plays romantically, and enjoy an absolute freedom. All the fabulous god or fairies are characterized with extraordinary qualities as tender skin like ice or snow, youthful looks like a child, a kind heart like Buddha and extreme abilities such as riding the clouds, driving the flying dragon, and roaming the whole world.

If we use modern language to describe "gods" or "supernatural beings," then they are the most ideal super-humans without ageing or death, omniscient, omnipotent, extremely benevolent, beautiful and living an extremely happy life all day long. Becoming "gods" means becoming the most ideal super-humans, which is the greatest aspiration of our nation.

Just as the Taiwanese scholar Gao Dapeng said: *"If we must use one word to sum up our responsibility and dream, the best word may be 'aspiration'."* As the old saying goes, *"Poetry expresses aspiration"*. Are'nt poetic fairy tales the best expression of Chinese aspirations? The meaning of 'god' should be understood and appreciated just from this 'aspiration'... It expresses the Chinese people's aspiration or ideal. It shows the most profound and highest aspiration of the Chinese nation. Since it is the Chinese nation's most profound and supreme aspiration, our mission is to realize it. Is it possible for us to turn this most profound and highest aspiration from an ancient myth into a reality of our time?

Yes!

So long as we gather all the technological powers of the world for the pursuit of eternal life, of omniscience, of supreme benevolence, beauty, and of a paradise, this myth will become a reality. Present technological progress alone has shaken the "natural law" of life and death of mankind, by declaring that we can make great breakthroughs in the natural limit of our life span in the near future. We already possess miraculous technologies that are enough to make Qin Shihuang (First Emperor of Qin Dynasty), Emperor Wu of Han Dynasty, countless monarchs and brilliant scholars bear resentment toward death if they could come to life again; gene technology, clone technology, stem cell technology, refrigeration technology, nanometer technology, information technology to name a few. These technologies have already offered us concrete approaches to become immortal through science. For instance, by the gene technology we can do all kinds of gene forecasts, gene diagnosis, gene therapy and gene reconstruction. By stem cell technology, we can make organ transplantation very simple and easy. Anyone can get bad parts, in his or her body, changed, or any organ can be repaired or replaced. The discovery of 'cell telomere' has given us the power to control the 'life clock'. By using all kinds of high and new technologies like birth improvement, health improvement and education improvement technologies we can take the straight way to become "gods" or can take the curved way to become "gods" through the "body preserving technology" and the "soul preserving technology"... All these are opening up an extremely bright prospect of living forever and becoming gods.

The present-day miracles created by science and technology, are already remarkable. It goes without saying the future prospect of science and technology will be more so. Who could have imagined that the extremely bulky and costly computers a few decades ago could suddenly become a day-to-day commodity filling every nook and corner of our society? Who could have imagined that Internet technology and information technology could change the whole world so profoundly? Especially our present gene science and life sciences, which have already undergone quantum jumps, will bring the revolution of mankind's own fate into an unprecedented era. Just as the futurist Naisbitt said, *"the Internet only enables us to do our past work more conveniently, but has not produced anything new. On the other hand, gene engineering may change the human race and its evolutionary process..."*

What is important is that we can use scientific research to transform the old way of working alone into international large-scale cooperation. We can unite all the natural and social scientists in the world, transcend the national boundary and invest huge amounts of money to carry out unprecedented grand-scale research globally. Then the breakthroughs we shall make in all kinds of high and new technologies will be thousands of or even millions of times more than the present success. Do we need to worry that we cannot realize the dream "of becoming immortal" very soon?

If we say the fabulous dream of "becoming immortal" is a "wishful thinking", then by relying on science, the dream has become a "thinkful wishing". The former was bound to come to nought, but the latter is sure to come true.

MUTUAL FIGHTING BETWEEN BROTHERS AND SHARING THE SAME BOAT BETWEEN ENEMIES

Everyone in the world comes for profit; Everyone in the world goes for profit.

——Records of the Historian

Throughout history, the two most shocking things are: "Mutual Fighting between Brothers", and "Sharing of the Same Boat between Enemies". The two sharply contradictory phenomena have appeared repeatedly in all dynasties and all eras.

Mutual fighting between brothers is quite common in history. In the period of the Three Kingdoms, Cao Zhi was forced by his elder brother Cao Pi (both of them were sons of Cao Cao, king of Kingdom Wei) to compose a poem within seven steps of walking (or else he would kill Cao Zhi), and the poem came out like this (in meaning): *"When you cook the beans by burning the beanstalks, the beans in the cauldron are crying to the beanstalks: We were born of the same root, why are you burning us so cruelly?"* The reciting of the sad poem seems to be still lingering in our mind. The great wise ruler of the Tang Dynasty, Li Shimin, killed two of his brothers in the incident of Xuanwumen Gate in a struggle for the crown. In the late Kangxi Period of the Qing Dynasty, fights among princes for the crown became white-hot, mutual conflicting and killing were too horrible to see...

An opposite example is the story of sharing the same boat by State Wu and State Yue. In *Sunzi's Strategics*, the story is like this: Those who are good at fighting are like the snake of Changshan Mountain. When you strike its head, its tail will come to help; when you strike its tail, its head will come to help; when you strike its middle, both its head and tail will come to help. Can we use the army like the snake of Changshan? Yes. The people of State Wu and the people of State Yue were enemies to each other, but when they took the same boat, they helped each other, so when a storm came, they saved each other, just like the two hands of our body. In other words they could reach the high degree of integration and harmony like the snake of Changshan. That is the classical allusion of "Sharing the Same Boat by Wu and Yue".

Mutual fighting between brothers and helping each other by former enemies are very common in modern time, too. For example; China and USSR, China and Viet Nam, used to be the closest "brothers", but later they suddenly became enemies and killed each other brutally. On the contrary; the cooperation between the Communist Party of China and the National Party of China, and the Sino-American alliance are examples of sharing the same boat between former enemies. When China and the USA came together, China had just stopped cursing "Yankee imperialism" in its newspapers and periodicals. When Nixon landed in China, his first sentence was *"I have come for the interest of the USA."* This common interest turned enemies into friends, and turned fighting into cooperation.

Everything is determined by interests. When the interests are different, even blood brothers will fight against each other. When the interest is same, even enemies like State Wu and State Yue can share the same boat.

The difference or similarity of the interests determines the splitting or joining of the world or society, determine fighting between brothers, or sharing of the same boat by former enemies.

This is the greatest secret history has given to us.

Today, threats for mankind do not come from the outside, but from within. Mankind's current major contradiction is not with Mother Nature, but within itself. We spend huge amounts of manpower, material resources, financial resources and energy in competition with each other, in intriguing against each other, in dog-eat-dog struggles, even up to the extent of perishing together. This is the root of all our world crises and human tragedies.

Kaal Dojicj, a scholar of the science of international relations, has pointed out: *"If human civilization is throttled within the coming 30 years, the killer will not be famine nor will it be plague, but foreign policy and international relations."* For many years, human beings have been on the jig all the time, living under the great terror of nuclear warfare, and this terror is still getting deeper and denser. The fast technological progress has not stopped the terror, but on the contrary, it has made the world even more horrible. While the nuke shadow is expanding continually, biological weapons with the same destructive power or even more terrible are appearing. Once these "demons" are mastered by terrorist organizations, it is very likely that humanity will be wiped out. The "9\11" attack was just a rehearsal. Humanity has already lost the least sense of security.

By what can we save this tottering world?

Only by mankind's common interest.

The splitting of the human world is the result of split human interests. Each nation is struggling for its own interests and rights. Hence, only by finding the common interests of mankind, setting the common goal for realization of mankind's common interest and making mankind aware of the common interest and common goal, can mankind go from splitting to alliance and harmony. This will eliminate all terrors, calamities and tragedies in the human world.

Becoming immortal together or evolving into immortals is mankind's greatest common interest and common goal.

Whether you are an oriental or an occidental, whether you are a yellow person or a black person, whether you are a Chinese, an American or a European, whether you are a Jew, an Arab or a German... we are all humans and our common enemy, real enemy and greatest enemy is just one, ...that is, death.

Faced with death, there is no escape for anyone.

Pascall once described the condition of all people like this: *"Let's imagine, some people are executed under the very nose of their fellows. Those who have survived see their own plight as their fellows' fate, so they are full of grief and just look at each other in speechless despair waiting for their own death. That is just a miniature of mankind's condition."*

Death is not the tragic fate of any unfortunate nation, but the common miserable fate of all mankind. The fear of death is not a special mentality of any nation but the common mentality of all mankind.

Nowadays, high and new science and technology have already offered the practical possibility to conquer death gradually and get rid of the common tragedy of mankind. Yet, only by joining together and making concerted efforts can mankind defeat this enemy and achieve the limitless freedom of life and realize the highest interest of us all. A small interest will submit to a great interest naturally. When we are aware of the greatest self interest, and mankind's common interest, we shall certainly shake off the limitations of narrow national interests, and certainly move from splitting to joining together, and shall be of one mind and certainly make concerted efforts to become "gods".

Today, we are very rich in material. In any Western developed country, one big automobile factory's annual output can meet the need of all countries in the world for a year. But we are still very poor in spirit, which has put the whole society into an unprecedented spiritual crisis. Even in the USA, the most developed nation in the world, Nixon was also deeply worried about its spiritual crisis, *"Today, our enemy is just inside us. One and a half centuries ago Alexis De Tocqueville already warned us that most of the people in the USA were addicted to materialism, lacking a lasting social cohesion force, the religions and philosophic thinking were shallow, and all these have resulted in a 'new absolutism', namely; mediocre, selfish and lack of ambition. Today's USA will be threatened by a new absolutism, unless it finds a common goal with a brand-new meaning again."*

The USA is so and the whole world is no exception, either. A spiritual crisis can only be solved by a spiritual method. Only by helping mankind to find a "common goal with a brand-new meaning", that is becoming "gods" together, can we get mankind out of the mental state characterized by mediocre, selfish, lack of ambition and change into a new mankind that can create a new world.

A crisis often leads to a favorable turn. Today's great crisis can become the greatest turning point for mankind. So long as we understand our common interest and common goal, people of all countries can unite together and work wonders for the future ...even if they are enemies like State Wu and State Yue.

WHEN A GREAT IDEA BEFALLS

When a great idea as an evangel befalls this world, it may become an offence to the common people bonded by outmoded conventions and customs, and it seems to be a foolery in the eyes of those who have read many books but do not have true profound thinking.

—Goethe

Using modern high technology and the great power of world union to help people to become "gods" together and help mankind evolve into immortals is an unparalleled great idea that may bring happiness to the whole world. It is not hard to imagine that its fate may also be the same as mentioned by Goethe. To many people, it is "an offence" to be considered as "a rebellion against orthodoxy" and in the eyes of those "masterminds" or "scholars" who are well learned but actually do not have real wisdom, it may also be regarded as a "folly" or "fallacy".

In human history, when a truth was discovered, at the outset, it was often incompatible with traditions like water and fire. Civilization lies in innovation, and innovation is often a synonym of rebellion against orthodoxy. The same is true of ideological innovation.

Whether it is an innovation in the theory of the humanities, in the theory of natural science, or a technology innovation, it is often treated as "strange thinking" or "fallacy" by those "scholars" or the "general public" confined by traditional ideas. From the "heliocentric theory" in astronomy and the "continental drift hypothesis" in geology to the "genetic gene theory" in biology all of them were once criticized as "fallacies to deceive people" or "absolutely preposterous". However, at last, all those "fallacies" and "absurdities" became mental wealth that has brought great benefit to mankind.

Originally Darwin's evolutionism was opposed strongly by many people, as they thought listing mankind together with beasts was a great profanity to man's dignity, which angered most of the philosophers, scientists and all the religionists of that time. When the bishop's wife, Mrs Wurster, saw Darwin's *Origin of Species* published in 1859, she said: *"Let's hope it is not true. Even if it is true, let's pray it will not be generally accepted!"* But now, *Origin of Species* has come into the classroom of American churches.

In 1919, when the rocketry inventor Goddard expressed his viewpoint that the rocket could fly out of the gravitational attraction to reach the moon and outer space, it also aroused a great hiss, even the *New York Times* mocked him as "the man on the moon". But it goes without saying Goddard's contribution to scientific development is obvious to all of us.

When cyberneticist Wina proposed the bold idea of putting the "robot" into production in management areas, it suffered ruthless criticism from philosophers.

When a puzzle pending for hundreds of years, in the mathematical world was solved suddenly by the mathematic paper of Abel, a little figure at that time, the then "King of Mathematics" Gauss cried out in surprise, *"Too terrible! He should have worked out such a foolish thing."*

The young French mathematician Galois also met with the same fate; He handed over his mathematics paper three times, but it was denounced by the authorities twice and got lost once. Galois commented on those "authorities", *"These people have lagged behind for one hundred years."*

Malthus' "population theory" was seen as an evil saying and was attacked severely by many people. When Madam Marguerite Sanger proposed and advocated family planning in USA, she was thrown into prison many times...

Not only are all kinds of ideological innovation and theoretic innovation often under fire, but the conservative forces are often very stubborn too, refusing any change. So, just as Planck, the great German physicist who brought up the quantum theory and sounded the clarion of technological revolution of the 20th century, said: *"It is impossible for a new scientific truth to win by convincing its opponents and making them understand it, but its winning is chiefly because its opponents have died at last and the new generation familiar with it has grown up."* Such a sad story is recurrent and hard to evade in history.

These recurring stories have reminded me of a saying by Hegel: "What we have learned from history is only that Mankind has never *learned anything from history."* However, I believe that since mankind has already entered the 21st century, and especially the whole country of China is advocating liberating ideology, seeking truth from facts and keeping abreast with the times, such a historic event will not recur.

Napoleon said, *"in the world there are two most powerful things: one is the sword, the other is thought and thought is more powerful than the sword".*

When the thinking is upright, the people will be upright; when the people are upright, the world will be upright. To value innovation, first of all, we should value innovation in thinking. To treat innovation in the right manner, first of all, we should treat ideological innovation in the right manner.

Laozi (ancient saint of China) said, *"When the upper man learns the way, he follows it with diligence; when the middle man learns the way, he follows it on occasion; when the lower man learns the way, he laughs out loud; If it is not laughed at, it will not deserve to be called 'the Way'."*

When a wise person hears the truth, he will practice it earnestly; when an ordinary person hears the truth, he will hesitate, half-believing and half-doubting, but when a foolish person hears the truth he will have a good laugh. If it is not laughed at, it will not deserve to be called "the Way" (great truth or law). No great way or great truth can avoid being mocked by foolish people, even being laughed at loudly.

But I believe more people are "upper people", not "middle people", or "lower people".

PART I - WHERE MANKIND IS GOING?

夢星

INTRODUCTION

April 14, 2039 is a historic day.

Who could have imagined, by taking out a cell from the hair to clone and with the help of magical artificial intelligence system, we can "revive" a great dead man? Recently, the top life scientists and artificial intelligence experts in the global village revived many great personalities, such as Darwin, Bacon, Hegel, Marx, Socrates, Nietzschean, Condorcet, Mao Zedong, Qin Shihuang (First Emperor of the Qin Dynasty), Confucius, Li Bai, Cao Cao, Cao Zhi, Zhuge Liang, Su Shi...

On an early morning full of radiant and enchanting sunlight, at the Great Hall of Science in Beijing, China, a unique seminar is being held on the subject "Where Are Human Beings Going?" The freshly revived great personalities gather to discuss the future of every member of the global village.

In Beijing's Great Hall of Science, the atmosphere is fervent with a lot of excited people. Although, any person in the global village can talk directly with the great people via his/her personal satellite system, yet, the great fans of Mao Zedong, Darwin, Confucius, Bacon, Li Bai etc, are willing to pay an extremely high price to see these great personalities with their own eyes. My neighbor Lao Gu is one of them. He has taken out all his savings and borrowed 3 million Yuan from Minsheng Bank, just for attending this grand seminar. The mere thought of seeing his idol Darwin triggers a boiling passion in his heart.

The first speaker at the meeting is Socrates. In his time, if he had paid the fine or run away, he might have saved his life, but he preferred to die without reluctance, which had angered so many people. He raises his right hand habitually and when the people have calmed down, he says: "At that time, I thought, I could still discuss questions after death, and see the people I admired, such as Homer, Thales and so on, but I never thought one could know nothing after death. I am lucky enough to be revived by Mr. Liwinson and other scientists, as I can see the world again."

He hardly finishes his speech when Confucius says: "In those days, when my students used to ask me questions about death, I used to play a petty trick, and taok a shy-away policy." Then, he utters that famous classical saying: *"Since we do not known enough about life, how can we know about death."* "In fact, I did not want to see people suffering from the pain of dying, so I found out a put-off. Since there is no way to solve the problem, we had better avoid it. Now conditions have got much better and I have found the chance to revise my saying at last."

The great poet Li Bai recites his poem aloud: *"Can you not see; people are sorrowful about their white hair before the mirror, which was jet-black in the morning but is as white as snow in the eve."* After reciting, he gives a deep sigh, saying: "Everyone says I am bold and unconstrained but nobody knows my misery. Lofty sentiments cannot hold up against the sadness of life and death. From the age of 15, I began to stuck on the religion for becoming god. I often worked together with Taoists in my life, longing for an eternal life. Now, the gene technology and life techno-logy have such a miraculous power. How grand it is! How grand it is!

Qin Shihuang (First Emperor of Qin Dynasty) is still that domineering, with a voice that sounds like a large bell: "It was me who suffered the most. Because of my eagerness for longevity, I was fooled by the people headed by Xu Fu and I became a laughingstock... But in the present time, Life science and technology have progressed upto such a level which was not possible in my time. Isn't the 1000-year project already on the verge of final success? Hey-hey I will be benefiting from it soon."

Looking at Emperor Qin Shihuang, Sakyamuni shakes his head, "You are so selfish, you are only considering yourself, why don't you consider all living creatures?" After twisting the beads before his breast, he adds: "Great! These life scientists are really great people, much higher than me. At that time, I gave up everything to seek a way for all living creatures to get out the sufferings of birth, ageing, illness and death, but I did not get anywhere. At last this great problem can be solved now. Mankind will enjoy an eternal life soon and the global village is just the paradise. It is a great blessing for all lives indeed."

Condorcet says triumphantly: "I said a long time ago, mankind would use science to conquer death, and then no one would die. Plato could only prove in theory that it is a foolish thinking to say that soul would become immortal after death. "

Now, it is the turn of Mao Zedong. His Hunan accent is still that strong, but luckily there is a perfect intelligent translation system. "Mr. Condorcet you are even greater than our Confucius but you might not know, at the age of 20, I wrote a poem saying *Believing a life of two hundred years, and thrashing the water for three thousand miles*. This shows my thinking was very advanced. Has the '200-Year Project' sponsored by some scientists already been completed successfully?"

At this moment Darwin gives a light cough, which draws the attention of all others. "Ladies and Gentlemen, let's return from the digression, as the people of the global village have great and earnest hopes from us. In that year, I put forward the evolutionism, making people aware that mankind has evolved from apes. Today's high technology amazes me. Gene technology, nanometer technology, artificial intelligence technology, new material technology, new energy technology, technology for living in the sea, technology for emigration to outer space... So many kinds of high and new technologies! Where will they lead mankind to?"

These words set everyone thinking. Every attendant is pondering: Since mankind has evolved from apes, where will it evolve to in the future?

Mao Zedong is the first to break the silence, saying frankly that he has already thought a lot about mankind's future: "As early as seventy years ago, I told Yu Guangyuan, Zhou Peiyuan and other scientists that Mankind would die because there will be more advanced things to replace mankind, and that will represent a higher stage of development. What will be those things more advanced than mankind? I did not give a clear answer, but now I have found, the best expression is the most vivid, most ideal and most accurate term in the traditional culture of our Chinese nation, that is, of course, *xian*, or *xianlei*!"

Li Bai is the first to clap hands for joy: "It suits my way of thinking! The tigers are playing the se (a musical instrument) and the phoenixes are flying back. Gods are coming one after another. I call myself 'god in wine'. How can we find any other word better than 'god'?"

Cao Zhi speaks in the best of his spirits: "I once wrote 'Poem of Roaming in Fairyland'. In my poem, I described the features of god in three aspects; first is living forever, second is boundless freedom and third is limitless happiness. Does that just tally with the direction of mankind's evolution? From my point of view it is just to the point."

Qu Yuan (poet in the Warring States Period), Su Shi (poet in the Song Dynasty) and so on nod in agreement.

"I disagree!" Nietzschean gives a shrug nervously, and begins to speak slowly: "'God'? No. In my opinion, we had better use my term, 'super-human'!"

Heidegger stands up: "Gods, after all, are things in your oriental culture, and the concept of ape is raised by our occidental people. Now, that we are facing the problem of suggesting a new term for mankind's goal of evolution, in my opinion, the best choice is 'superhuman'." Darwin nods a little. But Leibnits, Joseph Lee, Toynbee, Russell and others do not agree, as they think, before the resplendent ancient oriental civilization, that narrow-minded superiority complex of Western civilization is derisive.

Now, the meeting gets into a tumult, the people fall into two camps distinctly: Some support the term "xianlei(gods)" while others argue in support of "superhuman". Each sticks to his own stand, so, the emcee has to announce a decision by the majority through Internet-based voting by all the people of the global village.

One hour later the answer concerning mankind's fate comes out: A majority of the votes are for "gods".

Finally, the great people attending the seminar of "Where Mankind Is Going To" reach consensus at last, offering a satisfactory and complete answer to the people of the global village.

Mankind has evolved from apes.

Will certainly evolve into gods (immortals).

I. APES, MANKIND AND IMMORTALS

Of all kinds of understanding, the most important is the understanding of oneself.

In understanding of oneself, the most important is the understanding of one's own future.

"Understanding yourself", that is the famous inscription engraved in Delphi Bethel of ancient Greece. It is regarded as more valuable than all the ethnic writings of saints of all ages, and the wisdom flame reflected by it is like a beacon light for mankind's progress.

Mankind has been groping very hard on the way to understand itself. Mankind's origin is an enigma that has troubled people for a long time. People have employed all available means to find out that where they came from. In ancient China, people thought humans were made by Nvwa using yellow soil. In the West, people believed God created Adam and Eve, who gave birth to children, thus, mankind came into being. It was not till 1859, that the right answer was found by Darwin: Mankind has evolved from apes.

Thus for the first time, mankind gave a scientific answer to the question. That was the first declaration of mankind.

When Darwin declared this answer to the world, it was like a huge lightning, it shocked the whole world. Old theories and old forces regarded it as a great scourge.

Although Huxley won in his verbal battle with the archbishop, there occurred a noted "ape case" in the USA even in 1925. A middle school teacher was judged as "guilty" by the court and was fined for 100 US dollars, just because he taught Darwin's evolutionism. It is only now, that the idea of "mankind has evolved from apes" has gone deep into the hearts of the people.

In retrospect to history it is easy to see, the process of understanding oneself is so hard.

Of all kinds of understandings, the most important is the understanding of oneself. In understanding of oneself, the most important is the understanding of one's own future.

Untill mankind's future is revealed correctly, people just cannot help doing things in a rush, be eager for quick success and instant benefits. As we are not clear about our goal or direction, the result might be: either we fall into the trap dug by ourselves or be busy for nothing in a muddle-headed way, wasting huge amount of energy.

Only by finding out the right answer to the direction of mankind we can find the true way out and can do the greatest benefit for mankind.

As materialist dialectics tells us: anything has a process of origination, development and extinction. Mankind is no exception. Engels says: *"Fourier and his contemporary Hegel mastered dialectics in a subtle way... Just as Kant proposed in his writings on natural science the idea that the earth would go extinct. Fourier, in his study of history, proposed the idea that mankind will go extinct."*

Mao Zedong took a step further, clearly stating: *"Mankind came into being by birth, hence mankind will die too.... the earth will also die. But the death of mankind or the earth mentioned by us is not like the Doomsday in Christianity. We say mankind will die, or the earth will die, because there will be more advanced things to replace mankind, which will represent a higher stage of development."*

Here Mao Zedong not only pointed out clearly that mankind will perish but also clearly stated that mankind's death is just its leap forward, *"as there will be more advanced things to replace mankind, which will represent a higher stage of development."*

What is it, that is not human but higher than human?

We call it "race of gods". Just as mankind is a collective concept, xian-lei (race of gods) refers to the group of gods, being a collective concept.

The word "god" represents the finest and the most ideal future of mankind. There are no "gods" now, there were no "gods" in the past either, only in the future will there be "gods". Hence the "god" here is different from the God in religion or myths. Of course there are some similarities between them, as both are considered to be limitlessly wise, limitlessly fine, limitlessly superior, and in the supreme position in the universe.

Mankind will leap to be a non-human race higher than mankind, "immortals" which is a foregone conclusion in mankind's evolution.

Darwin's evolutionism has revealed that mankind has evolved from apes. And for its future direction also, we should find the answer from evolutionism.

In a sense, the study of mankind's future is a study of mankind's evolution. Since mankind has grown to the present state, mankind's power in science and technology has been great, enough to change its natural evolution into artificial evolution and passive evolution into active evolution. Thus, the depth and extent of evolution may be far beyond our imagination. In general, mankind's evolution falls into microevolution and macroevolution. Microevolution is mankind's individual evolution whereas macroevolution refers to mankind's collective evolution. Microevolution consists of body evolution, intelligence evolution and attribute evolution. What is more noteworthy than mankind's microevolution is its macroevolution. Macroevolution consists of society evolution, culture evolution and tool evolution. Lately some famous anthropologists in the world have come up to this idea: *The evolutional speed of mankind as a biologic organism is slowing down, but the evolutional speed of it as a "super organism", that is mankind's collective evolutionary rate, is advancing by leaps and bounds.*

Mankind's evolution includes three major microevolutions and three major macroevolutions, and in general, its evolutional rate is being accelerated as well. From the foremost primate to the ancient ape of forest, it took about 30,000,000 years to 46,000,000. From the ancient ape of forest to the australopithecine (ancient ape in the south), it took about 16,900,000 years to 20,000,000. From the australopithecine to homo erectus (Peking man), it took about 3,800,000 to 100,000 years. From the primitive men to the beginning of the slave society (21BC-475 BC), it took about 100,000 to 7,000 years. The slave society took about 15,000 years. The feudal society took about 2500 years. The capitalist society took only 300 years. The socialist society has lasted less than 100 years.

Mankind's evolution at such an accelerated speed may be regarded as "evolution of the evolutionary process itself". Moreover it just refers to the evolution without aim and without consciousness. If it is a purposeful evolution with self-domination, that is, if we change from spontaneous evolution to self-conscious evolution, make the best of high and new science and technology and unite all the energies of mankind, then mankind will have a greater power to further speed up the evolution at a tremendous speed. It will be more probable to evolve into the extremely high level and the boundless beautiful world very soon.

Eminent scientist of the former USSR, B N Hifuloff said: *"In the past, the inorganic natural world completed its transition to the life natural world: With silicon, granite and sand, it was possible for living cells to appear. It was the first leap. The second Leap was the transition from a life of simple biology to intelligence, thus, humans appeared. Now, we are living at the beginning of the third leap."*

"The third leap" indicates something that is not mankind but higher than mankind, and that is just what we call "xianlei(immortals)". "We are now at the beginning of the third leap" that is to say, today, we are just at the great turning point of evolution from mankind to immortals.

At first, there was no human beings on the earth. From apes to humans, it was a great leap on the earth. Several million years ago, the ancient apes were just monkeys jumping about tree branches, who were in the danger of being swallowed by fierce beasts at any moment or place... But now, from the caves and thatched cottages, people have evolved to such a state where they have built up skyscrapers one after another, spider-web-like dense highways, crisscross overpass bridges, what they use for travel are automobiles, trains and planes, in communication they use beep-pagers, mobile phones, videophones and so on, from boring wood to get fire to controlling the atomic energy that can destroy mankind for score of times... Today's mankind cannot be named on the same day with our ancestors. In comparison, how petty the apes were and how great mankind is! Could the apes and primitive men have imagined today's condition of mankind? The people of several hundred years ago or even one hundred years ago could not have imagined today's picture. If the already realized evolution is so hard to imagine, then the would-be evolution, the evolution of mankind to immortals, will be very hard to imagine for today's people, but is very possible at the same time.

The genetic differences between mankind and apes are only about 2%. But it is just this small difference of 2% that makes mankind different from apes in intelligence, behavior, mentality etc, as much as that between heaven and earth. The genetic differences between mankind and immortals may be also about 2% only. Today, by using the huge power of science and technology, the "God" in the new era(we) can not only solve the 2% of the puzzling problems, but also completely solve the rest of the problems that are much more than this 2%.

Today mankind's wisdom, power and conditions are thousands or millions of times greater than its ancestors, and we are fully able to achieve another leap, a great leap controlled by ourselves.

Now, automatically, a saying by Huxley comes to my mind: *"Once those who are good at thinking are liberated from the blinding traditional bias, they will find the best proof of mankind's great power different from mankind's low-grade ancestors. And from mankind's long process of evolution, they will find the reasonable ground for mankind's confidence for reaching upto grander future."*

In retrospect to mankind's evolution history our answer is: Mankind has evolved from apes, and it will certainly evolve in to immortals. It is an answer resulting from the earnest, in-depth and careful thinking about mankind's past and future.

Mankind will evolve into immortals, and we should consciously accelerate such an evolution. This idea may trigger a greater shock and more argument than what Darwin's idea about man's evolution from apes generated. Its opponents are deep-rooted traditional thinking and ideological shackles and what it offends are uncountable old theories, old principles and old forces. However, the power of truth is invincible. It will certainly have a great impact on our future life, academic career, occupation, wealth, happiness and life-span...

Everything will be changed by it.

Mankind will evolve into gods...

Let mankind make its own second declaration solemnly!

Now, we can say: How great the future is! How beautiful the future is!

II. LET "IMMORTALS" STEP DOWN THE ALTAR

Labor created mankind and technology creates immortals

The legend on immortals is of long standing in China.

What were "immortals" in the eyes of the ancient people? How did they live? Qu Yuan, a great poet in the Warring States Period, wrote in his poem "Travel Faraway", *What "immortals" eat is the dark yellow qi (air) of heaven and earth, what they drink is the clear dew and jade-like juice, every one of them looks bright and charming, their bodies are sturdy, their souls are unperturbed and extremely happy, they ride the great air flying to Heaven, and live a very free life.* In the writings of the romantic Zhuangzi (great philosopher about 2000 years ago), *immortals ride the clouds, and drive the sun and the moon, roam the whole world, they are not only free of the trouble of life and death, riding the stars and flying everywhere, but also look tender, charming and graceful, without any worldly worry.* Cao Zhi, the great poet in the Three Kingdoms Period, wrote like this: *"Immortals play chess on Mount Tai. Accompanied by fairies playing the lyre and sheng (a reed pipe instrument), they fly for ten thousand miles in an instant rising to the great void easily, Flying across clouds..."* In his eyes, immortals can do anything and possess anything. They can take anything from the universe, riding the phoenix, reining the tiger or lion, looking very majestic... So to speak, immortals in the ancients' eyes are extremely romantic, completely free and happy forever. In today's words, they are absolutely ideal persons, and their life is the most admirable. Hence, all the ancients in China dreamed to become "immortal".

The ancients' dream of becoming immortal has lasted for several thousand years. With regard to such a dream, most people think it can only be a dream, being beautiful but incredible.

Today, we must declare to the world that by relying on the divine power of high technology mankind will evolve into being immortals.

To get a deeper understanding of this bold prophesy, first of all, we must take off the mystic fog of thousands of years from immortals, analyze this concept in a scientific way, redefine it, so as to let immortals come down from the altar, thus changing the dream into a reality.

As a scientific term, "immortal" has the following four major attributes:

(1)Eternal Life

The first major attribute is living without ageing, transcending the time limitation, being completely free from the abyss of misery of "birth, ageing, illness and death", getting limitless freedom and eternal youth.

A person has only one life, and a life is for once. Only by possessing life can we possess everything else: career, family, love and happiness. Whoever you are, what you crave for the most is the extension of your life.

We not only need "a long life", but also need "to avoid ageing". If after enjoying a short youth you fall into the state of senility, lingering on in a steadily worsening condition, then it will constitute the greatest suffering. In *Gulliver's Travels*, Swift wrote, *"in an island country, those who have lived to the age of 80 will become anile and useless".* Such a long life is indeed a great misfortune.

"Immortals" never get old, they remain young forever. The only difficulty lies in "forever". Shakespeare said: *"Time may pierce the color finishing on the surface of youth, may dig trenches and slots on the forehead of a beauty and may eat up a bird of wonder! Nothing can escape from his sweeping sickle."* Realization of eternal life and keeping youth forever is the complete victory in mankind's conquest of time.

"Everyone must die". However the great development of modern technology has shaken this "natural law" that has always been regarded as unshakable.

Today, "living forever" has already begun to become possible. Some scientists have already predicted that by virtue of the deciphering the gene code, mankind's life span can be greatly extended. So far the optimistic estimate by some scientists is 1200 years and the most pessimistic estimate is not less than 500 years. Haris, a member of the British

Committee of Human Gene, pointed out, if we keep on studying, then theoretically speaking, it is not a hard matter to make mankind's average lifetime reach 1200 years. The discovery of the gene code alone has greatly changed the traditional lifetime limitation in the scientific circle, from one hundred years or so to a thousand years or so. We can imagine, along with the continual breakthrough in science, people's understanding of mankind's lifetime limitation will change greatly all the time, from one thousand years to five thousand years, ten thousands years, until an eternal life.

Living forever, as the most important attribute of an "immortal", is no longer a mere illusion now, but there is a possibility that it can be realized step by step through science. Mankind has already begun to possess this ability, and such ability will get stronger and stronger.

(II)Omniscience and Omnipotence

Omniscience and omnipotence are the attributes of God, if a person becomes "immortal", he will reach the summit in "knowledge" and "ability".

Omniscience and omnipotence are quite possible. Its foundation is eternal life. If a person can live upto 2000 years, and keeps learning and struggling all the time, he will equal a hundred doctors and his knowledge and ability can surpass hundreds of doctors.

What is more important, future science and technology may make omniscient and omnipotent "gods". Man-machine integration, brain-machine integration, gene engineering and so on, will help people to move towards omniscience and omnipotence. Once the microcosm is completely conquered, the science and technology findings in this aspect (such as high energy physics, super-high energy physics, molecular biology, quantum biology and so on) will enable the human body structure and functions to change fundamentally, and graduall move closer to the goal of omniscience and omnipotence.

For example, if we implant a computer chip into the brain, or use bio-protein to make an integrated circuit so as to realize man-machine networking, we will have no need to spend time cost and energy cost of ten years or even a few decades in memorization of knowledge. Scientists have predicted, mankind will get a huge amount of wisdom in a moment, as easily as copying a software on to a computer. If a 5-year-old child wants to learn English, the scientist will download the English language information into the child's brain. A whole set of *Encyclopedia Britannica* may be turned into a mini memory circuit on a mini chip, which can be stored into the human brain at once. So long as we realize the brain-computer alliance, we may help people to grow towards "divine ability to know the unknown" and "omniscience and omnipotence", finally progress up to the level of "godly brain".

Meanwhile, the man-machine integration in many ways may also greatly accelerate the process of "omnipotence". We can not only implant the chip into the brain to build up our intelligence greatly, but also connect such a chip to the human nervous system, so that we can control the chip and all kinds of micro-machine units in our body, so as to able to see things that can be seen only through X rays, ultraviolet rays or infrared rays. Then, the so-called "peculiar functions" such as see-through function, remote-seeing function, flying function and so on will become everyone's abilities. Then, the person will no longer be an ordinary person, but an omnipotent "superhuman".

(III)Supreme Kindness and Beauty

We have good reason to believe that mankind can make substantial progress towards omniscience and omnipotence in near future.

People often use "gods of Buddha's heart" to describe great virtue of benevolence. Every god pities on all mankind, having a kind heart like Emperor Yao and Emperor Shun (ideal emperors in the legend of ancient China).

The supreme aim is the highest virtue, that is, having the supreme moral character and best quality. After becoming "immortal", the "man" will become a saint or mahatma with super terrestrial qualities.

For the progress of social civilization and for mankind's progress forward, everyone must have a noble moral character. This is also for the sake of an individuals' well-being. The happiest person will be someone with a noble moral character. Those who love others are always loved by others. Those who benefit the world are always benefited from the world.

For the most ideal person, acquiring a saintly moral character is the most essential requirement. To achieve this, on one side we need the development of social civilization and on the other hand, we also need the power of science. A person's moral character is also affected by the genes. As the saying goes, *"You can change mountains and rivers but not a person's nature"*. For example, according to some scientists, the adventurous character of the Kennedy family may be related to their genes. If we work on every aspect, from genetics to systematic education and social civilization, we may cultivate the highest moral quality for everyone.

Meanwhile, immortals also have the best appearance. People always use "goddess" to describe a beautiful look, often say "pretty as a fairy", and always regard "immortals" as the most beautiful. The ancient philosopher, Zhuangzi, described gods as *"having the skin like ice and snow, graceful as a virgin"*, every one of them is the most nice-looking person.

Everyone loves beauty, and everyone wishes to become an eye-catcher. When someone asked the famous wife of Roosevelt, Mrs. Elinor Roosevelt, what was biggest regret she had in her life, her answer was: not being prettier. It is a sentence uttered after a careful thinking from one of the most worshiped and beloved lady. And such a wish does not belong to females only. The world-famous writer Leo Tolstoy wrote: *"What else can affect a person's prospects than his\her appearance? There is nothing more than a person's appearance that can determine whether the person is lovable or annoying."* British scientists have found, men with a handsome look find better jobs than those lacking attraction, and earn more money. The researcher in London's Gilde Hall College, Killings, points out that Secretaries with ordinary looks are likely to have an income 15% less than the pretty ones. Some researchers have also found, that men considered to be lacking attraction are likely to have an income 15% less than the more handsome ones. Females with ordinary looks are also likely to have an income 11% less than the prettier ones.

Having an appearance like a goddess, was only a dream in the past. Before long from now, by using modern science and technology, everyone may have a beautiful countenance beyond match.

The gene engineering will uncover the gene related to appearance, then we will be able to control our own appearance, height, skin color, hair quality, and the development of bones, nerves and so on. By taking a pill, you may turn your hoary hair into black hair, in this way "hoary hair" will become a term of the past. Even the innate eyeball color can also be changed. You can select your skin color according to your wish.

Cosmetics will be "made on demand", and you may have the unique face refining cream of all kinds, designed according to your own skin quality. You can even stay at home and ask the Internet-based beauty treatment masters to serve you. Cosmetic technology has advanced by leaps and bounds and it is very common to use artificial skin, artificial bones and even the five sense organs. Thus, if you want to change your appearance, it will be as easy as writing a cheque. A health and beauty expert working in Canada, Sindhi Theo, said: *"It is no longer something concerning only face, It is like improving unsatisfactory parts of your body by using the technology available."*

Do you wish to have the beautiful hair of Julia Roberts, the nose of Cleopatra, the eyes of Zhao Wei, the lips of Demi Moore, the stature of Marilyn Monroe, and the wisdom of Athene? You can realize this at your will. Even if you are as ugly as Quasimodo, you don't need to feel sad. By a little adjustment of the gene, you may shape yourself into Tom Cruz.

Beauty is no longer the monopoly of a few, but has become a possibility for everyone. It will bring about another emancipation of women, and they will no longer need to worry about their age as a disadvantageous factor while competing with young girls. In year 2000, 65-year-old Sophia Roland defeated many young and pretty super models to become the "Belle of the Millennium". Such "grandma-typed belles" will be found everywhere in the future society. Anyone, regardless of the age, can make himself (herself) very attractive, and can change the look many times in a life, so as to be the ideal beauty all the time.

(IV) Daylong Ecstasy

Whatever we do, finally, it is for our maximum happiness. The essence of happiness lies in joy. In all ages, mankind takes the pursuit of limitless happiness and joy as the highest cause. With regard to the world in which this highest cause has been realized, Christianity calls it "Paradise" or "Land of Promise", Buddhism calls it "Pure Land", or "Elysium", Islam calls it "Heavenly Garden" or "Fairyland". No matter what it is called, it refers to the highest prospect with incomparable joy.

If living is a pain, and it is impossible to see the joy of tomorrow, people will choose to end their life themselves. Can you not see, there are so many people going to the lower world at their own will in each country, each nationality, each year, each day, each hour and each moment? A long time ago, Li Dazhao (1889-1927, famous communist and martyr) uttered this truth, that is simple but many people have not uttered yet: *"Only the joy-seeking life-philosophy is the natural and true life-philosophy."* Even if a person can live for a thousand years or two thousand years, but if he is in suffering all the time, I think, there will be no one wishing so. Why has euthanasia been legalized in many nations and has been accepted by many people? Why has helping others to end their life become a very humane and loving act, when we are fully certain there is no way to change a painful life into a joyful life? Why do some philosophers say *"Those who invent euthanasia are great, those who are willing to accept euthanasia are great and those who help people with euthanasia are also great"*. Isn't this enough to prove "the joy-seeking life philosophy" is a true-life philosophy?

As a proverb says, *"As happy as a god"*. "A god" is incomparably and daylong joyful. That is just what future mankind should attain, that is ecstasy 24 hours a day. By technology, we may cut down the sleep time. Using the gene map, we can find out the gene that makes people feel happy and optimistic, and can implant the optimism gene or joy gene into a pessimist. We can also use all kinds of technology, for example, by finding out the part of the brain that produces joy, we may stimulate it continually, so as to be joyful all the time.

Above all things, the daylong ecstasy must be gained from creative work. When work has become the primary necessity in life, and when everyone can do the most interesting work, it will become the greatest headspring of joy. In future society, all the dull and heavy jobs will be given to the robots, and all mankind needs to do will be challenging, creative and entertaining jobs.

"Lyre, chess, painting and calligraphy, poetry, wine and flowers:;All of them were indispensable in my life. But now all have changed; I have to deal with firewood, grain, oil, salt, soy, vinegar and tea instead". It is doggerel to describe the trouble of trivial housework. From now on, all the niggling such as "firewood, grain, oil, salt, soy, vinegar and tea" will be trusted to the robot, and everyone can enjoy the pleasure of "lyre, chess, painting and calligraphy, poetry, wine and flowers".

Meanwhile, education reform will show more concern for people's pleasure, helping them to learn how to make joy, how to be happy, how to remain optimistic, especially to teach them the techniques to develop an optimistic and joyful character. Having a joyous character and way of thinking may also contribute to daylong ecstasy.

Once mankind acquires daylong ecstasy, they will make a complete turn from the sad state of "It is hard to have a chance to laugh happily", the extreme bitterness of life will become an unprecedented ecstasy, and the world of bitterness will become the real Elysium.

In conclusion, "immortal" represents a leap of mankind, being the most ideal superhuman, without ageing or death, omniscient and omnipotent, extremely benevolent and beautiful, living an extremely happy life all day long. If we use a formula to express it, it is like this:

"Immortal" = living forever + omniscient and omnipotent + supreme virtue and beauty + daylong ecstasy (ultimate goal)

"Immortal" = super longevity + super high intelligence + super kindness + super beauty + super joy (present goal)

If we want mankind to evolve into immortals, first of all, we must let everyone grow in a super way in all aspects. First, we should realize "Everyone is like Peng Zu (in longevity)", "Everyone is like Kong Ming (in wisdom)", "Everyone is like Yao and Shun (in virtue)", "Everyone is like Xishi (in beauty)", and "Everyone is like Meitreya (in virtue and beauty)". Of course, "super" is a relative concept. For example, in "super longevity", let alone 800 years as Peng Zu, if we reach the "200 years of life" mentioned by Mao Zedong, it will be the first key step in mankind's evolution into immortals.

The transition from apes to mankind was a result of natural evolution through competition and natural selection, but the transition from mankind to immortals will be a result of evolution dominated by mankind itself through science and technology.

As we know, living forever and becoming "a god" has been a fantasy of all ages. But we know clearly, science and fantasy are interrelated to each other as fantasy may activate science while science may realize fantasy. Fantasy is a midwife of science, and fantasy can be realized only with the help of science. Let's see How many of mythic fantasies in *Fengshen Yanyi (Tales of Gods and Ghosts)* and *Pilgrimage to the West* have not been realized by means of science? Isn't the present world just a mythic world of the old time? *Qitian Dasheng* (Mahatma As Higher As Heaven) Sun Wukong could fly 108,000 miles in one somersault, but now an ordinary American citizen has been able to ride the rocket to roam the outer space, and soon everyone may take a "trip to the outer space". Sun Wukong could get into the belly of Iron-fan Princess, similarly the nanometer technology will be able to make a micro robot very soon that can get into your blood vessel to do operation for you...

Who is so powerful so as to let "gods" walk down the altar? Who can enable mankind evolve into immortals? No doubt, it is technology!

If we say labor created mankind, then we can say technology will create immortals. However, what is more important than technology is human beings. The reason is: "To achieve all these, the key does not lie in technology, but lies in the humans, lies in whether we humans will work this way. Technology is like the kite, and it can fly higher and higher, but we should not forget that the holder of the line is man, and only he is the real master of the kite. Finally, technology will return to the hands of humans."

We must be very clear and specially clarify this point: in achieving the eternal living, omniscience and omnipotence, supreme kindness, supreme beauty, and daylong ecstasy, the key is not technology (technology is already moving to that goal), but it is "whether man will work towards the goal". So long as we make efforts towards that goal, "gods" will certainly step down the altar.

III. GREAT VENTURE AND GREAT TRAGEDY

Faced with the sorrow of "When can life return after death?" and "Enjoy wine and song while we can, for life is short" (famous poem lines in ancient China), of course, great people like First Emperor of Qin Dynasty, Emperor Wu of Han Dynasty, Second Emperor of Tang Dynasty and so on, were not willing to succumb to the fate of "A man seldom lives to be seventy years old".

Unlike Westerners, in pinning their hope on the other world and on the immortality of the soul, the Chinese have always believed in eternal life and strived to gain it directly through their own efforts. Ancient Chinese were crazy about the pursuit of longevity, seeking the eternal existence of both flesh and soul, and seeking to be immortal.

The movement of seeking immortality began at least from the *Spring and Autumn Annals Warring States (476-221 BC)* period. According to *Records of Warring States*, "A guest offered a drug of immortality to the King of Jing." At that time there were already people engaged in seeking and making the drug for immortality. According to *Records of the Historian*, many kings, such as King Wei of State Qi, King Xuan of State Qi, and King Zhao of State Yan, sent people to enter the sea to look for the three godly mountains of Penglai, Fangzhang and Yingzhou, as the legend said there were many gods in the mountains and there were drugs for immortality. Thus, a great movement of seeking gods and seeking immortality began.

In the time of Qin Shihuang, the movement of seeking immortality grew to its heyday. Qin Shihuang paid a high price in order to seek an eternal life. He gathered many necromancers who were skillful in the art of immortality and sent them to all parts of the country to look for drugs of immortality. He was fooled by some of his subjects such as Lu Sheng, Xu Fu, but he never stopped the struggle for searching the drug of immortality, but finally all his efforts went in vain. In the period of Emperor Wu of Han Dynasty, people all over the country were active in looking for the god mountain and seeking immortality. After all such movements of looking for the god mountain overseas failed completely, and there was hardly anyone dreaming about it. People began to doubt this way to become gods, but started taking another way, that is, relying on their own efforts to become immortal. Thus, they began the process of making cinnabar by melting gold, firmly believing that taking the gold cinnabar will help people to become immortal. From then on, gold refining or alchemy became a lasting movement in Chinese history. The first emperor who took the gold cinnabar was Li Shimin, the very famous second emperor of the Tang Dynasty, but he lost his life because of it...

Qin Shihuang (First Emperor of Qin Dynasty), Emperor Wu of Han Dynasty and Second Emperor of Tang Dynasty were among the few hero emperors in China's history, and Mao Zedong commented on them in his famous poem "Qinyuanchun, Snow" as *emperors appearing once in a thousand years*. None of them were fatuous, all of them had great talent and bold vision, but they set their supreme goal at "becoming immortal", and did no spare any price for it.

Everyone fears death, everyone longs for a boundless life and a boundless youth. Whether you are a high official or a humble lackey. Whether you are a statesman, entrepreneur, artist, or an ordinary person, no one is an exception. Only that emperors' demand for this was more urgent. The reason was also very simple: their life was more worth treasuring than that of the average person. They thought that the value of "becoming immortal" was greater and higher than being an emperor.

While longing for longevity, in the face of sad feelings of "When can life return after death?" and "Enjoy wine and song whenever we can as life is short", such great persons as First Emperor of Qin Dynasty, Emperor Wu of Han Dynasty and Second Emperor of Tang Dynasty, showed their superior spirit of "My life lies in me, not in Heaven" and their revolutionary thinking of "Man can conquer nature". Hence, they did not succumb to the fate of "A person seldom lives to be seventy years old".

Besides First Emperor of Qin Dynasty, Emperor Wu of Han Dynasty and Second Emperor of Tang Dynasty, there were innumerable immortality seekers among the emperors of all ages, one stepping into the breach as another fell.

Among the emperors of the Tang Dynasty alone, there were five who lost life because of taking the longevity drug! All heroes followed suit in seeking immortality. According to *Biography of Liu Xiang* in *Records of Han Dynasty*, the famous Confucianist Liu Xiang was also keen on alchemy, but he did not succeed, and almost lost his life for it. The famous poet Li Bai's story of trying to become "a god" by taking the cinnabar is well known of course. The Four Brilliant Scholars in the beginning of the Tang Dynasty; Wang Bo once studied the classics on "gods", Lu Zhaolin almost lost his life because of taking cinnabar, Bai Juyi (great poet in the Tang Dynasty) learned the art of alchemy together with Yuan Zhen, but because the timing was not proper or the vessel was not sealed well enough, all the drug flew away, making the poet feel very sorry.

People were killed by the cinnabar one after another. They learned at last that the mixture of gold, lead and mercury should never be eaten, so, the external art of alchemy failed completely.

After this fiasco, people turned to the internal art of alchemy, that is, by training the vital essence within the body, to get the spiritual "panacea" for immortality. Actually, it was a qigong exercise. The internal art of alchemy became the mainstream in the science of immortality, and it was greatly developed after the Tang Dynasty.

Today, when mentioning god-seeking history, many people cannot help sneering at it. Indeed, in today's point of view, their practice was really laughable. Yet, we might as well give a cool and careful analysis to what should be sneered at, their means, or their purpose to "become immortal"? By becoming "immortal", we could not only get boundless joy and happiness, but also avoid the tragedy of "dying in action before the final victory", so that many people might have sufficient time and sufficient conditions to make good. Becoming "immortal" could enable people to get the greatest happiness, and to achieve the greatest success so as to make the greatest contribution. Should the struggle for becoming "immortal" be sneered at? As a matter of fact, they just represented the ancient Chinese's pioneering spirit, revolutionary spirit and exploring spirit.

Is that a scientific exploration? Is that a scientific spirit?

"Scientific correctness" has become an ingrained and unbreakable idea for several centuries. However, science also involves failure. In the 606 times, it failed for 605 times, and it succeeded only in the last attempt. When Edison invented the electric bulb, he failed for more then ten thousand times, and finally he brought brightness to the world. Were the previous failures not scientific explorations? Did they not represent a scientific spirit? In order to realize a great goal, we often need to try for ten times, a hundred times, a thousand times, ten thousand times... and finally we succeed. The ancients in China never stopped their effort to be immortal, and their road was so tortuous: from the three large-scale adventures to the three overseas godly mountains to seek gods, from Emperor Wei of State Qin to Emperor Wu of Han Dynasty, to the later deaths of many kings because of taking of cinnabar, to the turn to internal practice of alchemy... no one made a great coup in the struggle.

Nevertheless, although the seeking of immortality failed, in the process of exploration, the external and internal arts of alchemy led to the ancient science. The external art of alchemy combined many branches of science into one, such as chemistry, mineralogy, botany, zoology, medicament, astronomy and so on, becoming the source of the oriental science that was earlier than the Western science (in Europe, until the 16th century, science was still called "natural magic"). It is just because of this, that as time went on, many subjects grew out of the science of immortality, such as chemistry, health preserving science and so on.

Since the ancients failed again and again in their pursuit of immortality, people began to look at "immortality" with a cold and critical eye, feeling that it was naive and absurd. However, can we neglect this point: Haven't many wishful ideas of the ancients such as the "clairvoyance", "clairaudient" been already turned into reality by the invincible might of science?

Mankind has already entered the space age, and landed on the moon. But five hundred years ago, in the Ming Dynasty, there was a man called "Wan Hu", who bound himself onto 47 crude rockets, wishing to fly to the sky. He was bombed into pieces in a great explosion. The sad finale of "smashing his body to pieces" could not cover up the brilliance of Wan Hu's dream. Later, astronomers used Wan Hu's name to name a ring mountain on the moon.

Many of the god seekers in history also ended up in death. They were like "Wan Hu", as stirring and tragic as Wan Hu. Maybe in modern people's eyes, their doing was just like Wan Hu's binding himself with 47 crude rockets, being so ridiculous. Many people laughed at Qin Shihuang and Emperor Wu's god seeking. They laughed at Li Bai and Su Shi's indulgence in immortality, and laughed at the alchemy art... However, in this laughter, they did not know that they are neglecting what we could learn from the ancients. Laughing or sneering at someone or something may blur people's vision, affect people's wise perspective, and lead people away from in-depth thinking. Did the ancients' pursuit contain some wisdom and value? Especially in the present time, does it deserve our revaluation?

As for Wan Hu's act, people of those days would never think it was a great exploration, but only when the landing on the moon became a reality can people revaluate Wan Hu. Likewise, in treating those god-seekers of ancient China, perhaps, only when people have conquered the limitation of time and realized immortality can they revise their thinking and revaluate their opinion.

Since the tragic end of Wan Hu's "smashing his body to pieces" has already been changed into the great victory of mankind's flying into the blue sky and conquest of outer space, how can we have no reason to believe that the ancient god-seekers' sad stories will be converted into the fine reality of everyone becoming a god?

The ancients' god-seeking process was a process of mankind's struggling to conquer the natural limitations and transcending nature, that is, the process of marching from the greatest realm of necessity to the greatest realm of freedom. We should not conclude that their destination was wrong just because their road was wrong. Although they failed again and again, their ideal was mankind's greatest ideal. Their hope was mankind's greatest hope. Their exploration was mankind's greatest exploration.

Mankind's history is one from the realm of necessity to the realm of freedom. When we recall history, we can say: The ancient god-seekers were the spearhead in mankind's struggle to transcend nature, transcend necessity and march from the realm of necessity to the realm of freedom. Their explorations were great, and we should cherish their memories. They were not foolish, nor lamentable. Today, if we do not realize this great ideal while we can, we will be the most foolish and the most lamentable.

IV. PEOPLE ARE PRISONERS OF THEIR IDEOLOGY

What is surprising is, whatever you forecast, the truth is always underrated. Even the most extreme forecast lags behind the reality.

—Negroponte

Today, unprecedented great leaps have been made in technology, which has made it possible for us to realize the great ideal of leaping into immortals. However, corresponding great leap has not occurred in people's thinking and ideology.

Technology always walks ahead of ideology. Before any new idea comes into being, a new technology has already begun to roar like an angry wave flapping the coast.

Technological revolution is changing our life. Today's house has almost become a world of all kinds of household electrical appliances: at the door we have the electronic bell and the monitoring video camera. In the guest chamber we have a TV and VCD set. In the study we have computer, telephone and facsimile equipment. In the bedroom we have cordless telephone and air conditioner. In the kitchen we have electronic rice cooker, electronic oven and microwave oven. In the toilet, we have the electronic closestool and shower... The newly born Internet has connected thousands upon thousands of households together, and the electrification and the electronic revolution have put us into a mythic world unthinkable by the ancient people. Automobiles are galloping on the cobweb-like dense highways, high-speed trains are dashing on the vast tracks, 10,000-ton steamships are sailing in the boundless sea, sky-masters are flying in the immeasurable sky, and we can reach any place in the world instantly. All kinds of modern means of transport are playing a immeasurable role in mankind's expansion of activity in space, in saving time and in raising of work efficiency. The human society is no longer developing at snail pace, now life's changing speed has reached a very surprising level.

Today, we have ushered in another technological revolution that is even greater and even more profound, bringing about shocking changes to everything. A single gene can extend the life-time of a drosophila by one or several times, and same can be done with mankind's gene in future. In the future, when we inject an elixir pill or a longevity gene into the human body, it may rejuvenate the body that was growing old and about to die soon. Scientists can also use a completely new heart or liver cultivated by stem cell technology, thus, we may repair or replace any "part" of the human body at will. By adjusting our telomeres, we may get the right to set our life clock backwards. All activities on the earth and all the temporal and spatial variations of the earth's environment may be loaded into a computer, the computer chip will be as cheap as a draft paper, piling up around mankind, you may talk to any tool or electric appliance and all of them can give wise answers like the "Snow

White" in the fairy tale...

All those are just a few examples, and the more stirring revolutions will come next. Bill Gates said in his *Future Speed per Hour*, *"people often overestimate things to happen in the coming two years, and underrate the development trend of the coming decade"*. When you are galloping forward by riding the wind of technology, the distance between fantasy and reality becomes very vague, in fact, maybe there is only one sheet of paper in between. When interviewed by a reporter, Negroponte stressed repeatedly: *"What is surprising is: whatever you forecast, the truth is always underrated. Even the most extreme forecast lags behind the reality."*

Technological advances have strengthened mankind greatly, technological leaps have broken up the myth that everything was created by God. Maybe, it is just because of the technological revolution that Nietzschean cried, *"God has died"*.

Technological impediments are not so hard to overcome or break through, and in contrast, it is harder to break through or cross over the gulf in people's ideology. The biggest prison in the world is the prison of the brain. People are prisoners of their ideology. The true revolution occurs in people's brain, and that is the ideological revolution. An ideological revolution is a complete re-judgment on the established values, it is to liberate people's spirit from the prison of old ideas, and enter a brand-new world. If there is only technological revolution without ideological revolution, people's sufferings might increase automatically. In a sense, technology and ideology are just like people's feet and shoes. If the feet grow at an accelerated speed but the shoes remain unchanged, there will be a sharp conflict between technology and ideology, and such a conflict will be very painful for the people, like a shoe pinching the foot. If we cling to the traditional idea and restrict or cut down technological development, it will be like cutting the feet to fit the shoes, thus leading to a great loss for mankind.

In history, ideological revolution was always unable to keep up with technological revolution. When mankind's first test-tube baby "Brown" was born in 1978 in Britain, it caused a great ethical argument, and many people looked at this "extraterrestrial guest" with a strange eye, but now test-tube babies all over the world have exceeded 300,000. Even the blood transfusion technology met with a strong resistance at the beginning, and even now there are still some people refusing blood transfusion treatment in western countries. The process of the acceptance of the contraceptive drug was also a process of struggle with mankind's traditional concept. Until 1869, at the annual meeting of medical science in Britain, someone still condemned it: *"All kinds of invention for restriction on the quantity of childbearing are indecent"*. In 1916, Madam Sanger was thrown into prison because she advocated birth control in the USA, and the entire world regarded it as a great scourge. It was only until 1965, one year before her death, that US Supreme Court cancelled the law of prohibition of married couples' use of contraceptives. In 1972, both sexes of unmarried people got the same right; in 1973, women's abortion right also got the recognition from the law. It was only then that birth control technology got its rights of existence, use and development finally.

Whether a technological progress can be accepted by mankind or not, ultimately depends on whether it can bring happiness to mankind, and not on whether it tallies with the moral standards, legal provisions and traditional ideas of a certain historical period. It is a universal truth, being correct forever.

The truth is constant but the ideology is changeable. If ideological revolution cannot keep up with technological revolution, it will impede or restrict technological development. If ideological revolution conforms to technological revolution, it will be able to promote and boost technological advances. The difference between an old ideology that impedes technological revolution and a new ideology that boosts technological revolution is not a difference between 1 and 2, but a difference between total disaster and total happiness of the world.

The present time, characterized by uncountable technologies and unprecedented revolutions, is forcing us to make a choice. What new ideology should we choose to face the new technology? If we still cling to the stale and wrong ideas, we will either lose the chance or become the victim of our own mistake.

Once an ideology is formed, it will control people's mind stubbornly, controlling people's viewing angle in observation, thinking mode, value orientation and pattern of behavior, thereby controlling everything. If you have been poisoned by a wrong idea, no matter how high your intelligence quotient is or how rich your knowledge is, it will be of no use. When the famous, upright and incorruptible official of the Qing Dynasty, Hai Rui, saw his little daughter taking some refection to eat from a manservant, he flew into a rage and forced the small child to fast to death. In today's view, what a foolish and ruthless conduct it was! But should it have happened to Hai Rui, a saint with rich knowledge, high intelligence and high morality. A wrong idea is like a devil: Once you have been "hypnotized" by it, intelligence or morality will vanish completely.

A wrong ideology may cause a learned wise man to conduct the greatest foolery. Similarly, when a wrong personal ideology becomes the wrong ideology of a group of people, it will form a stubborn habitual ideology and forced habit of the society. It becomes the most terrible evil force.

Women's footbinding was a custom lasting nearly one thousand years in China. The so-called "three-inch golden lotus" was regarded as a symbol of civilization, education, rules of etiquette and beauty, whereas the natural and strong foot was ridiculed as a symbol of barbarism, ugliness, and breach of morality, thus, women with such feet could find no husband, and they felt too ashamed to live in the world as well. Abolishing the footbinding was "harder than going to Heaven"! "The footbinding process was so terrible that the child used to cry in pain, but the mother used to beat her daughter mercilessly, forcing the tearful daughter to bear the great suffering. Even on-lookers could not help feeling sorry for it, but the parents were dead to all feelings, showing no pity at all"! As the saying goes, "A pair of little feet, a vat of tears", and such a suffering lasted a thousand years. Nevertheless, at that time, millions of people showed no objection to such a "self-hurting" foolery, supporting it one hundred percent. Did this ugly conduct that might let modern people feel deeply ashamed resulted from a low intelligence or low education? No! It only resulted from the ideology. Social ideology can tear into pieces, those trying to challenge it, and the people themselves often judge any thought violating it as evil.

Before the force of habit resulting from social ideology, even the emperor holding the highest power had to put his hands up. When the famous general Zuo Zongtang returned to the court in triumph, the emperor was very happy, and issued an order to call him in. Zuo

Zongtang received a warm welcome on his way. But he was detained by the gatekeeper. It turned out that, in the Qing Dynasty, any official coming from outside to the court must give some gift to the gatekeeper for entry according to the custom. According to Zuo Zongtang's status the gift money was as high as 80,000 liang silver. But Zuo Zongtang did not have so much money, so he could not enter and had to settle down outside the gate for the time being. When the emperor learned about it, he praised Zuo Zongtang's incorruptness, but also felt very puzzled, as it was the "routine of the court". Finally, the empress took the lead by offering half the amount of money. As for the rest, Zuo Zongtang's followers raised the money by all means, thus they collected suffecient gate gift at last. Zuo Zongtang's illustrious battle achievements (he had just cracked down the Taiping Heavenly Kingdom rebellion) and the emperor's invincible might could not even resist a small gatekeeper. The gatekeeper had no special ability and what he based himself on was just the rule "This is the routine of our court, and no one can be excused from it". How great the force of habit formed by ideology was! Such a habitual ideology has put mankind into a very absurd mind-set. So long as it is a habitual thing, it will be regarded as natural, without need of explanation or argument, it is right of course, no matter how absurd or ridiculous, it cannot be changed.

A person's wrong thinking may result in a personal tragedy, a nation's wrong thinking may lead to a national tragedy, and mankind's wrong thinking may cause mankind's tragedy. Mankind may pay a disastrous price for its wrong ideas, but at the same time, may also attain great happiness because of its correct ideas.

For thousands of years, the deep-rooted ideas in the human mind are "It is hard for a person to live to be one hundred years old" and "Everyone must die"... We have taken such ideas for granted, believing them to be true without exception. Even though science has already proved mankind can break through the natural limit of life span, and can "realize the dream to live a thousand years before 2030", they do not want to do so. They are afraid to do so, just because it is against the traditional concept. If someone has broken the traditional concept, he will often be labeled as "absurd", "prattling", "terrible" and so on. If you say, we can not only live to be a thousand years old, but also live forever ultimately, and mankind can evolve into immortals finally..., you will be sneered at or laughed at by many people. For mankind's mind-set: Once we have built up a fixed concept, then if a new idea meeting the new condition is against this old concept, we do not reflect whether the old concept is right or wrong, but use the old concept to judge whether the new one is right or wrong.

Almost all the sufferings and calamities of mankind are caused by mankind itself and are results of mankind's own wrong ideology. Only by changing the wrong ideology can we change the wrong ways of thinking, wrong patterns of behavior, wrong ways of life and wrong ways of survival resulting from it. Only by establishing the right ideology and getting rid of the ideology impeding mankind's happiness and blocking mankind's life development, can we get a brighter future.

All that we can control is, ourselves, that is, our ideology. Tagore said: *"Mankind's perpetual happiness does not lie in the acquisition of anything, but lies in giving himself to something greater than himself, giving to a concept that is greater than his individual life... Such a concept may make it easier for mankind to give up everything he has, even his own life. Before mankind finds out a great concept, his existence is unfortunate and pitiful, and only such a great concept may liberate him."* The crux of the matter is whether we have found such a great concept.

Today, what kind of concepts should we build up to be suitable for the technological revolution: boost the technological revolution and speed up development for enhancement of mankind's happiness? We should build up a concept: "Everything centers around the human beings, and the human being centers around his life" to treat life sciences and gene technology study, so that we may enjoy the limitless happiness brought about by technological revolution. We should set the concept "Treasure peace and harmony, and seek mutual benefit and winning together", so that we may prevent the tragedies of constant appearance of powerful and new weapons, constant escalation of arms race, continual aggravating of terrorism and continual approaching of mankind's self-destruction, and develop towards mankind's pulling together and helping each other so as to create a bright future together. We should set the concept "The world is changing, and everything is changing", so that we may prevent conservative and outmoded ethic ideas from impeding technological revolution, so that mankind may keep abreast with the times and keep a most tolerant and most active state of mind in boosting technical changes continually.

In the West, the industrial revolution and technological revolution took place only after the Renaissance, which remolded the Europeans thoroughly, causing great technological, economic, political and social changes in Europe. The Renaissance was just a great ideological revolution whose greatest value was in the realization of a great liberation in people's ideology and spirit. Now, we need a new great ideological revolution that is, a new Renaissance. Just as the great scientist Qian Xuesen said: *"Now what we face is the Second Renaissance in fact which will be a great event. The First Renaissance occurred in the 16th century which broke the fatuity of the medieval time and opened up the road of development to modern science. What I mean is just that there are still many things that are enslaving our mind. Those who are bonded in mind seem very fatuous. What we should do is just to break the bondage, so as to bring about a greater total leap in our understanding and reconstruction of the objective world. Isnt it a second Renaissance?"*

In the 21st century, mankind needs a second Renaissance that is a stormy ideological revolution. Only after the second Renaissance has come can an even greater technological revolution follow.

Technological revolution and ideological revolution are the twin wings for mankind's flight. Only when both wings are used together can mankind soar to the incomparable fine world it is longing for.

V. GREATEST IDEALISM VS GREATEST REALISM

So long as we can dream, we can turn it true.

—mankind's oath engraved on the gate of Kennedy Space Navigation Center, USA

Hawthorne says: *"Ideal is the master of the world."*

For a ship without a definite course, wind from any direction is a dead wind. A person without ideal is just like someone who has lost the way, he is not only in the dark about where to go and what to do tomorrow, but is even unclear about what to do today. Deng Xiaoping says in the "Four Cardinal Requirements for the New Generation", the first is "Have an ideal". It is so for an individual and the same is true of all mankind. He Lin (an educator) points out: *"The reason why mankind is different from animals and the reason why a great person is different from an ordinary person just lies in whether they have an ideal and whether the ideal is high or not."* Thus, it can be seen that the ideal is very important. In mankind's ideal an error as small as the breadth of a single hair can lead you to astray by a thousand miles. It is not important whether a man works hard or not, but what is most harmful is losing the macro direction or going to the opposite direction. For us, a lofty ideal is never a dispensable ornament, but the soul of our life and the moving force of our life. That is why Doctoevckii (Russian writer) said: *"Both in the case of mankind and in the case of a nation without a lofty ideal, it cannot survive."*

Mankind's leap to being immortals is the greatest and loftiest ideal of all the ideals of mankind. This means, all future "humans" will be able to become "gods" enjoying an eternal life, will be omniscient and omnipotent, will be extremely kind and happy, and have absolute freedom.

Becoming "gods" is a dream characteristic of the Chinese nation. The "god of poetry" Li Bai once wrote: *"I long for a five-colored rainbow serving as a long bridge to Heaven. If gods love me, they will raise their hands to invite me".* Not only Li Bai yearned for a god land, but uncountable Chinese were yearning for that obsessively, such as First Emperor of Qin Dynasty, Emperor Wu of Han Dynasty, Second Emperor of Tang Dynasty, Qu Yuan, Cao Zhi, Chen Zi'ang, Li Mi, Li He, Su Shi and so on. The word "god" is an integration of all fine wishes and prospects of mankind. Mankind's greatest sorrow lies in the shortness of life. Everyone is bound to die, and the soul will perish with the flesh. When a person has died, there is nothing left, just a mound of yellow soil at the mercy of rains and storms. But a "god" can live forever, transcending the finiteness and temporality of life, so as to be able to exist eternally, have eternal freedom and eternal joy.

Mao Zedong once wrote: *"Believing a life of two hundred years, and thrashing water about for three thousand miles".* It shows his idealist and romantic mind. But in comparison a "god" not only has a "life of 200 years" but lives forever and is merry forever, which is idealism much greater and more romantic than Mao Zedong's "life of 200 years".

An ideal is a dream, and greatness often results from dreams. Just think if mankind loses its dreams what will the world be like? In viewing the history of science and the history of mankind we could find such a truth. Without dream there would be no scientific development and mankind's progress.

On the gate of the Kennedy Space Navigation Center, USA, there is such a line of mankind's oath: *"So long as we can dream, we can turn it true."* Since Wan Hu's dream to fly in the sky has already been realized, why can we not realize the dream of "becoming gods"?

A romantic dream could generate a great ideal. The ancients' dream about gods can be a great ideal for mankind today. It will be the greatest, most splendid, most glorious and most precious ideal contributed by the Chinese nation to the people of the whole world.

In the 21st century, the most important thing is neither money, nor resources or even technology, but the common ideal that can unite all the people in the world, namely, the ideal for mankinds progress to being immortals. This is the greatest contribution of the Chinese nation to the world, and this contribution is much more valuable than the "Four Great Inventions" that we Chinese are proud of.

It is not enough to have a great ideal only but we must also stand on solid ground. Whether an ideal can be realized or not depends on whether its link to reality is tough or not. Only by closely combining the ideal with the real world can the ideal bear fruit. Thus, while crying up idealism, we should not forget the other pole, realism.

Mankind's progress to being immortals is the greatest ideal in the world and nothing is more perfect than it. Meanwhile it is also deeply rooted in reality, and deeply rooted in the hard soil of science that values reality and practices it the most. In the past, mankind's dream of being immortal was a fantasy and daydream, and was unpractical. Today, for the first time in history, mankind's highly developed modern technology has made it possible for mankind to be immortals. The economic globalization and global integration has made it probable to unite all the forces of mankind for realizing the great common goal of mankind: progressing to be immortals.

Idealism is great, and realism is great too, the combination of both is even greater. The ideal is the other shore and the reality is this shore and only by marrying up ideal and reality we can sail from this shore to the other shore. The late premier Zhou Enlai said: *"An ideal is necessary, as it is our direction of progress. Only when the reality is guided by an ideal it can become promising. On the other hand, an ideal can be realized only in realistic efforts."* Idealism and realism are the two basic forces of mankind's progress. In all activities of mankind, these two forces are restraining each other and acting on each other, but a real great progress results from the combined action of these two great forces. Becoming "gods" together and mankind's progress to immortals is just the maximum combination between idealism and realism.

If we say that becoming "gods" is a great ideal, it is easier to be accepted by people, but it is hard to get it across when we talk about its feasibility. The most common mistake people make is: their eyes are covered up by the past impracticality.

Now, I am going to tell a story and an experiment.

An elephant Lin Wang is a protagonist in the zoo. He is tied hard and fast by a small iron chain to a small cement pole. Lin Wang is both tall and strong weighing several hundred kilograms, but he has been tied tamely through out his life, only moving about the pole, one circle after another. Doesn't he want to see the world beyond this small area? Yes, he wants to do so very much. Is he unable to break off the iron chain? No, he has the strength. Yet Lin Wang does not have that idea. Lin Wang also thought about breaking off, but that was when he was very small. In those days Lin Wang hoped to break off the iron chain very much to have a glimpse of the outer world, as there were more beautiful meadows and a bluer sky. He tried his best to break loose desperately. But he failed. He tried for the second time, third time, tenth time ... N^{th} time to break away from the iron chain, but he failed again and again.

"No, I totally do not have the ability to break away from the chain". Uncountable failures made Lin Wang come to such a definite conclusion. Thus, Lin Wang gave up the idea to break loose from the iron chain. In this way, Lin Wang grew up day by day. So, Lin Wang has never left the small circle of the elephant hall. He has even come to such a belief that his ability is just as limited as he can only enjoy such a small world and spend his life the same way.

As many failures in his young age have printed the idea "I cannot",

"Impossible" firmly in his mind, he thinks he can only spend his life this way. The same is true for mankind. The uncountable failures in the ancient times in the pursuit of an eternal life have printed deeply the idea of "We cannot" and "Impossible" in the social consciousness and collective unconsciousness of mankind. Mankind has already come to the conclusion that humans can only live in this way, and can only have such a short life.

Lin Wang thinks he still has that little ability as in the past, and his wrong estimation of his own power has ruined his freedom and happiness. Today in many problems doesn't mankind also think that its own power is still at the level of the ancients? Totally wrong! Mankind, please open your eyes widely, look at what level your own ability has reached; we have the "clairvoyance (videophone) and the clairaudient (mobile phone), we go up to Heaven (landing on the moon, spacecraft), go down into earth (underground city ocean-floor palace), the Internet enables all mankind to share wisdom without distance, the miraculous robot can walk into the common household, the clone technology has made it possible to produce artificial skin, artificial muscles, artificial bones, artificial internal organs and so on, 96% of the organs of the human body can be changed, the gene technology can design the "perfect" later generations... But do we really know our own power? Have we really been aware of our present power?

Many failures in the past have made us believe that some things are totally "impossible" or "impractical" even now. Just as the "spiritual chain" resulting from the iron chain has "locked the elephant forever", this spiritual chain of "impractical" has bonded us severely, putting us into a great self-restriction. But how many of us have woken up?

Also, there was such an experiment: someone put a cabrilla in a fishbowl. In the fishbowl, there were many tiny fish but a piece of glass was used to separate the cabrilla from the other tiny fish. When the cabrilla was hungry, it tried desperately to catch the tiny fish but it failed each time, and each time, it only bumped against the glass. Hunger forced the cabrilla to strike the glass again and again, but it failed each time. Later, the glass was taken off, the tiny fish swam by the cabrilla all the time, but the cabrilla never wanted to catch them. Thus, this cabrilla actually died in the fishbowl with rich food!

We are not the cabrilla. However, if we do not awaken, we will have the same tragic end as of the cabrilla.

At the mention of becoming gods or living forever, many people will think it is absolutely impossible. Becoming "gods" and living forever have already become tokens of waggery and absurdity. Yes, it was totally impossible in the past. Nevertheless, should we still use the old eye to look upon today?

Today earthshaking changes have taken place in every aspect of our life, that "glass" blocking us for thousands of years is being smashed gradually by the mighty high technology. But mankind is not aware of the disappearance of that "glass". A beautiful great world is just at hand, within our reach.

Everything can be changed. What could not be achieved in ancient times does not represent an impossibility now. What we could not do yesterday may become a reality tomorrow. When mankind uses the past "impossibility" to restrict today's actions, it means we are making the greatest tragedy for ourselves.

The cabrilla died of hunger amid the rich food. It is the greatest tragedy for the cabrilla. Similarly, when we already possess the ability to change our fate completely and evolve into immortals, but we choose to perish, is it not our greatest tragedy?

Milton cried in his *Paradise Lost*: *"Wake up, wake up, or else we shall sink without end!"* Mankind wake up, hurry!

When mankind has really awakened, it will be aware of its direction: it is evolving into immortals. It is a great ideal without comparison and meanwhile it is also a scientific prediction based on the dialectical materialism philosophy characterized by "neither only believing the authorities, nor only believing books but only believing the practice", and resulting from modern science and technology and their development trends. Hence, it is both the greatest idealism and the greatest realism in the world, being a perfect combination between the greatest realism and the greatest idealism in the world, Such a perfect combination is awaiting the wise mankind to struggle for it.

VI. NEW CULTURE, NEW MANKIND AND NEW WORLD

Today, we possess information but do not possess culture. We possess knowledge but do not possess ideals. We possess technology but do not possess goals.

The present era of mankind is just like what Dickens described at the beginning of his book *A Tale of Two Cities*: *"It was the best of times, it was the worst of times, it was the age of wisdom, it was the age of foolishness, it was the epoch of belief, it was the epoch of incredulity, it was the season of light, it was the season of darkness, it was the spring of hope, it was the winter of despair, we had everything before us, we had nothing before us, we were all going direct to Heaven, we were all going direct the other way…"*

Mankind possesses an unprecedented surprising great power. Nevertheless, mankind's fate is still an unknown variable. On one side is the Heaven and the other side is the Hell. We may "go up directly to Heaven", but may also "go down directly to Hell".

In the current world, there are countless contradictions, people are falling apart crises are hidden everywhere and war is going on continuously. In the millennium that has just passed, how many wars were there in our world? Owing to the differences in concept and statistical standard, it is very hard to come to a precise number. US scholar Quincy Wright has a very rough statistic in his *A Study of War*. In the 838 years from AD 1100 to 1938, in Europe alone, there were as many as 4452 "fairly great wars". In the first half of the 20th century, two world wars broke out unprecedentedly. In the First World War, 33 countries with 1.5 billion people took part and in the Second World War, 80 countries with 2 billion people were involved. Finally, the USA used the atomic bomb on Japan, thus, our world has entered an unprecedented terrible "nuclear world", which is a world in which it is possible to destroy all mankind in an instant. After the cold war the economic development worldwide has lost its balance, the gap between the rich and the poor is increasingly expanding , all kinds of sharp conflicts among nations and religions are intensifying, and what is worse is terrorism, this "malignant tumor" has stemmed from the conflicts. Nowadays, terrorism has already developed from kidnapping, assassination, bombing, skyjacking, taking of hostages and small-scaled raids to "human flesh bomb", suicide bombing and the astounding "9\11" event. If things go on like this, not only a peaceful life will no longer exist, but this world will become a horrible hell.

Such a world is never the ideal world we wish for.

Nobody likes to have an old world that will "go down direct to Hell", and nobody is unwilling to choose a new world that may "go up direct to Heaven".

The New World is a uniform world, a harmonious world, a completely peaceful and a highly cooperative world. In other words, the New World is a globally integrated world taking the greatest interest of all mankind as its soul. In the New World we shall not only have global economic integration but also have global cultural integration and global political integration. In the New World all mankind will be of one mind, pull together, work hand in hand with each other, and realize the ideal of "all mankind be one family".

The world is composed of people. Only a new mankind can create a new world. The present world where we are living in is just the result of the present mankind.

What is the present mankind like? Diderot (French philosopher) has such a description: *"Mankind is both strong and weak, both humble and noble. It is able to penetrate minute things but often turns a blind eye to obvious matters."* In many ways, mankind is so powerful that it can turn the whole world upside down; but at the same time, it is so weak that by a mere touch of the finger on the nuclear button, all mankind can be destroyed in an instant. It can be so penetrating in thinking that it can study the minute things in the outer world, but it often turns a blind eye to the greatest common interest within mankind. Such a mankind, of course, only deserves a world of mutual struggle, mutual killing, continuous war and countless violent events.

Only a new mankind will deserve a new world. What will the new mankind be like? It will not be weak in heart, will not be humble in manner, will not turn a blind eye to important matters, but will be much stronger, nobler, and have a greater insight into things. It will be able to see clearly the common interest and long-term interest of mankind, and be able to penetrate the self interest and immediate interest. In other words, the new mankind will be a mankind that will have a clear understanding of the best common interest of all mankind and make all-out efforts for it. It will be the mankind that can create the new world in the best and fastest way.

The new mankind will be just the "new-styled person" as mentioned by Floom, *"I believe that only when a new-styled mankind appears a united world can exist in the real sense."*

Such a new-styled man has "appeared from the age-old bondage of soil and blood, he feels he is the son of mankind, is a citizen of the world... He loves all mankind, and his judgment will not be distorted by his national loyalty."

To cultivate such a new mankind, we can only depend on a new culture. Culture is the bottom most software system composed of mankind's or a nation's world outlook, life philosophy, values, ways of thinking, patterns of behavior and so on. **Culture is the DNA of mankind or a nation. Culture is the second DNA of every person.** The key to the change of human behavior lies in the changing of the ideology. The key to the changing of ideology lies in the changing of the culture. **When the person is upright, the world is upright. When the mind is upright, the person is upright automatically. To have an upright mind, we must have an upright culture.**

Only the appearance of the new mankind can lead to the appearance of a new world. Only the appearance of the new culture can lead to the appearance of the new mankind. So long as you are a human, you share the same humanity, the same demand and same interest with other humans. We are in dire need of construction of a new culture rooted in common humanity and common interests. The new culture that is able to generate the new mankind and create the new world is not the culture of any nation or country, but is a culture that has integrated the pith of the cultures of all nations in the world, and is able to represent the common core values and common best interests of the whole mankind.

Mankind needs such an advanced culture: it has integrated the pith of scientific culture and religious culture, integrated the pith of modern culture and ancient culture, and integrated the pith of occidental culture and oriental culture. The new culture composed of the "human-centered culture", "life culture", "ecological culture", "peace and cooperation culture", "mutual benefit culture" and "Great success (great accomplishment) culture, is such a culture.

When "the killing and conquering culture", "militarism" and "Darwinism" become the mainstream culture, it will produce a war-loving old mankind and a war-stricken old world. On the contrary, only the "new culture of mankind" composed of the "human-centered culture", "life culture", "ecological culture", "peace and cooperation culture", "mutual benefit culture" and "Great success culture (culture of great accomplishment)" can produce a peace-loving and human-loving new mankind, and can generate a new world characterized by harmony, unity and global integration.

Human-centered culture: Among the three levels: individual-centered culture, people-centered culture and mankind-centered culture, mankind-centered culture represents the highest level. In this culture, everything is judged according to mankind's interest and mankind's happiness is the supreme yardstick. Everyone has the conviction that "Everyone's real nationality should be mankind", and that "The greatest happiness of mankind is the foundation of ethic and law-making" (Jeremy Benthan), takes a wide look at the whole mankind, cares about the whole mankind, and has the cultural sentiment of *tianxia zhuyi* (ideal for the whole world)" of the Chinese culture and the saint-like passion of "caring about the whole world". It does not mean giving up one's own nation, but only means that one can, first of all, have the whole mankind in mind in dealing with international relations and all problems. Once we have had such a human-centered culture, we will be able to eradicate the source of war and terrorism, namely ultranationalism, will take mankind's interest as the starting point and ending point in doing everything, and will find the best and greatest way out for mankind.

Life culture: It develops and expands the value of life, helps all mankind set up the idea of respecting, treasuring as well as esteeming life, letting everyone have the idea "One life is more valuable than a solar system", and makes everyone believe that only treasuring life is the most humane choice. Only such a life culture can make it possible to eliminate mankind's old contempt and overlooking of life, and arouse mankind's boundless love for life. With a life culture that treasures life without comparison, we will be able to open up a broad way for the priority development and rapid development of life science, life technology and life economy, and promote world peace profoundly and extensively, so as to build up a new world without war.

Ecological culture: It seeks "Unity of Heaven and Mankind", esteems nature and values the harmonious coexistence and harmonious development of mankind and nature. Protection of the ecological environment is never an issue just for any nation or any region, but the common issue of all mankind. With such an ecological culture, everyone will value the ecological environment, love our common homestead—the earth, regard the protection of the whole earth as our duty, and it will help to cultivate the concept of "We are in the same boat, so we must help each other", which may not only prevent the destruction of the environment by war, but also promote the alliance of mankind.

Peace and cooperation culture: It helps to bring about reconciliation, peace, harmony and unison between individuals, nations, mankind and nature. Pacification with nature and within the same species is a call of the time, and an urgent need of the time. The peace and cooperation culture helps all people to unite and cooperate, even between conflicting and contradicting parties. The peace and cooperation culture will greatly promote internationalization, globalize great alliance and great merging, which is just what the New World needs the most.

Mutual benefit culture: It stresses the wisdom of "mutual benefit", seeks mankind's common interest, keeps a foothold in the common interest, seeks win-win relationships, and pursues the prospect of "Everyone is happy". The mutual benefit culture stresses the combination between self-interest and others' interest so as to benefit all people. The mutual benefit culture stresses that the common interest is higher than everything, helps to change conflicts of interest into unity and harmony of interest, so as to help the world tend to unity and unison.

Great success culture: It implies the spirit to strive and excel by intensive effort, and implies the spirit to try to make continuous progress and achieve sustainable development. The sense of great success means the sense of becoming top-grade and outstanding, and the sense of seeking satisfaction and perfection. Seeking the top great success means seeking great achievement together, and helping the whole mankind to make great achievements. The common great achievement of mankind will be their becoming immortals together. Human life span jumping to infinite life from the transient ordinary life. Human development going towards full-range genius-leveled development from the ordinary overall development. That is the full manifestation of the pursuit of excellence and perfection characterized by the Great success culture. With such a Great success culture mankind will cultivate a broad mind, not be bewildered by narrow and shallow interests, but will pull together and struggle together, so as to reach the consummate prospect of mankind together.

In short, we must use the new culture in place of the old culture in order to replace the old world with a new world and the old mankind with a new mankind. All the above-mentioned six cultures belong to the new culture promoting the unity of the world and global integration, and all of them will help to create a new world, that is; a uniform, harmonious, peaceful and cooperative world. Only by the extensive spreading of these six cultures can we produce a new mankind and walk up to the New World.

Here, what should be stressed specially is: Culture is never knowledge. Knowledge is just the surface color of the soul, whereas culture can penetrate your soul, melt into your blood, and get down to your marrow. If mankind can really possess the advanced human culture that is composed of the six major cultures, then the present mankind may become the brand-new mankind and build up the brand-new world.

The close relationship of the new culture, new mankind and new world is illustrated as the following picture:

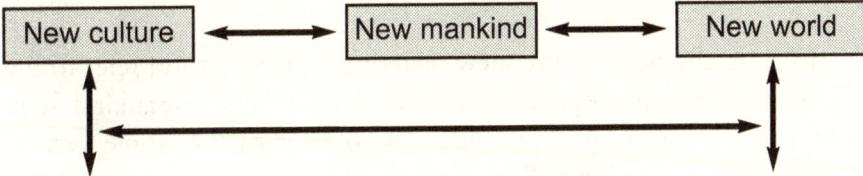

On one hand, the new culture brings up the new mankind, the new mankind builds up the new world, and the new world boosts the growth and development of the new culture.

On the other hand, the new world helps the growth of the new mankind, the new mankind can create and develop the new culture, and the new culture can guide and promote the new world directly.

The new culture, new mankind and new world together constitute a benign circulatory system of interaction and inter-promotion.

Today, we possess information but do not possess culture. We possess knowledge but do not possess ideals. We possess technology but do not possess goals. We are still uneasy, for—the world's science and technology developing too fast. We are still anxious, for—we worry that our renewal of concepts is too slow. We are still impetuous, for—there are too many chances before mankind. We are still thinking, for—we are confident that the fate is controlled by mankind itself, forand we are longing for the birth of a fine new mankind, and longing as well as for a fine new world.

The eagle flies lower than the chicken sometimes, but the chicken can never fly as high as the eagle. Mankind sometimes flies as low as the chicken, or even crawls like the cockroach, but once mankind is awake, it will soar up as high as an eagle.

VII. NEW MANKIND, HIGH MANKIND AND SUPER MANKIND

The day when God created man, He regretted. Mankind is the first creature that tries to change its own evolutionary process.

If we say the new culture, new mankind and new world refer to the whole horizontal development of mankind, then the new mankind, high mankind and super mankind (immortals) refer to the whole vertical development of mankind. The coincidence point of the horizontal and vertical coordinates of human development is the new mankind (as is shown in the Fig.), that is, the new mankind is both the center of the horizontal development of mankind and the paramount key point of the three-step vertical development of mankind.

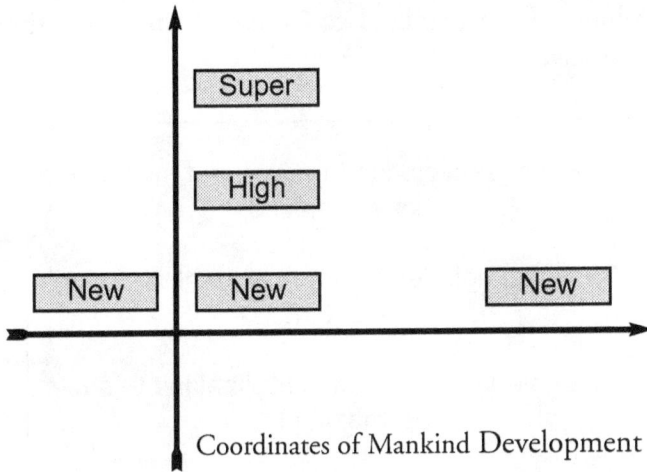

Coordinates of Mankind Development

In retrospect, we have found that mankind has already experienced a quite long stagnant state in it's evolution history. Scientists also think that the impact of natural evolution on mankind is weakening. In the past, the evolution from apes to mankind was a result of natural selection, chiefly because of the pressure from the natural environment. Now, if mankind wants to have a second mutation or leap, it must have the ability to dominate its own evolution and design its own evolution.

Everything depends on the design. When the design is good, the space shuttle can fly to the sky. If the design is not good, even the children's plane in the park may drop, causing an "air crash". The terrorists' skyjacking succeeded because the terrorists' design of the skyjacking was better than the security design of the airline company. The counter-terror experts succeeded in their anti-skyjacking action, because the counter-terror design was better than the skyjacking design. Whether mankind can evolve to the supreme prospect at the highest rate also depends on mankind's wise design characterized by carefully arranged steps or stages.

By one step we cannot reach Heaven. But if we keep a down-to-earth spirit, put up the Heaven ladder well, and go up step by step, we may reach Heaven at last. Mankind's evolution to immortals is not easy matter, so it must take some steps and needs a practical spirit. We should not only stride towards that ultimate objective, but also think over the whole process and design the practical steps.

The evolution from mankind to immortals may take three major steps or three stages—

> First stage: evolution from present mankind to new mankind.
>
> Second stage: leap from new mankind to high mankind.
>
> Third stage: flight from high mankind to super mankind.

Present mankind—new mankind—high mankind—super mankind. The starting point is the present mankind (mankind in today's reality) and the end point is super mankind (immortals).

A journey of a thousand miles begins with a single step. The most important is the first step. Whatever difficulty we meet, so long as we can take the first step that seems to be very reasonable indeed, we will find that what follows will be very easy, whether to determine the second step or actually take the second step. The leap from the present mankind to the new mankind is the crucial first step in becoming immortal. This step may mean the difference between success and failure. With the first step taken, we will be able to carry all before one.

Then, what is the new mankind? **The new mankind is a mankind that sincerely believes in the supremacy of mankind's common interest, believes in the top value of life, and advocates harmony between nature and man and cooperation among all people.**

Today's mankind is often confined to the narrow national interest or the interest of any small group, without the notion and consciousness of the great value of mankind's common interest. That is why some rich countries shift off their environmental crisis to backward countries. On one hand, they stress their vanguard role in environmental protection, but on the other hand, they transport the poisonous garbage to those countries that have no processing capacity for it. It is just because of the mere consideration of their country's interest that such things happen.

Regardless of the heavy opposition of the international community, President Bush goes his own way, continues to carry out the NMD and refuses to sign the *Kyoto Protocol* for protection of global environment. It is just because of the absence of the notion of the top value of mankind's common interest, that the USA posing itself as a democratic country only stresses democracy within its own country but runs counter to democracy in the international community. But the new mankind means that everyone in the world can "go beyond the old tradition of narrow loyalty to the tribe or country, and be loyal to all mankind".

In today's mankind, their low-esteemed and overlooked value of life is astounding, especially in our great populous country educated by the ancient tenet of "dying to achieve virtue". On the early morning of June 23, 2005, a 40-year-old farmer migrant worker from Gansu Province, Wang Guoqing, was detained by three law enforcement officers. They asked him to pay 10 Yuan as the water charge, but he did not have money with him, so the officers began to strike and kick him. What was more heart-striking is, when several residents of Wang Guoqing went to the municipal government to complain, the desk clerk said: "Don't damage our reputation". Because of such an indifference, because of lack of money at that time, and because of no timely treatment, Wang Guoqing died the next day. Just for the sake of 10-Yuan water charge, a vigorous life was lost!

Yao Li, a woman clerk in Daqing Branch of Construction Bank of China, was attacked by two armed gangsters. Her raising an alarm got no responce. So in order to avoid all the money from being robbed away, she was forced to hand out more than 10,000 Yuan to the gangsters. Although she made up for the loss with her own money immediately, her working institution dismissed her from her party membership and her public position. In the eyes of Daqing Branch of Construction Bank, a precious life = 13568.46 Yuan RMB. Her institution had good reasons: The national interest is higher than anything else. In a crucial moment how can you give up the money to the gangsters? In a crucial moment, Yao Li, as a communist, should have offered up everything of hers, even the most precious life. Is the 10,000-Yuan money more valuable than the life of a person? For the sake of the 10,000 Yuan, a weak woman should have used her flesh and blood to withstand the 5-pound hammer, should have died a martyr, rather than giving up a penny of the country's property. Only that can be considered as "great" and "noble". But she did not "have a life-and-death struggle" with the gangsters, so she had to lose her party membership and even her job, so she should be cursed as "coward" and "renegade".

In foreign mass media, we often see moving reports on stories of self saving, such as a child escaped from a disaster, an old woman saved herself in a disaster, and all those who saved themselves successfully in distress are regarded as heroes and praised. However, in our country's mass media, laudatory words are only given to those who save others, whereas successful self-savers are never regarded as heroes. The reason why our public opinion does not praise a person who saved himself successfully is, only that self saving is just for keeping the individual's own life, but does not have such noble significances as protection of the collective property, maintenance of the national interest and so on. Hasn't our culture been overlooking the life of a person? Under all kinds of beautiful banners, we are willfully trampling, humiliating or even throttling life. Isn't it the most lamentable?

The root of mankind's war and all tragedies is just the fact that we lack the values of "Life Is Supreme". "Life Is Supreme" does not mean only "my" life is the most precious, but any living person's life is equally precious, and everyone's life should be respected and treasured. Only when all mankind regards "Life Is Supreme" as the highest guideline can we exterminate war, the evil demon.

In the meantime, deterioration of the ecological environment and deterioration of the international relations are also the greatest calamities confronted by mankind. Only the new mankind armed with the ecological culture of "harmony between nature and man" and the peace and cooperation culture characterized by "harmony among all people" can eliminate the greatest problem by the root, and clear up the disastrous consequences mankind may have to suffer.

The new mankind not only believes in the greatest value of mankind and life sincerely as well as advocates harmony between nature and man and among all people, but also they are new people who meet the four cardinal requirements: "having ideals, having a high moral standard, having a good culture and having great abilities".

"Having ideals" means having the ideal to make mankind happiest, having the ideal of "Mankind will certainly evolve into immortals" and having the ideal to realize full globalization. The present globalization is only a single-sided globalization, a globalization of the developed countries, a globalization of multinational corporations, and a globalization of rich only; but what we need is a full globalization, that is, a globalization of all countries and all people, and a globalization that may bring the greatest happiness to all mankind. That is the "ideal" of the new mankind.

"Having a high moral standard" means having the most universal moralities of mankind, "love" and "righteousness". The most essential power of morality comes from love. As the saying goes, *The greatest secret of morality is love*" (Shelley). The new mankind should love themselves, love others, and have a universal love. Also, they should be just, kind, and faithful. Those may become mankind's universal moral virtues.

"Having a good culture" means having the six major cultures of mankind's new culture, namely, human-centered culture, ecological culture, life culture, peace and cooperation culture, mutual benefit culture, and Great success culture. Culture means a thorough breeding from tip to toe and from inside to outside. People with such a culture will be far better than those who are only literate, but will be the really lettered people in the new era, that is, people needed for realization of the ideal of "becoming immortal together".

"Having abilities" means having the abilities needed for realization of the ideal of "becoming immortal together". Of all kinds of panic, the most terrible is ability panic. In the book *Ability Panic*, I listed ten major abilities in the "Ability Menu", and all of them are necessary for the new mankind, especially the core abilities: learning ability, thinking ability and innovating ability. The new mankind should be far stronger than the present mankind especially in these core abilities, that is, they should have a super learning ability, a super thinking ability and a super innovating ability.

The period from the present mankind to the new mankind is the key stage and key step in the evolution from mankind to immortals. To achieve a decisive victory in this stage, we must rely on the new culture and new education, we must rely on cultural innovation and educational innovation, and we must rely on cultural revolution and educational revolution.

We need a high-efficiency globalized education system to turn out the new mankind that "sincerely believes in the top value of mankind and life, and advocates harmony between nature and man and among all people", and is characterized by the four requirements of "having ideals, having a high moral standard, having a good culture and having great abilities".

What is the high mankind? The high mankind is a mankind with high qualities in all aspects, a mankind that has attained a high level in all aspects, such as health, intelligence, ability, virtue, beauty, happiness and so on. To realize the leap from the new mankind to the high mankind, we do not only need a high and new culture, and a high and new education, but also need the power of high and new science and technology. In "Part Three" afterwards, we shall detail "birth optimization through technology", "education optimization through technology" and "health optimization through technology", so as to use technology to optimize mankind in all aspects, raise mankind's integral quality, and enable the high mankind to possess the high qualities needed for the leap to the super mankind.

The super mankind is the race of immortals, being the ultimate objective we are trying to reach. The super mankind also falls into three stages—the initial super mankind, the medium super mankind and the high-leveled super mankind. The initial super mankind's life will reach 1000 years. The medium super mankind's life will reach 10,000 years. The high-leveled super mankind will jump to eternal life, living as long as Heaven. Out of these three stages the most important is to realize the dream of the initial stage, that means everyone can live up to 1000 years, and can have corresponding development in the other respects. Scientists predict, this can be realized before 2030, and this prophesy is based on the progress of gene deciphering. If we can improve mankind's quality from now on, make efforts consciously towards this goal, then we can not only ensure to realize this dream in 2030, but also possibly achieve this goal much earlier. So long as we can guarantee to reach the stage of the initial super mankind, that is, making everyone live up to the age of one thousand, it will not be hard to realize the dreams about the medium super mankind and the high-leveled super mankind. The reason is like this: so long as everyone can ensure a life of 1,000 years, need we be afraid that we cannot attain the medium super mankind goal within a thousand years? Maybe, it will not need a thousand years, only need 100 years, or 200 years... Then, mankind will be more confident and daring in the evolution into the high-leveled super mankind of eternal life.

Scientists have told us long ago: *"Just on the day when God created man, HE regretted. Mankind is the first creature that tries to change its own evolutionary process."* Indeed! Mankind has already possessed the ability to extend its own lifetime and change its own fate and track. So long as all mankind unites as one, help each other in the same boat, we shall certainly progress from the present mankind to the new mankind, from the new mankind to the high mankind, and from the high mankind to the super mankind—immortals!

PART II - SUBVERSION OF TRADITION AND FETAL MOVEMENT OF FUTURE

夢星

INTRODUCTION

Today is Wen Wen's 321st birthday and many of her friends have come from various stars to congratulate her.

Gaoqiao Xue has come from Mars, and just three minutes ago, she was consulting data in Damin History Library, she is quite satisfied with the speed of the interstellar instant transport. Ciolla has sent a 10,000-year-old coral reef to Wen Wen from the undersea apartment house of Hawaii, Xiao Feng, who is living in the Himalayas, has also come out of the cave palace to see her. Fresel from Paris, took a private airship and used 15 minutes to reach Wen Wen's home on Mercury. A group of friends have gathered and the atmosphere is very hot. The chief reason for them to become friends is they have the same interest—study of ancient relics.

"I have just browsed the history around the year 2000. Please guess: In those days, what was the hottest topic in their mass media?"

Wen Wen smiled, with her arch eyebrows sticking up: "Little Xue, don't keep us guessing, just tell us!"

The 267-year-old Gaoqiao Xue is a novelist and all her writings are based on the ancient history. Although she can consult the historical accounts of hundreds of years ago on the computer, she still likes browsing the newspapers of hundreds of years ago in Damin Library, which are all out of print now as they have stopped using printed matter for several centuries. In the early morning, after drinking the cocoa, Gaoqiao Xue opened the screen on the wall as a rule, and clicked for the interesting "news" around the year 2000. What appeared most frequently should be such a term: "jiedi lian" (love between an older female and a younger male).

Wang Fei and Xie Tingfeng (famous singers) got into the crisis of "jiedi lian"...

Can our "jiedi lian" last long?...

Opinions vary about "jiedi lian"...

Some refuse "jiedi lian"......

Oh! it turns out that, although people around the year 2000 were much freer than those in the first few centuries AD, in love affairs, yet, most of them still felt "jiedi lian" was against common sense, and they could not accept this kind of love between a younger man and an older woman.

"Wang Fei was only 12 years older than Xie Tingfeng." Gaoqiao Xue talks to himself, "Can it be called 'news'? People of those days were too laughable. If they knew my husband is 120 years younger than me, would they fall, and faint? But it is just as ordinary as our going undersea to take a holiday every week." At the thought of yesterday's story that a 17-year-old little boy showed love to her although she is 267 years old now, she cannot help smiling, which makes her gem-like face more charming. Then, she finds a piece of "news" on the eve of the year 2000: in order to celebrate the beginning of the millennium, people held a beauty contest, and a 60-year-old Sofia Roland came out first, as a "grandma-leveled belle". She gives a laugh disapprovingly. How can a 60-year-old Sofia Roland be called a "grandma-leveled belle"? Judging from the photo, she looked so old and haggard! Should such a woman be called a "belle"? Then, what about me? Haha, I would be a belle in the cubic level of a grandma?

At Wen Wen's birthday party for the age of 321, Gaoqiao Xue can not wait to share these thoughts of hers with her good friends.

"Hey, do you believe? Most people around the year 2000 did not think high of 'jiedi lian', especially when the age difference is above 5 years!" Gaoqiao Xue raised her voice in excitement.

"Only a 5-year difference or above? Nothing to speak of."

"But they did not think so. In those days, if a man was old, so long as he had power and money, his marriage with a young woman was accepted by the society, but what about a woman? If BF was a chap younger than her, even if she was in her peak years, even if it was out of a true love, there would surely be a young and pretty girl to replace her position in future." Gaoqiao Xue was indignant in saying so.

"Youth is gone in an instant, and beauty is lost in a moment!" Xiao Feng chimes in: "I heard this was the sentence most often uttered by women with a deep sigh in those days."

"What a pity! So to speak, countless beautiful girls were disfeatured by the sharp knife of time. But in those days there was indeed no force to save them." Gaoqiao Xue, Wen Min, Qiao Er and Lemon get into meditation. They sigh at the fate of the past women and also feel so happy about the present people who can not only enjoy a long life but also keep young forever. Among them, the oldest is the 667-year-old Qiao Er, and among those admiring Qiao Er are persons ranking from 19 to 37, 57, 97, 177, 277, 477 and 877 in age...

"All right, Wen Wen, do you have any wish to be realized today?" Fresel changes the topic.

"Nothing special. I only want to change a look and a shape today. I have already designed a look for myself for the coming year. The other day, I submitted my letter of application to the Physiognomy Bureau, and already got their approval. OK, I will have a change right now. Please wait a moment."

When Wen Wen reappears before her friends, her blond hair has already been turned into soft and brilliant black hair like silk, flowing down just like a black waterfall; her angelical cheek, elegant legs... Wen

Wen's new look shocks everyone present there.

Gaoqiao Xue cannot help crying: "My God! Our 321-year-old Wen Wen has surpassed their youngest and pretties star in the year 2000. Who will say 'Beauty is gone in a moment'? We ought to tell them, all mankind is completely able to—

Keep beauty forever,

Keep young forever.

VIII. QUESTIONING LIFE

When the shadow of Azrael approach, All people in a sudden retrospect will find what they have been striving for in all his life is worthless, and only the life that can never be recovered is so worth treasuring!

Old mankind has repeated the same fate for thousands upon thousands and millions upon millions of times. Fresh lives have turned into bones of the dead one after another. All thinkers, philosophers, statesmen, scientists, entrepreneurs, artists... No matter how great they were, none could escape from such a tragedy. It reminds me of the deploring of Wang Xizhi, the great calligrapher of the Jin Dynasty, in his "Preface of Collection of Writings at the Orchid Pavilion": *"Great indeed is the issue of life and death. Ah! How sad it is!"* Whenever I think here, I can find no expression for the grief in my heart.

Nevertheless, as far as the man in the street is concerned, life and death has become a common sense and they take it for granted.

People often turn a blind eye to what is taken for granted, even if it is absolutely important. Just like air, people have taken air for granted so they often forget it. People have taken the short life for granted, so they often forget to think about life.

Everything in the world goes around life, and everything in the world originates from life. Mr. Lin Yutang (a famous scholar) said: *"Science is nothing more than the curiosity about life, religion is the veneration of life, literature is the praise and deploring of life, art is the appreciation of life..."*

If we make a philosophical thinking on life, we will find the triple absoluteness of life; absolutely essential, absolutely once, and absolutely precious.

Absolute Essentiality of Life

Life is the fundamental precondition of existence of all living creatures. "Since it is a living creature, its first important matter is, of course, the life. The reason for being a living creature just lies in the life. Otherwise, it will lose the meaning of the living creature."(Lu Xun) Human beings, as a living creature, is no exception, no matter how special he is. A human being can be called a human only when his life exists. In the existence of a human race in the world, the absolutely principal question is how to maintain his (her) life.

The absolute essentiality of life means—with the life, there is everything; without the life, everything is lost. As the old saying goes: *"Where there is life, there is hope."* So long as the life is existent, there is a hope to get anything that life can produce. Any need can be met, any loss can be recovered, and even a failure may become a success. On the contrary, death is the greatest misfortune, for it not only destroys all hopes, but also makes anything already gained lose its meaning. With life lost, even if you have money, power, love, friendship or anything, all will be lost together with it.

Although the ancients realized long ago that "Wealth and properties are external things besides the body", yet, among the people today, there are still numerous people who "risk their life" for these external things. When they have obtained the external things besides the body, such as money and power, they have completely lost their capital, their own life! A rich man becomes the Chairman of the Board of Directors, he has assets worth millions of dollars, he has risen from abject poverty, because of "desperate struggle" for more than one decade. But now he is lying on the sickbed, on the verge of leaving this world. When some young adorers visited him, he said: "I would rather be a pauper now. If I could exchange, I would use my family property worth millions of dollars in exchange for life..."

In this world, some are cudgeling their brains for money, some are bucking bitterly for power and some are shedding tears and blood for position... Money, power, position and life, which is the most important? Once life perishes, anything attached to it, such as wealth, power, status, honor, love and so on, will disappear instantly like water bubbles.

All fine things in human life have their meaning only when they are bound to life. Even the greatest statesman, greatest scientist and greatest thinker—for example Einstein: when we talk about him now, he is only a name, or symbol and it is meaningless to the Einstein when alive and kicking.

Human development must be an overall development. An overall development chiefly includes three aspects namely—morality, intelligence and body; the "body" is the most fundamental. Mao Zedong hit the point with one sentence: *"Without the body, there will be no morality and intelligence."* Non-existence of the body equals non-existence of the life, and all things will return to nothingness. However high a person is in morality and intelligence, when he dies, all his morality and intelligence will be lost completely with it. Thus, we can see the absolute essentiality of life.

Judging from the micro aspect, that is, as far as the individual is concerned, life is of absolute essentiality. Judging from the macro aspect, that is, as far as the society is concerned, life is of absolute essentiality, too. Marx put it in this way: *"In any human history, the first precondition is undoubtedly the existence of living individuals."* The first precondition is naturally the most essential precondition, being of absolute essentiality. Any human history or any human society must take "the existence of living individuals" as the precondition absolutely. As a matter of fact any individual has his existence in life; once the life is lost, the individual will no longer exist. Hence, in the whole society or the whole world the most fundamental and absolutely essential thing is only life, that is, everyone's life. Speaking from the most basic point, what makes the human society go on endlessly and go up vigorously is neither power, nor wealth, or productive force, but life, everyone's life!

Life's absolute essentiality is life's first essential attribute.

Absolute "Once" Character of Life

Ostrowski (Russian writer) has a widely spread, well-known saying: *"For a man, the most precious thing is life. Life belongs to man only once."* As for the "once" character of life, the Chinese were aware of it as early as in the ancient times.

"Everyone has got one life and it cannot be regenerated" (*Taiping Jing*). *"After death, there is no regeneration"* (Ruan Ji, 210-263, thinker and writer of the Three Kingdoms Period). Life's "non-regeneration" is absolute and there is no exception for anyone, at all times and in all countries. Life is extremely precious, just because it can never be regenerated.

Christianity has created an illusory Paradise, and designed a world on the other shore, offering people a "check that can never be cashed". Buddhism has fabricated the theory of "Metempsychosis (transmigration)", making people believe that life is not "this time" or "this generation", letting them fancy "the great Beyond" or "eternal life after death". In fact, Socrates could have been exempted from punishment or run away if he had paid the fine, but he was willing to die, because he believed after death he could still discuss things, and could still see those he admired, such as Homer, Thales. The samurais in Japan wielded their broadsword to paunch themselves "heroically", because they thought they could transmigrate and be great warriors again...

All of them have been proved as untrue and ridiculous by today's science. China's ancient thinkers, especially Taoists, got the profound understanding of the "once" nature of life long ago.

The first classic book of Taoism, *Taiping Jing*, stated clearly— *"The death of anyone under the sun is no small matter. After death, he would no longer be able to see the world again, and his veins and bones would become soil. Death is the greatest loss. Everyone lives between heaven and earth, everyone gets only one life, and no life can be recovered."*

They affirmed the "once" nature of life completely: To everyone, death means the end of all, the total destruction, eternal Finis and eternal ruination. Just because the Chinese nation has a more profound understanding of the absolute "once" character of life, the Chinese ancients showed an exceptional thirst for "longevity" and "eternal life".

Only when you have realized the absolute "once" character of life and realized that death signifies a sinking into the abyss beyond redemption, can you treasure life without comparison. Only when you have known profoundly "All things will vanish after death" can you really understand the core essence of life.

An old saying is well said: *"Both Confucius and Robber Zhi have become mote"*. Confucius was the greatest mahatma while Robber Zhi was the most notorious robber. Whether a mahatma or a notorious robber, neither could escape death, and both turned to dust and ashes. Hence, death is mankind's common enemy. Only by a thorough understanding of death can we have a profound understanding of life. Let everyone of us remember firmly—Life is only once, without return! Our common greatest enemy is none other than Death!

Absolute Preciousness of Life

Life's absolute essentiality and absolute "once" character determine the absolute preciousness of life.

There is such a story—Someone ran across a successful person and complained about his abject poverty to the latter and asked for advice.

The successful man only smiled, and then asked him: "If I use ten thousand dollars to buy a hand of yours, will you agree?"

"If I lack one hand, how can I work?" He shook his head.

"Now, what if I offer a hundred thousand dollars to buy one foot of yours?"

"A thousand-mile journey starts with the first step of my feet. I can get one hundred thousand dollars all right, but my journey from now on will be a terrible zero!" The successful man nodded the head, then, in a way of bargaining in the marketplace, he said: "Then, I offer one million dollars to buy your head, OK?"

The man said sulkily: "Without the head, there will be no life. Without life, what use will money have? Even if your money can keep my life, I will lose my thinking and lose my creative power."

"OK," the successful man said gravely, "You possess sturdy hands and feet, a flexible brain, and a healthy life. That shows you own at least one million dollars of wealth. How can you say you have nothing?"

Of all things, the most precious is a human being and in a human being the most precious is the life. For a nation, the most precious thing is not the great deal of gold or foreign exchange it has reserved, nor its land or other resources, but the health of its people and the life of its people. Americans seek "zero injury and death" so, for the sake of zero injury and death, they carry out a "luxurious war", which is not worthwhile in the eyes of some Chinese. A Chinese senior colonel mocked at the seemingly business-minded American army in his *Ultralimit War*, saying they should put tonnages of dollars for the "zero injury and death", US soldiers were as precious as jade porcelain vases that were very fragile. He questioned, "Whether such a practice is worth while". Just because of the tenet to save the life of an ordinary soldier, we have the film *Saving Private Ryan* directed by American film director Spielberg. They send a team of soldiers to bring back a son to his heartbroken mother. It could be possible only when we have understood the absolute preciousness of a person's life.

A person's life is the person's total most essential interest. People's life is the people's total most essential interest. Treasuring life limitlessly is absolutely necessary. Treasuring life the most is the greatest virtue while treasuring life the least is the greatest evil.

In education, the most precious is life education; in science, the most precious is life science; in culture, the most precious is life culture.

Those who struggle for mankind's life, for the prolonging and well-being of mankind's life, are the most real, the most kind and the most beautiful people, and their contributions are the greatest contributions, so, they will win the greatest respect for certain.

There is nothing more important than human life, and life is priceless. This is a common sense. The greatest truth is often the most ordinary common sense, the plainest truth. Many people and many nations often neglect it or overlook it unknowingly because this truth is the plainest and commonest. Montaigne (French thinker) said: *"Of all the weak points we have, the rudest one is the contempt of our survival."* Since human civilization has progressed to the 21st century, we should overcome this rudest weak point resolutely and completely.

Life is priceless. This is a common sense, a great common sense our society and our world should highlight as much as possible. In the present time, characterized by high technology, high risks and high fickleness, if we are still muddleheaded about this most fatal "common sense", there will be no reason to excuse in any case.

IX. REVALUATION OF LONGEVITY

Time is the domain for ability and all development.

—Marx

In talking about longevity, people often think it is a matter that should be considered and cared about only by the "hoary-haired generation". And as a matter of fact—usually, those who are interested in longevity are also old people in the evening of their life. As they do not have many days left, their concern for longevity is just for getting more days in their natural span of life. For ordinary younglings, only when they have reached old age do they begin to have some sense of longevity.

The chief reason why ordinary people lack the sense of longevity is that they do not quite understand the great value of longevity. In the eyes of ordinary people, longevity is nothing but a longer period of old age, but in fact, the value of longevity is far from that.

Judging from many aspects, the value of longevity is extremely great and inestimable.

First, judging from human's need. Engels pointed out long ago—"*A person first needs survival, secondly needs enjoyment and thirdly needs development*". A person must try to survive, or else nothing can be talked about. Second, he needs enjoyment and should not be content with survival only. Third, there must be development; development of character, development of intelligence and development of capability... Longevity is the essentially decisive factor determining the three general needs. Life span is the individual's survival time. Survival is a person's general fundamental need; the level of longevity concerns directly how much this fundamental need has been met. Once life ends, the right of existence will be lost completely. Longevity also concerns the right of enjoyment. The longer your life is, the more you can enjoy life. The length of the life also determines the degree of enjoyment of the right of development. Fundamentally, both development of the person's own quality and development of his career depend on the length of life. Especially in an era characterized by life-long education and life-long learning, the length of the life is proportional to the level of ability and level of quality. The longer the life is, the higher the final ability and personal quality will be.

It is obvious that a long life determines the degree of a person's right to existence, his right to enjoyment and his right to development. It determines the degree of satisfaction of the three major needs, namely survival, enjoyment and development. It directly concerns his achievement in academic career, occupation career and everything in life. So we say that longevity is the total and general manifestation of the people's fundamental interest, and the per capita life is the highest indicator in judging the degree of realization of the whole fundamental interest of the people.

We can say longevity has a great value without comparison by judging not only from human need but also from great achivement. In my books *Secrets of Great Achievements* and *Ability Panic,* I mentioned the "Dacheng" (great achievement) concept many times. Dacheng refers to all-round great achievement namely—great achievement in academic career, great achievement in occupational career and great achievement in life. In all these three major aspects, longevity plays the most essential and decisive role.

Longevity and Great Achievement in Academic Career

Great achievement in academic career means great accomplishment in knowledge, ability and general quality. Confucius said: *"At fifteen, I had my mind bent on learning. At thirty, I stood firm. At forty, I had no doubts. At fifty, I knew the decrees of Heaven. At sixty, my ear was an obedient organ for the reception of truth. At seventy, I could follow what my heart desired without transgressing what was right."* Everything goes up along with the growth of age, scaling constantly towards the supreme prospect. Of course, the relationship between age and quality has no absolute proportion. However, through continual learning and continual training, age will be certainly proportional to the level of ability and overall quality. Marx says: *"Time is the domain for ability and all development"*. Generally speaking: the longer the life is, the more possible it is to raise the general quality of life and thus, the more probable it is to have great achievements. Even a mahatma like Confucius could reach the level of doing things freely without going against the rule only when he reached the age of seventy.

Judging from another point of view, it is also a very inspiring fact that at seventy, Confucius could follow what his heart desired, without transgressing what was right, then, if we train ourselves for another two, three or more "seventy years", will everyone be able to become a mahatma? Will everyone be able to reach the level of "following what his heart desires, without transgressing what is right"? If we live healthily up to 200 years, 500 years, or 1200 years old, do we need to worry that we cannot get 5 doctoral degree, 10 doctoral degrees, or even 20 doctoral degrees?

Longevity and Great Achievement in Career

Life span is the total amount of lifetime and working time an individual enjoys. Time is power and time signifies victory. By winning time we can get victory. Whatever undertaking you are involved in, time is the fundamental determinative factor of achievement. A scholar once said with emotion: *"A researcher could live in a slum, eat coarse meals, wear rags, could deny recognition from the society. Yet, as long as he has time, he could continue to devote himself to scientific research. Once he is deprived of his free time, he would be ruined completely, and will no longer make contributions to knowledge."* The longer a life is, the longer the time for tackling key problems will be, the more possible it will be to make outstanding contributions and achieve great accomplishments in career.

Since ancient times, countless heroes have bowed for longevity or gasped in admiration of gods. It seems to be very strange, but in fact it is not strange at all. The more aspiring a great person is, the deeper he will feel about the transience of life, thus, the stronger his fear about the perishing of life will be, and the greater his anxiety about time will be. That's why some great people with heroic spirit as high as heaven, such as Qin Shihuang, Emperor Wu of Han Dynasty, Cao Cao and so on, had such a great desire for eternal life. And that's why Mao Zedong uttered his bold manifesto *"Believing a life of 200 years, and thrashing water about for three thousand miles".*

Longevity can make greatness even greater and make splendor even more splendid. The most splendid achievement of Deng Xiaoping, the general designer of China's reform, was made after his venerable age of 73. Without his long life, there would be no brilliant achievement of today, and even a tragedy like Zhuge Liang's "death before the victory" might have occurred.

Longevity and Great Achievement in Life

Great achievement in life means getting the greatest happiness, and it refers to the total amount of the greatest happiness in one life.

In calculation of a lifetime's total amount of happiness, we may use the following formula—

Total amount of happiness in a lifetime (magnitude of happiness in a lifetime) = average amount of happiness each year × life span

When the annual amount of happiness is certain, the length of the life span will be the only multiplier determining the total amount of happiness in one lifetime, that is, it is the only determinative factor of the magnitude of happiness in one life.

The ancients of our country already had a clear understanding of the truth that life span is the main factor of the magnitude of happiness. China's earliest Confucianist classic *Shang Shu* says: *"The five aspects of happiness—The first is longevity, the second is wealth, the third is health, the fourth is virtue and the fifth is living up to the heavenly mission."* In the ancients' thinking, "longevity" is the first in the "five aspects of happiness".

Since ancient times, people of our country have always used "Double Felicity of Happiness and Longevity" to describe the highest level of great achievements in life. They have often used "Happiness as boundless as the East Sea and Longevity as high as the South Mountain" to describe the highest goal in life. Double felicity of happiness and longevity means in our whole life we should lay equal stress on longevity and happiness. Thus we can see the close relationship between longevity and the happiness of the whole life, and the high position of longevity in great achievement of life.

Time determines everything. The longevity problem is a matter of the total time of a life, that is, the length of the total survival time, enjoyment time and development time in a life. Ordinary people lack a strategic mind, have a short sight—only know the short time at hand, but do not value the long time of the whole life. If it is not the greatest folly, at least, it is the greatest impolicy.

Feng Zikai (famous painter) said: *"Usually, people's comprehension of time seems to be only enough to control a short period such as taking a ship or a bus. With regard to the centenary long-term life span, they are unable to control it, as they are often bewildered by the part, neglecting the whole."* It is really the common fault and a serious fault of ordinary people. We ought to advocate cherishing time like gold all right, but in cherishing time, the more important thing is cherishing life. Thinking little of longevity and making no efforts to seek longevity are the same as not treasuring life, the same as not treasuring time in the real sense, at least, mean those people do not know how to treasure time on the whole. In this way, they can never get the greatest achievements in academic career, work and life.

A long life can expand a person's right of existence, right of development and right of enjoyment limitlessly, and can also make it possible for him to achieve the greatest success in academic career, jobs and life.

When someone asked Buffett, the famous US investor, "Once you become the richest man in the U. S., what other goals will you have?" He answered without hesitation: *"to be the person with the longest life in the U.S!"*

We must point out the long life mentioned here only refers to the long life with a good health and youth as the preconditions. I am very interested in the new concept "healthy longevity" set forth by the World Health Organization. The length of "healthy longevity" refers to the average number of years which a citizen lives, and it results by subtracing the time of health damage because of illness, injury and so on. We seek not only a long life but also a healthy life at the same time.

Moreover, we should try not only to get a "healthy longevity", but also "a youthful longevity". That is to say, we should seek not only a healthy long life, but also a youthful long life. A new branch of learning on keeping youth, namely—"science of youth", has come into being for a long time. The science of youth is just for keeping a person's working capacity and vitality in all his life, extending and retaining his youth, even making youth stay forever. In the pursuit of a youthful long life, on one hand we must fight against ageing physically, and on the other hand we must also fight against ageing psychologically. Fighting against ageing psychologically is even more necessary. If all oldsters have the state of mind of "The old horse in the stable is yet dreaming of heroic exploits"(poem by Cao Cao), we may say they are not old mentally. An American author Samuel Erman wrote an essay "Youth", which has been regarded as a talisman for prolonging youth by many great figures. The famous General MacArthur inlaid "Youth" in his picture frame and kept it on his desk all the time. The founder of Panasonic Electric Company of Japan, Matsushita, remarked with deep feeling: *"For 20 years, 'Youth' has always been accompanying me, and it is my motto."* In "Youth", there are such sayings— *"Youth is not a time of life; it is a state of mind; it is not a matter of rosy cheeks, red lips and supple knees; it is a matter of the will, a quality of the imagination, a vigor of the emotions; it is the freshness of the deep spring of life." "Youth means a temperamental predominance of courage over timidity, of the appetite for adventure over the love of ease. This often exists in a man of 60 more than a boy of 20."*

Only such a person can enjoy a youthful long life, being even more youthful than a 20-year-old chap. Such a person is bound to make great achievements in academic career, in jobs and in life.

A long life has a matchlessly huge value, not only to the individual, but also to the society.

If we make an analysis of the life periods, a life can fall into the production period and the consumption period and so on. Production period—consumption period—contribution period. For instance, if an average life is 80 years, we may have the following picture: A person goes through the kindergarten, the primary school, the middle school, the college, until doctorate study and postdoctoral study, he is 30 years old then; he retires at 60 (now it is even earlier). So, he only works for 30 years. He passes away at 80, and from 60 to 80 he needs support from the society again. His production period for the society (working period) is only 30 years, far shorter than the consumption period of 50 years, thus, the contribution period has become a negative. Although a person's actual contribution to the society depend on the length of the working period and contribution period, yet, the too short working period and contribution period is a great obstacle to social progress indeed. If we can live a long life and it is a healthy and youthful long life, then along with the extension of the lifetime and the corresponding extension of the working period, we may increase our contribution to the society greatly. When the lifetime increases from 80 to 150, even if we study until 30 and begin to work then, and retire at 130, the working period is as long as 100 years, and the consumption period is less than 50 years. So, the contribution period will increase greatly (from a negative of 20 years to a positive of 50 years). Hence, the society will be able to accumulate wealth much more and at a faster rate, so that it may progress much faster. When people's life span reaches 500, or 1200 years, people's working period and contribution period will increase by many times. So, the social accumulation and progress will greatly speed up.

The ratio between personnel's service period and personnel's training period is of decisive significance to economic and social development. For any product, people lay a great stress on its service life, and the same is true of personnel. Personnel are different from products, the products lose their functions with the increase of service time whereas personnel increase their abilities with the increase of service time (working time). The longer the working years are, the more competent the person will be; the older the age is, the more skillful the person will become. Along with the growth in age, a person will accumulate more wisdom, experience, knowledge, and skill. Some neurology researchers have found, after the age of 70, more inter-connecting "branches" will grow out among the brain cells, which can enhance the depth of one's knowledge. Psychological study has also proved, people's mental mechanism will become healthier and more perfect gradually along with the increase of age. It is obvious that oldsters have many advantages that are incomparable to those in other age groups. Our society has neither made the best of old people's strong points and advantages nor has it extended the working lives of all kinds of qualified personnel greatly. If we can prolong the life of all kinds of qualified personnel, especially their healthy and youthful life, we may greatly increase their working years and contributing years, thereby greatly increase their contribution to the society and accelerate social progress.

The famous physicist Seman said with emotion: *"If scientists' life can be prolonged for another 10 years, the civilization progress of the world will never be like the present state."* By prolonging scientists' life for mere 10 years we can exert such a great effect on the civilization progress of the world. If we can prolong their life by 50 years, 100 years, or even 1000 years, what a huge effect it will have on world civilization and mankind's progress!

Schopenhauer said: *"When a genius is alive, people don't estimate his greatness. One century later, the world will recognize his greatness. It will long for his return."* However, no matter how great and outstanding the person is, how we long for his return, he will never come back. The moment he dies, his great wisdom will vanish completely.

Edison, Einstein, Curie, Helen Keller, Roosevelt, Lu Xun, Chen Jingrun... If they could return or have been living till now, how much material wealth and mental wealth they would have brought to the society, and what a great role they would play for human civilization! Saint-Simon wrote: *"If France suddenly lost 50 excellent physicists, 50 excellent chemists, 50 excellent writers, 50 excellent engineers... France would become a corpse at once. For these people are doing jobs most helpful to the whole nation, boosting the civilization progress of France. If we traine such a group of people again, we would need at least the time of a whole generation."* If all these most excellent people can have a healthy long life or even an eternal life, it will undoubtedly be a great blessing to the progress of all mankind.

Let's put aside the fact that the time in which these great people make contributions to the society is only a few decades and what a great pity it is. Even every ordinary person has huge potentialities, and such huge potentialities are often restricted by the shortness of the life, so they cannot be discovered and exploited fully. The old woman painter Moses was lucky enough to find her drawing talent at 80, but many people did not discover many kinds of outstanding potentials they had even until death, to say nothing about making good use of such outstanding potentials. Floom said: *"Everyone has all the potentialities of mankind, nevertheless, the shortness of life does not allow people to realize their potentialities, even in the most favorable environment.*" If a person's life can reach 300 years, 1000 years or even longer, then the degree of exploitation of all the potentialities of everybody will be one thousand or ten thousand times more than the present, and their contributions to the society will be one thousand or ten thousand times more, too. What a great significance it will be for the increase of social development and mankind's progress!

Hence, whether judging from the individual or from the society, we need to reconsider the value of a long life. We have a special need to review the idea that "Longevity is the most valuable" uttered by our ancient sages in *Taiping Jing*.

"Will you tell me what is the most valuable of all things between heaven and earth? Of myriad things in the universe, longevity is the most valuable."

X. REFLECTION ON ETERNAL LIVING

Freedom is a very moving word. Freedom is as light as a cloud and also as heavy as Mother Earth. Pursuing eternal life means pursuing the infinitely great "primary freedom", that is, the most important and most essential is freedom of everyone.

Eternal life, just as its name implies, is endless survival or limitless prolonging of life, that is, no death.

When First Emperor of Qin Dynasty and Emperor Wu of Han Dynasty were racking their brains in pursuit of immortality, how grand they looked. But all their fine dreams were dashed to the ground, and they became laughingstocks of the world. In quest of eternal life, the ancients failed again and again, but tried again and again, ending up in nothing. The countless failures have forced people to give up this idea, so, present-day people usually take eternal life as laughable. Moreover, some people feel the pursuit of eternal life reflects a too much clinging to life, and they are afraid of being blamed for preferring life to dishonor, so they don't feel assured and justified whenever talking about eternal life.

But in my opinion, we should rehabilitate the reputation for eternal life.

Seeking life without death is human's natural propensity. In the bottom of the heart, everyone is longing for living forever. The famous writer Ba Jin said: *"Living" is beautiful, and liking "to live" is human nature. An animal does not fear death because it does not know there is something called "death". Dreading death and loving life is the most normal and most essential human nature. Longing for living forever should not be reproved for any reason.*

In the pursuit of eternal life, the precondition is affirming life is priceless. Life is absolutely essential, absolutely "once" and absolutely precious, having an unrivaled value. We admit that life is priceless, then seeking longevity and seeking perpetuation should be regarded as pursuit of the most priceless thing in the world. The longevity-seeking undertaking should be seen as the most meaningful and most valuable undertaking.

The value of a long life is in the limitless expansion and increment of the value of longevity. As aforesaid, a long life has a very huge value, as the ancient saying goes, *"Longevity is the most valuable"*. Longevity determines not only the degree of a person's right of existence, right of enjoyment and right of development, his achievement in academic career, achievement in occupation career and achievement in life, but it also determines the degree of exploitation of human resources, the degree of utilization of human resources, social development and mankind's progress. Living long is the limitless augment of life span in "quantity". Not only all sorts of values of life span are inherent in living long, but by living long these values of life span may be amplified by a thousand times, ten thousand times, millions of times... Thus, human's right of existence, right of enjoyment and right of development will be expanded limitlessly. As the right of development is infinitely great, a person can develop endlessly, so, the aforesaid ideal prospect of gods, such as omniscience and omnipotence, highest virtue and highest beauty, will also be certainly realized.

Living without death means an endless life and an endless time for development and progress, so it enables people to get limitless achievements in academic career, occupational career and life, and enjoy boundless happiness. Because of boundless achievements in academic career and working career, they can make infinitely great contribution to the society. Hence, of myriad things in the universe, a long life is the most valuable.

Seeking immortality is seeking the infinitely great "meta-interest". In the academic world, there is a concept called "metanotion". Metanotion is "the notion on the common nature of mankind", "the most important or most fundamental notion in the entire scope of mankind (Zhang Shaohua). "Meta-interest" is the interest rooted in the common nature of mankind, being the most important and most fundamental interest within the entire scope of mankind. What interest is more fundamental than the interest of life or the interest of survival? Seeking immortality is seeking the infinite meta-interest. As far as mankind is concerned, there is no greater pursuit than this in the world.

Pursuing immortality is the same as pursuing the infinitely great "meta-freedom", that is, the most important and most fundamental freedom within the entire scope of mankind. Freedom is a very moving word. Freedom is as light as a cloud and also as heavy as the Mother Earth. The whole history of mankind is a history of the pursuit of freedom, and man's degree of freedom is always changing. Primitive men had the least freedom and they could not break away from the enslavement of nature: a feral animal or a rainstorm might end their life. Only after they mastered some natural laws did they gain some controlling power over their own survival. Slaves in the slave society had no freedom, but social progress could not tolerate cruel oppression, so, slaves got their personal freedom at last. In modern society, people have begun to make unremitting efforts to seek freedom of thought, freedom of speech and even their work is becoming freer and freer, as home-based office work and liberal profession have become the trends for future... However, none of these struggles and efforts for freedom has touched the "meta-freedom"—freedom of life. Everyone has a life and every young person has a youth, but everyone's youth and life will be taken back without exception: it is really uncontrollable by anyone. That is the greatest non-freedom of mankind. **Once the freedom of life is lost, all freedom will be lost with it. Hence, after mankind gets one freedom after another, seeking eternal life, this "meta-freedom" means seeking the greatest and most thorough going freedom of mankind.**

Seeking eternal life means struggling against death. Death is mankind's common enemy, threat and fear. How to uproot this threat or overcome this fear, is mankind's greatest problem. People have tried to use philosophical or religious methods to solve this problem. Montaigne said: *"Studying philosophy is just studying how to die."* Feuerbach said: *"The tomb is the headstream of religion."* First Emperor of Qin Dynasty and Emperor Wu of Han Dynasty made great efforts to avoid death, but the result was still "cold corpses in gold coffins buried underground". Alchemists and Taoists also sought "to get rid of the difference between life and death"; even if by living in mountain forest one could live longer, yet, life lost its meaning because of emptiness and monotony. The failure of China's ancients in seeking immortality does not mean the effort to avoid death is wrong, but only that the methods they adopted could not reach the aim. If we neither use the method of First Emperor of Qin Dynasty and Emperor Wu of Han Dynasty, nor use the method of the alchemists and Taoists, but find out a correct way to get out of the maze of death, then, the efforts for eternal life will not be criticized again, but will be regarded as just and nice.

One hundred years ago, the enlightenment scholar in the French Revolution, Condorcet, wrote his last work, hiding himself in an attic near Paris, regardless of the threat of death sentence. Condorcet thought: history and science are struggles against death. His book was written for the conquest of death. So the last words he wrote were: *"Science is meant to conquer death, then, there will be no person dying in future."*

The way out found by this great scholar was not fantasy, but science. He had the conviction that science could conquer death, and bring the happiness of eternal life to mankind. Although he himself took poison and left the world forever, he was more eager than anyone else to bring eternal life to all mankind. But two centuries later, few scientists or thinkers have such a noble ambition as Condorcet's, even though we are more and more approaching the realization of that ambition. Today, we can roam in the outer space, can make artificial celestial bodies, can make artificial lives, can clone mankind itself, but why cannot all the mankind face our common greatest threat, unite together and use the divine power of science that can compare beauty with God, to seek eternal life and solve the greatest problem of mankind in concerted efforts?

Normally, we should be wiser than the ancients, and more able to think about longevity and seek eternal life. Yet, why could the ancients rise to fight against death whereas we are more likely to submit to the will of heaven and resist change? The reason may have something to do with the "thinking set" resulting from the countless failures, that is, the numerous failures have killed the great aspiration for eternal life completely. On the other hand, the fast-paced and high-blundering life has made modern people rotate at a fast rate like tops, without time to consider the long life or death problem. Slowly modern people have forgot the matchlessly beautiful dream of eternal life.

As a matter of fact, we already have good conditions to conquer death and realize this dream. Even long before, the great writer Ba Jin believed: *"Some day in future, science will conquer death indeed"*. This "day" has already appeared in our schedule. It is high time now. Mankind should have such a belief firmly, and utter such a great aspiration. Science can turn the dream of eternal life into reality, and can conquer death. We must use science to realize man's dream of eternal life and to conquer death.

In short, eternal life not only can be pursued, but can also be realized, and we should rehabilitate its reputation.

Living without death is very precious and living without ageing is even more precious. Living without death means: infinitely prolonging human life time, enabling people to enjoy an infinite life. Living without ageing means: infinitely prolonging human's "youthful life", getting rid of decrepitude. Death is mankind's enemy and decrepitude is mankind's another enemy. Now, human's youthful life is only twenty to thirty years. If we can prolong it to 100 years, 1,000 years, or 10,000 years, how wonderful it will be! It is obvious that living without ageing is more ideal and more attractive than living without death, and it is the absolutely perfect prospect. We should strive for living without death, all right, but should strive for living without ageing all the more.

The change from youth to anility is a gradual process, anility is not something that befalls suddenly, but approaches us "silently". I am 21 years this year, enjoying my youth lightheartedly, but I never feel it is a great enjoyment, and never think about the fineness of youth. If anyone talks about anility or death, it will sound like telling a remote and old story of other people. I believe many younglings have the same state of mind as I have, we never have time to think about such things that seem to be irrelevant to us, and only after twenty or thirty will we feel that the "remote" prospect is so near to us.

Mr. Li Yonghao, the richest man for the year 2001 in Chinese Mainland rated by *Forbes*, said: "If possible, I am ready to give all my wealth to swap age with a youngster". You see, even the wealth of billions of dollars can not win against the transient and beautiful youth.

I have been existing in this world for 20 years now. 20 years passed in the twinkling of an eye, and my memory of the picture at 5 or 8 is still close at hand, so vivid. In this way, another 20 years will also flash over very soon. When I was jumping with joy for China's success in winning the hosting right of the Olympics, I thought, when the Olympic Games will be held in 2008, I would be 28 years old, nearly 30, so, I could not help feeling a chill. It suddenly occurred to me as most of the protagonists in novels are forever young boys and girls at their flowery age. Why is each of Jin Yong's novel so touching and soul-stirring? Just because all of them are romantic tales about bread-and-butter boys and misses. In our mind, Huang Rong is forever so charming and lovely, Guo Jing is forever so handsome and sturdy, and they are youthful forever. We have never seen what the anile Huang Rong and Guo Jing will look like. In *A Dream of Red Mansions*, when Baoyu heard "Once the spring is over and youth is gone, flowers fall and people pass away, both going to the world beyond" in "Song for Funeral of Flowers" by Daiyu, he could not help thinking that Lin Daiyu's flower-like beauty would vanish to somewhere that no one could find, how will he control his heartbreak? Since Daiyu's pretty face of youth would have to disappear in the end and could no longer be found back, the other pretty girls, such as Baochai, Xiangling, Xiren and so on, would also fall into the same fate. Not only beautiful looks would vanish at last, even the existence of the persons themselves would also change into "a mound of yellow soil".

The novelist's wish could keep the protagonist's young and pretty face forever, but people in real life have no means to fix their life to youth forever. **Whether a hero or a flunky, despot or bully, phoenix or unparalleled genius, all will get old, become ugly and die.**

At the beginning of the *Princekin*, there is such a saying: *"Every adult used to be a child."* More than that—every old woman used to be a maiden, every old man used to be a lad. One generation after another, all people are repeating this terrible fate. A lad full of animal spirits and with a heroic bearing will become a stooping, staggering old man full of wrinkles on the face. A maiden looking like a slender flower of jade will become an ugly crone. It only takes several decades.

Girls are unwilling to admit that all the present crones have resulted from the flower-like maidens. Yes, they are evading the fact. When they see that their own parents had the youth and beauty as themselves, they are even amazed. Were you ever so youthful too? And, what they will face will be the same fate, "in a fillip of the finger", they will become the same, too. The face and hands are full of senior plaques, almost all the teeth have dropped, the mouth is shivering and drooling, the skin is so loose that you can pull it up for one foot, the walking is very unsteady... Simon says in her *Second Nature*: *"The mother-in-law is a vivid portrayal of anility, and when she gave birth to her daughter, she already faced the doom of anility. Her adiposity and wrinkles are reminding people that adiposity and wrinkles will also befall youthful brides, which foreshows the future of the bride sadly"*. The man was full of resentment for his mother-in-law's mocking because he was very clear, his wife's fate was also everyone's fate, also his own fate.

Look, this is the fate of all of us as well as the fate of all mankind!

We need determination, willpower, imagination and creativity to change the fate of all of us and to change the fate of all mankind, and stride from this shore of anility and death towards the other shore of eternal life.

The people of this generation should shoulder the mission to end this greatest tragedy of mankind and realize the highest ideal of mankind that has been cherished by us for uncountable generations. When we have such a power but we do not make the corresponding efforts, how foolish, how cowardly, how pitiful, how lamentable and how deplorable it will be!

XI. LEARN FROM SUN WUKONG

Anybody is a prisoner sentenced to stay of execution, and the death sentence is bound to be executed sooner or later.

—Hugo

In the classical novel *Pilgrimage to the West*, the most attractive, most impressive and most lovable figure is Sun Wukong.

Sun Wukong started as a little stone monkey who rushed into the Dragon Palace to get the Gold-Hooped Pole, rebelled against Heaven, raised a great turmoil from Heaven to earth, even to the extent of forcing God to grant him the title of "Qitian Dasheng" (Mahatma as High as Heaven). The great turn for him was his finding of a great master and acquirement of the matchless super ability. Why did Sun Wukong want to find a master to learn abilities? What super ability did he get?

In the first chapter of *Pilgrimage to the West*, the story goes like this—

While the Monkey King Sun Wukong was enjoying the feast with his monkey soldiers, he felt a spell of sorrow and shed tears suddenly. (Let's see, what the hell worried this hero monkey so much so as to make him shed tears?)

The monkeys bowed hurriedly and said: "Why are you so worried, Your Majesty?"

The monkey king said: "I am merry and gay now, but when I take a long view, I cannot help worrying." (What on earth did he worry about?)

The monkeys said with a smile: "Our king is so insatiable! We gather happily everyday, living in godly mountains and the Elysium, with age-old caves and divine land. We are neither ruled by the kylin, nor by the phoenix, nor are we trammeled by any throne in the human world, enjoying limitless freedom and boundless happiness. What long view can worry you?"

The monkey king said: "Although we are neither controlled by the human law, nor threatened by fierce beasts, yet, with the passing time, we shall get old, and King of Hell will govern us in the dark. Once we die, won't everything be in vain?"

Hearing the words, all the monkeys hid their faces and began to moan, conquered by the fear of death. (Even monkeys with a little intelligence could "moan", let alone the lord of creation.)

Sun Wukong is a provident monkey indeed: While enjoying the great ease, he can take account of the tragedy of "Once we die...". But Sun Wukong is never someone who just sits back and waits for the doom. He took action at once: he wandered about the whole world to look for good masters, and spent about eight or nine years, just for "getting a way to live forever". After so many painstaking efforts, he found his ideal master at last. Here, his dialogue with his master was very interesting. We might as well excerpt it as follows—

The master said: "Let me teach you something in the branch of '*shu* (divination)'. OK?" Wukong said: "What is it?" The master said: "In this branch of learning, you will know how to practice divination, so you will be able to go towards good luck and evade bad luck." **Wukong said: "By this, can we live forever?" The master said: "No! No! We cannot." Wukong said: "I won't learn that! I won't!"**

The master said: "Let me teach you something in the branch of '*liu* (schools of philosophy)'. OK?" Wukong asked: "What can I learn from it?" The master said: "In this branch of learning, there are Confucianism, Buddhism, Taoism, the science of Yin Yang, Mohist School, medical science and so on. You can study all kinds **of philosophic thinking, or pray to Buddha and get help from gods." Wukong said: "By this, can we live forever?" The master said: "It doesn't look so.."...Wukong said: "In that case, it doesn't have a long value, either. I won't learn it! I won't!"**

The master said: "Let me teach you something in the branch of '*jing* (silence)'. OK?" Wukong said: "What can I get from it?" The master said: "It is about quietism, about fasting, about sitting in meditation to understand the dhyana, about abstaining from speech and keeping indoors, doing all kinds of qigong exercise, such as qigong in sleep, qigong in standing, and so on." **Wukong said: "By this, can we live forever?" The master said: "It doesn't seem so, either."...Wukong said: "No eternal meaning, either. I won't learn it! I won't!"**

The master said: "Let me teach you something in the branch of '*dong* (motion)'. OK?" Wukong said: "What does it mean?" The master said: "It is about taking active actions in life, such as taking in the *qi* of *yin* to tonify yang, bending the bow and treading the crossbow, rubbing the navel for qigong, processing miraculous drugs, making ding (an ancient cooking vessel), swallowing red lead, smelting the autumn stone, and so on." **Wukong said: "By these, can we live forever?" The master said: "No that hope, either."...Wukong said: "I won't learn it, either! I won't!"**

Sun Wukong refused all those marvelous arts, only being bent on the way to live forever. After getting the favor from his master, what he asked for was still the greatest wish of his.

Sun Wukong is exceptionally clever indeed, extremely intelligent and farsighted. He knows which is the most precious, which ability is the most worthy and most worth seeking. Out of all things in the world, what is the most precious? Life! Out of all abilities in the world, which is the greatest?

Greatness comes from comparison. Chernyshevsky said: *"When we compare one thing with others, whatever seems much higher is the great one."* Of the educational thought of the great educationist Spencer, the most precious point is about the comparative value of knowledge: *"The most important question does not lie in whether that knowledge has value or not, but lies in its comparative value."* In knowledge, there is comparative value and in ability there is also comparative value. In Sun Wukong's mind, there is a very solid concept on comparison of abilities. What he was seeking despite so many hardships and such a high price was the most valuable ability, the ability to live forever.

However, in comparison to Sun Wukong, we are far more foolish. Even till now, we do not know what should we pursue the most, and what ability we should develop the most. Many countries are vying with each other in making the atomic bomb, many people are so zealous for developing all kinds of new manslaughtering weapons.NMD and TMD have come into being regardless of the cry against them throughout the world... What mankind is zealous about is to develop the ability to kill itself, and they are competing madly in the ability to destroy human race, their own species. Their pursuit of such abilities has reached a crowning extent: they owned the ability to destroy mankind by scores of times long ago, and some people are even planning to work out a weapon in the near future that is so powerful that "by one sneeze, a body of person could get cancer". Nevertheless, the most worth seeking ability, the ability to live long, or the science and technology on longevity, has never aroused a little attention from mankind.

There are thousands of branches of science in the world, but there is not a single science specializing in longevity. There are numerous academies of science all over the world, but there is not a single academy of science on longevity. There are uncountable institutions for developing nuclear weapons in the world, but there is not a single institution for research on longevity. The global military expenditures are as high as trillions of U.S. dollars, but there is not a half penny investment in the technology of longevity! Why can't we be as clever as Sun Wukong? Why can't we go all out to seek the ability of long living among all abilities and why can't we go all out to develop the science and technology of longevity among all kinds of science and technology?

All mankind, the whole world and the whole society should learn from Sun Wukong; to do what is most worth doing, and seek the ability that is most worth seeking. We should organize all the best brains of the world; the best biologists, the best medical scientists, the best nanometer technology experts ... turn the military expenditures of the whole world into expenditures for longevity research, and gather all the forces of the world to tackle key problems in this area. If so, the probability of breakthrough in longevity research will be 100%, and all that is left will be a matter of time.

In learning from Sun Wukong, we should not only learn from his firmness and over-absorption of mind in seeking the most valuable ability, but also learn from his thorough understanding of life and death and his longest view on life. Without such a long view, it will be impossible to have such over-absorption of mind and pursuit for eternal life. People usually do not have such a supernormal vision as Sun Wukong, they are only able to see very narrow and shallow things at hand. When a small pebble is very close to the eye, it may seem to be greater than the whole world. The huge sun is too far from us, so, to human eyes it looks like a small red ball that can be held in the hand. There is nothing more important and greater than living forever, but as it is too far away from our present life, it is often neglected. Usually, o rdinary persons lack Sun Wukong's vision and "long view", but only when death approaches they awake to the truth, but it is too late by then.

Human life seems to be written in water, years always pass in haste, just as the famous writer Zhu Ziqing wrote: *"When I wash my hands, time elapses through the basin. When I eat, time passes through the bowl. When I think alone, it passes through my fixed eyes. When I feel it going away hurriedly and reach out to hold it, it passes from my hand. At dark, when I lie in bed, it strides over my body cunningly, and fly over my feet. When I open my eyes and see the sun again, one day has already passed. I hide my face to sigh. But the newly coming day's shadow begins to flash over in my sigh."* One day, one week, one month... fly over one after another. Only when we look back, we will feel that things several years ago, 10 years ago or 20 years ago were just like what happens now. Only at this moment when you "count the time on your fingers" can you utter words full of deep feelings. Feng Zikai (famous painter) expressed the feelings like everyone out of us: only when he reached thirty he could have some sense of "life has come to 'the beginning of autumn'". In the past, my thought was so shallow that I felt spring could stay with us forever, and man could stay in youth forever, but never thought of death. Also, I thought the meaning of life only lay in living, and my life was most meaningful, as if I would never die. Until now... "Sun Wukong could worry about "long life" as the most valuable thing while he was "enjoying a merry-making festival every day". Everyone who is enjoying youth and life now should have such a long-term worry. Everyone should appreciate profoundly the saying by the eminent writer Victor Hugo: *"Anybody is a prisoner sentenced to stay of execution, and the death sentence is bound to be executed sooner or later."* Everyone will face the worry about anility, and no one can escape the fatality of death. However, science plus revolution can change everything. While we do not feel the worry about decrepitude and are not faced with the threat of death, we should pay a great attention to death and challenge it.

The famous British historian Doctor Toynbee uttered a sentence in a very serious tone when he was talking with Ikeda Otsukuri (famous thinker and writer of Japan) and Mr. Ikeda. Otsukuri thought that this sentence has sharply revealed a fact, that is, leaders of the world always shy away from "death", the fundamental question. Doctor Toynbee said: *"None of the statesmen and leaders of all walks of life has ever challenged this fundamental topic, but they only take an evading attitude to it. It is coward-ly, and also the most shameful."* Not challenging death is "cowardly" and "most shameful", then, challenging death and conquering death are undoubtedly the greatest and most heroic.

In *"Chapter of Tolerance"* by Zhuangzi , there is a story about Yellow Emperor going into the mountains to seek the Way. Yellow Emperor ruled China for nineteen years and his ruling was free from corruption. When he heard Guangchengzi (a sage) was living in Kongtong Mountain, he went to visit him. He said to Guangchengzi: "I heard that you know the highest Way, so I have come to consult you about the pith of the Way. I want to have a good harvest, let winds and rains come in their proper time forever, and let all my people live and work in peace and contentment. What should I do?" Guangchengzi said: "What you ask for are just some nonessential things, so I cannot tell the highest Way to you." After saying those words, Guangchengzi fell asleep at once, pay-ing no attention to him again. Yellow Emperor came again from the lee-ward, moving his steps little by little kneeling, made several big kowtows to Guangchengzi, and asked in a new way: "I heard you know the high-est Way. Would you please tell me how to live a long life?" Guangchengzi sat up suddenly and said: "That is right. You have come to the right track now. Come over, I will tell you..." And then Guangchengzi talked about the major principle on how to live a long life.

Zhuangzi, Guangchengzi and Yellow Emperor took longevity as the supreme Dao (Way), and thought, in comparison with longevity, poli-tics was only a nonessential matter. Just as Toynbee said, leaders of the present world always avoid the issue of "death", that is, avoiding the highest Way and the greatest thing. Even if we do not say they are cow-ardly or most shameful, at least, they are only doing nonessential things. They are either busying themselves with arms race and competition in national power, or through the bottle-bottom-like spectacle glass, watch-ing whether the growth rate has increased by one digit, so as to add to their political achievements... But the real great undertaking concerning people's livelihood, the real great matter of perpetual value, that is, prolonging the people's life, has been cast aside. In fact, only this under-taking can be considered as "representing the fundamental interest of the greatest majority of the people".

The most fundamental duty of a country should be to enhance its people's most fundamental interest. The most fundamental interest of the people is the "set" of the individuals' most fundamental interest, and all the individuals compose the collective body of the people. Protecting the most fundamental interest of the people is nothing else than protecting the most fundamental interest of the individuals that compose the people. The nation exists for its people, and there is no "national interest" that is higher than the people's interest, neither abstract "interest of the people" nor "national interest" that is irrelevant to the individuals' interest. As far as every individual is concerned, only the existence of his life is his most fundamental interest. So to speak, as far as a nation or government is really loving its people and working for the people is concerned, the most fundamental interest it protects and the most fundamental goal it seeks, should be to do the best to realize the most fundamental interest of each individual, that is, longevity, or even eternal life.

Ikeda Otsukuri's teacher Mr. Heta said: *"One century later, as many as 100 million people will no longer exist in Japan. Whenever I think of their fate, I cannot help being astounded."* In saying these words, his sad feeling was clearly seen on his(Ikeda Otsukuri) face. If we take things as they are, one century later in China or in the world, there will be far more than 100 million people dying. Yet, common people have become accustomed to indifference to birth, ageing, illness and death, and have got used to the end-result of death. Such a habit and indifference have become numbness gradually. Only those thinkers who ponder over and worry about the people's fundamental interest will feel sad and shed tears for a matter that most people are already accustomed to and are insensible to.

Today's mankind is still like the one whom Nietzschean criticized in his essay "An Anti-Era Review": *"Heavy habits, trivial and mean matters are congesting each corner of the world, and they form a dreary air over ground, hanging over all great matters, impeding and troubling great matters' road towards immortality, taunting them, smothering them, making them lose their breath."* People have got used to developing manslaughtering weapons, engaging in all types of war, indulging in beer and skittles, and seeking a lopsided high cost of living, which are mocking at and smothering great goals. Only by extricating ourselves from such lowdown and mean matters can we walk up onto the great and noble road.

Wake up, mankind! Tear off the worldly net of vulgarity, seek what is most worth seeking and develop what is most worth developing!

XII. NONSENSE OF EXPERTS

If some elderly scientist claims with confidence that something will never happen, please don't notice his words.

—Clark

The present time is characterized by more and more minute subdivision of disciplines, we give more and more credit to experts and worship them blindly. As long as anything is said by an expert, it is correct, one hundred percent doubtless. If a wrong viewpoint is uttered by a man in the street, people will think carefully to judge between right and wrong, but if it is from an expert, it will be very likely to be accepted by both the public and the society without thinking. The consequence resulting from this is especially terrible.

On one hand, we are immersed in "superstition about experts" and unable to extricate ourselves, but on the other hand, history tells us repeatedly that many experts or authorities talk nonsense. **Mr. Lu Xun said: "Experts often talk nonsense."**

Do experts really often talk nonsense?

Yes. Many cold facts in history have given the answer.

As bicycle repairmen, Wright brothers invented the plane, and this story is very thought provoking. At that time, many experts and authorities, for example, the famous French astronomer Legendre (the first person who used trigonometry to measure the distance between the moon and the earth), the great German inventor Siemens, the German scientist Helm Hotz (one of the finders of the principle of conservation of energy), the American astronomer Newcomb and so on, came out to assert that it was totally impossible to invent a plane, as they had already proved scientifically that a machine heavier than air could never fly. As soon as the experts said so, it was like removing burning wood from under the boiler, everything was over. Those who had wished to offer financial aid to the research took away their funds completely. Only Wright brothers, as they were ordinary repairmen acquiring knowledge by self-study, did not know the "great" nonsense by those "great" experts, relied on their bold imagination and practical efforts, and sent the plane into the sky at last. Suppose they also heard the nonsense from the experts, and believed their words blindly, realization of mankind's dream to fly into the sky would have been delayed for uncountable years, maybe not even till today.

There are too many similar examples. In 1943, Chairman of the Board of Directors of IBM Company, Thomas Volin told people with confidence: *"I think 5 computers are sufficient for the world market."*

Marshal Foch, President of Senior Military College of France, was very fond of the newly appearing airplane of that time, saying, *"The plane is a very interesting toy, but it has no military value at all."*

In 1899, a committeeman of United States Patent Office issued an order to cancel his office, and the reason was: *"All things that can be invented in the world have already been invented."*

Among those who made the above judgments, which one was not an expert? What's more, they were not ordinary experts, all of them could be called endsville great experts in their own specialized fields, but facts have proved that the judgments they made about the future were mere laughingstocks.

Just as Hegel said, *"the greatest thing mankind has learned from history is that we have never learned anything from history"*. Although history has reminded us again and again that experts may also talk nonsense, we still believe them blindly.

When interviewed by a reporter, the famous futurist Toffler said: *"In the reports of the New York Times, Today's U.S.A. and so on, you may see many questions derived from the past. In these reports, some scientist or expert said it is impossible"*. It has reminded me of the book *Future Profile* by Clark. In the starting chapter of the book, there is an eye-catching list, which enumerates many statements about things that were claimed to be impossible. All those things "that can never happen" have happened, and have been achieved by top-grade scientists of the time. Hence, Clark said: **"If some elderly scientist claims with confidence that something will never happen, please don't notice his words"**.

Today again we hear many experts' nonsense, uttered either lightly or seriously. Especially because of the participation by the mass media that exaggerate deliberately so as to create a sensation. The introduction and analysis of many latest research results have been distorted completely.

The senior specialists showing face frequently in news media tend to make the most terrible supposition about the consequences of high and new science and technology, and then criticize them severely. Take gene for example: we ought to jump for joy because of the birth of the first gene map of mankind and fancy endlessly from the great blessing this scientific progress has brought to mankind, but on the contrary, in introducing gene technology, our mass media are ceaselessly exaggerating the all sorts of terrible consequences such as gene weapons, gene pollution and so on. Some experts even said: *"Wipe out science, and kill all scientists for the bad things they have done."* It seems that gene technology will certainly become the most horrible demon of mankind. Maybe it is a result of some media's willful clipping, but anyhow I feel such experts do not have the minimum brain or minimum wisdom.

"Within 500 years, mankind's average lifetime will not reach 100 years" and this conclusion was made by two scientists from Chicago, Olshanski and Bruce Kanis. The reason given by two scientists is like this: The improvement in life expectancy of mankind in the last century resulted largely from the lowering of the infant mortality rate and the lowering of the lethality rate of infectious diseases (such as tuberculosis) which reduced the rate of premature death of young people. They declared it in a very positive tone: *"In these aspects, we have already exhausted our longevity resources, and you can never save a young life twice. Even if nobody dies before the age of 51, the average human lifetime can only re-increase by 3.5 years."*

Those words seem to be reasonable at first, and especially the scientist's title often makes you believe him without doubt. It is hard to comment after hearing the conclusion, but when you hear the reason, you might laugh out loudly. Their reason is the life-time dilation in the last century was short because of the conquest of some pertinacious diseases such as tuberculosis and so on. Everyone knows, tuberculosis of that time was an incurable disease just like cancer now, but it was conquered by medical science at last. Then in the new century, why does mankind stand no chance of conquering such incurable diseases as cancer, AIDS and so on? If mankind conquers cancer, AIDS and so on, can't our average lifetime increase much more again? Mankind's average lifetime has been growing all the time, and at present not only there is no trend of stagnation, but on the contrary the trend is growing faster and faster, especially in the developing countries that accounts for the majority of world population. "Can only increase by 3.5 years" does not agree with the actual condition clearly. The statement "we have already exhausted the longevity resources" is laughable beyond description. What facts can prove "longevity resources have already been exhausted"? None. The truth is just the opposite: life science and technology have brought about unprecedented and limitless longevity resources. Gene engineering technology, stem cell technology and so on can not only overcome "incurable diseases" such as cancer, AIDS, etc., but they have also brought about an unimaginably and limitlessly fine perspective for mankind's longevity and eternal life. Just the discovery and full use of the "ageless living" gene alone can lead to wonders that were impossible in the past. How can you say "longevity resources have already been exhausted"?

What is most worth thinking is not experts' nonsense itself, but that we still have such a blind faith in what experts say, without the least judging ability of our own. Just take this statement for example; "Mankind's life time cannot reach 100 years within 500 years" it is nonsense in the field of life sciences in fact. And, practice has shown that many contemporary experts have uttered similar pessimistic nonsense in the domain of life sciences.

Just about 20 years ago, in the book *Who Should Play the Role of God* written by Jeremy Rivkin and Howard, it talked about the good things the genetic engineering might bring to mankind and some related problems. After publication of the book, most of the molecular biology scientists, policy makers, media workers and science fiction writers in America regarded it as alarmism and pure imagination, thinking the science described in the book would appear at least 100 years later or perhaps several hundred years later. Even most scientists engaged in the research on genetic engineering also thought; that the description was only a "fictitious" future. However, in the 20 years after the publishing, species from transgenosis, clone, surrogacy, human organ making and human gene operation... each of the breakthroughs of science and technology predicted in the book has been realized. But according to the experts' assertion then, these should be the things of 100 years or several hundreds years later.

History has proved repeatedly "Experts often talk nonsense". Yet, we still believe them blindly. In many aspects and most of time, we just follow experts' lead. We just believe whatever experts say, and lack our own thinking ability and judging ability. It is a lesson we should learn.

The megatrend of development of the times has greatly reduced the number of experts thinking "Within 500 years, mankind's average lifetime won't reach 100 years", and most experts are very optimistic about the future life expectancy of mankind. Haris, member of British Committee of Human Gene, pointed out; *if we keep on studying, theoretically speaking, it is not a hard matter to let mankind's average lifetime reach 1200 years.* Shouldn't we make special efforts to develop such exhilarating high and new life technologies greatly? Some assertions made by experts in this respect are not necessarily correct, and moreover, it is unavoidable to have some nonsense uttered in the guise of "expert".

The famous U.S. moralist Daniel Karakhan wrote a book *Setting the Limit*, in which he opposes using modern medical technology to prolong life at "a high cost". According to his opinion, when a person lives to a certain age, we should not use complicated organ transplantation, heart surgical operation or other technical means to prolong his life unnaturally. The reason is nothing else than this: When our investment in health care is limited to a certain amount, the more we spend in "prolonging life by technology", the less the expenditures used for the majority of members of the society will become. Thus, it will lead to a state where: Only a handful of people will benefit, but the majority will suffer a great loss. It is clearly an unreasonable allocation of resources. Many experts have such a worry: "prolonging life by technology" and "designing more excellent and perfect people" can only widen the gap between the poor and the rich and aggravate the problem of polarization.

Nevertheless, we know to our cost, we should never use a still eye to look at a problem. Technology is most innovative and dynamic, and no one can do better than it, in controlling cost reduction. In the past, few people could afford to buy a computer, but now computers are becoming cheaper day by day, and its high speed in updating and depreciation is astounding. A few years ago, the mobile telephone was something only the rich could afford, but now even farmers carry it with them, and we often hear the ringing of mobiles ceaselessly on a bus. Now it seems that the cost of transplantation of the heart is quite high, but after the stem cell technology and clone technology have matured, the cost will certainly be reduced greatly and continually. If we increase our efforts in research and speed up our steps, it will surely become something that everyone can afford. If we give up our research efforts from now on because of the high cost, then the terrible price will frighten poor people back and only the rich will afford it. Moreover, the disproportionation of resources allocation is never a fault of development of high technology, but is a social problem. Owing to social and historical reasons, disproportionation of resources between poor countries and rich countries, and between poor people and rich people, is a quite common phenomenon. Today, some poor people can only eat wotou (steamed bread of corn, considered as the worst food in China) whereas the rich spend thousands of dollars just for beautifying a nail, or spend tens of thousands of dollars just for a bottle of imported wine. Such an inequality can be adjusted only through social reform and social policies. But in prolonging life by technology, we should never stop our research efforts just because of such inequality, on the contrary, we should go all out to speed up our research so as to reduce the cost at a fast rate. We are familiar with the "Rule of Moore", that is, the price of the integrated package of semiconductors will drop by one time every 18 months. The more progress is there in technology, the greater the price reduction will be. The same is true of high and new life technology: the more deep-going the research is, the greater the price cutting for life prolonging will be.

What is more important regarding "prolonging life by technology" as a costly matter, is a mistake in concept of value. Life is absolutely essential and absolutely precious, prolonging life is absolutely necessary, being the most fundamental interest of all the people in the world. Moreover, along with continual progress of technology, it will not bring a "burden" to the society, but will prolong each person's contribution period greatly, bringing a faster development of the whole society and a faster progress of all mankind.

Thus, it can be seen that specialists' viewpoints are not necessarily correct. After hearing specialists' assertion, we still need to think independently. **"Today, old-typed foolish citizens are many, but new-typed 'foolish citizens' have also reached a certain scale."** Old-typed foolish citizens are those illiterate people, but new-typed foolish citizens are those who follow authorities like sheep and believe experts' nonsense blindly.

In short, we should know that experts' talking nonsense is inevitable, and they may even abuse the credit the society and the public have placed on them, using the mass media to spread a lot of nonsense to mislead the public and the society. Here, if we have no independent thinking ability and we trust everything of experts, it will be very dangerous, and will ruin our great prospects.

XIII. TWILIGHT OF LIFE SCIENCES

Bill Gates' slogan "Let each desk have a personal computer" has already begun to get old-fashioned, now, let's put forward a new slogan that is full of revolutionary enthusiasm: "Let everyone be able to live forever!"

Fighting against death is mankind's perpetual mission, being the most revolutionary great struggle of mankind since ancient times. In order to seek eternal life, the ancients looked everywhere for health preserving arts, sought ways to cultivate themselves according to a religious doctrine, smelted red drugs, ransacked godly mountains, begged gods and Buddha... They have failed again and again, and the failures were very tragic.

After entering the 21st century, faced again with the ancient dream of "eternal life" that has spanned the whole history of mankind, we need to think in a new way; How on earth can we realize this matchlessly fine dream?

By begging gods or Buddha, we cannot realize it, and by relying on Heaven or Earth, we cannot realize it either. Only by science can we realize it.

Technological development, especially in the twilight of life sciences, has already frequently shown the great possibility of eternal life, although it is just at the beginning stage. For turning this possibility into reality, nothing remains but to make further efforts. Then, let's take a look at some great developments in life sciences.

The great achievement in the human genome project is very exciting.

Human genome refers to the totaling of all the genetic information determining the human body's birth, ageing, illness and death. The purpose of the international human genome project is just to make clear the entire gene ordering of human genomes and draw a picture for it. US scientist Dulbecco said: *"Mankind's DNA is mankind's most essential information. All things in the human world are closely linked to this ordering."* The "human genome project" costing as much as 3 billion US$ is a great project comparable to the "Manhattan Project" or "the Apollo Program". Today in the USA, even an ordinary taxi driver knows this grand plan of "using 3 billion US$ to subvert 3 billion nucleic acids". When the whole picture is completed, the core secrets of human life will be disclosed to the world.

I. Gene Research

The successful completion of the human genome project will be an unprecedentedly great scientific finding, and will bring about a great hope for eternal life. A parasitologist in the US State Institute of Health, Michael Gotlib said: *"We have always been path-finding under a street lamp. Now, we have seen all the problems clearly at last."*

The great breakthrough in gene research has enabled mankind to see through all the questions. As far as eternal life is concerned, the gene engineering has also brought about the most fundamental solution. The secret recipe for eternal life or eternal youth is neither hidden in any mystic place in the world, nor can it be granted by "God", but lies in everyone's cell, or to be accurate, it lies in human genes. The reason is: for any individual, the influence of gene on life time is the most fundamental. Some species group has a surprising long life (more than ten thousand years), but some species group has an unthinkable short life (less than one day), and the lifetime limitation of each species group is fixed, which shows that lifespan is determined by gene essentially. Why is the life of someone long while that of another is short? Here, gene also plays a basic decisive role. Hence, by revealing the secrets of gene, we may find the golden key to longevity or eternal life.

Although scientists have not been able to do large-scale study on the "longevity gene" in an organized and planned way, the bit-by-bit study has led to many great discoveries. For example, we have found the gene "I have not died", the "Methuselah gene" and so on.

A research group in European Tumor Research Institute announced in Milan: they have found a gene related to the lifespan of an organism. By restraining the effect of such a gene, we may prolong the life of an organism. The researchers have found: after getting rid of a gene called "P66SHC" in the body of the mouse for testing, or restraining its function, the mouse's resistance to disease was strengthened, and its lifespan was prolonged greatly.

Scientist Simmer Bense, in the US Fornea Technical College found a drosophila that lives longer than other drosophilas, because there is a special gene functioning in it. Now, by utilizing this gene they can prolong the life of drosophilas by 35%. Bense named this gene "Methuselah Gene". Methuselah is a figure in The Bible, and he lived to the age of 969. If such a gene in mankind can be found and made the best use of, we will be able to make man's life time exceed Peng Zu (a Chinese figure who lived to the age of 800) or Methuselah.

Doctor Stephanian Helfand and his colleagues in the Connecticut State University's Medical Centre chanced on a new "longevity" gene in their research project on drosophilas. It is another discovery after the "Methuselah Gene". The chief reason why such "longevity" gene can change the lifespan of drosophila is that it can control the energy absorption of the drosophila's cells, so that the drosophila's cells may "diet". Such a gene is distributed on the two chromosomes of the drosophila. If we only change the gene on one chromosome, the drosophila's life will extend by one time or so, but if we change the genes on both chromosomes, the drosophila will die of too much "diet". Helfand and his colleagues named this magical gene "I Have Not Died".

Scientists in the Massachusetts Institute of Technology reconstructed the gene of a nematode and prolonged its life by more than a half successfully. In the scientists' reconstruction of the genes of the nematode, they removed this gene or added a like gene. The result showed: If it has two SIR2.1 genes, the nematode that can only live for two weeks originally can live three weeks.

The discovery of such longevity genes has brought about a fine prospect for mankind's longevity. That is only research on animals, but scientists have also made breakthrough in the research on humans. The scientists in Scripps Academy of California did a research on the genes of some old people above 90, and they found, in more than 99% of the cases, the function of these old people's gene was completely normal, and their genes functioned almost like the genes of neonates. Then, why do we still get old? The researchers in Illinois University at Chicago have basically found the answer. They say, the secret of ageing might lie in a special gene called "p21".The molecule gene science professor Igory Robinson started up the p21 gene in human cells and he said: *"By starting up this gene we can cause significant changes in uncountable other genes, and all these genes have something to do with ageing and diseases relevant to ageing."*

So, it is obvious that there must be genes in the human body that are closely related to longevity and anility, and by a full and deep understanding of such genes and by regulating and modifying them properly, it is hopeful to reach the purpose of longevity or living forever.

Except this, the gene engineering also has a very specially important function. It can combine human genes, animal genes, vegetal genes or any species' genes into one individual life. It is an earthshaking change, which may trigger the most profound and most tremendous transformation including extension of human life. As we know, in animals there are long-living ones such as the tortoise that has a lifespan as long as a thousand years, and many unicellular animals can "live forever" so long as the environment is suitable; in plants, there are also long-living ones such as the ten-thousand-year-old ancient cypress, and even monarchs had to learn from them for keeping a long life. If we can find out the genes that help them live long or even live forever and then transplant or duplicate them for humans or develop drugs with the same effect, then we will open up a matchlessly fine prospect for prolonging people's life.

Gene engineering may trigger a series of revolutions; such as the magical genetic forecast, genetic diagnosis, gene therapy, gene reconstruction and so on, all of which have displayed an extremely bright prospect for the realization of eternal life. Most "incurable diseases" such as cancer will be eradicated and AIDS will not be worth mentioning. The surgeon will "work by the mouth without using the hand", as there will be advanced technology working instead of them, from predication to treatment and everything. We may even be able to do cytothesis (cell repair) for the dead cells in the spinal cord or in the brain, then Zhang Haidi (a famous female writer in China who is suffering from paraparesis) will also be able to walk so fast as if flying... Gene research has also shown that the chief causes of ageing may be related to the defects in the human body's "maintenance" and "repair" systems, and such systems can also be controlled by genes ultimately, thus, by transforming or "closing" some genes, we may control the ageing process of mankind. Mankind can also adjust their own life style according to the "gene map", so as to be in the best life environment all the time. Owing to these reasons Mr. Haris, the British scientist who is in charge of the genome project, asserts, it is not a hard thing for the average human lifetime to reach 1200 years or so.

The continual breakthrough in gene research can not only prolong human lifetime, but also make people ageless. William Haseltine, Chairman of the Board of Directors of Human Genome Scientific Research Company, points out: *"It is very likely for us to find out the headspring for keeping youth forever from our own genes. Cell replacement may be able to help people to be young and healthy forever."*

Besides the gene engineering, people have found the great value of the stem cell for ageless life. In December 1999, US journal "Science" announced the results of rating of the scientific progress of the world of the year, in which the research findings in the stem cell were listed in the top position in the ten major fields of scientific progress, before the human genome engineering, although the latter had attracted worldwide attention and cost a great capital outlay.

In "stem cell", "stem" meaning as "tree stem" or "origin", thus, a stem cell is a cell of origin (originating cell).

What contribution can the stem cell make to ageless life? Professor Chen Runsheng of Biophysics Institute of Chinese Academy of Sciences tells us about such a research. Harvard's researchers first taught two parrots to sing and then damaged the brain central nerve of one parrot, so this parrot lost its singing ability. Then, the researchers took the stem cell from another parrot and fixed it into this damaged parrot, and a wonder happened: the parrot that had lost its singing ability could sing again as before! After differentiation, the stem cell restored the damaged brain central nerve of the parrot. After the stem cell enters a foreign body, it can repair and integrate the original function of the damaged cell according to the receptor's information.

(II) Stem Cell Research

Because of the stem cells the concept of organ transplantation is being replaced by the concept of cellular transplantation gradually, and it has also solved the rejection problem of all transplantation. When your eyes become nearsighted or are injured, just by taking out the stem cell from your eyeball, you will be able to recover your eyes. When the heart, liver or lungs go wrong, is it still necessary to transplant a new organ? Just by planting your stem cell, you can reach the aim of organ transplantation, and there is no rejection reaction as it is pure self-repairing.

In the past, scientists always thought only the stem cell taken from the embryo had the potential to grow into different tissues. The new research progress has made it easy to take grown-ups' stem cells and there is no ethical problem, which has brought about a new hope for stem cell research. Along with the continual progress of life technology, in the measurable future, mankind will be able to use the self-produced "original parts" conveniently, thus, wherever an organ goes wrong, we can do self-repairing. In this way, we do not need to worry about the inability to "live forever"?

Moreover, the stem cell has changed "anti-ageing" into "removal of ageing", marking a new milestone for the conquest of ageing. The past anti-ageing method relies on hormone drugs, such as melatonin, growth hormone, thymic factor, dehydrogenated epiandrosterone (DHEA) and so on, plus some vitamins such as A, E, C etc. Common elements such as kalium, calcium, magnesium, iron, etc., and microelements such as selenium, germanium, etc., can also be used in the drugs. These drugs may have a little anti-ageing effect. But the stem cell technology can replace all kinds of stale parts very conveniently. If we take the human body as a machine, when we find that some cells, tissues or organs are not ideal, we may renew them, thus, we can remove aging completely from our body.

The research on cell telomere has opened up another door for eternal life. After splitting for 50 or 60 times, human cells will stop splitting, showing an ageing state. In the past, it was just according to the analysis of the number of times of cell division of the human body that scientists thought the number of times of human cell division was usually 50, and the average division cycle was 2.4 years. Hence, they reached the conclusion: Human longevity limit is 120 years.

Now, scientists have found a substance called "telomere", existing on the top of the chromosome, and a "telomerase" that can maintain the length of the telomere. Each time the cell splits, the telomere on top of the chromosome is shortened; when the telomere can no longer be shortened, the cell will not be able to keep splitting. Then, the cell begins to die. However, it is found that cancer cells can split again and again without limit, and never grow old. The non-ageing of the cancer cell has become an important clue for research on the normal cell. Once the genetic gene of the telomerase is generated in the human cell, the human cell will be able to split limitlessly like the cancer cell, then we will be able to prolong human life limitlessly.

California Gerun Biotech Company's researcher Cal Harley made an experiment to use the gene engineering method to change the cell telomere. Professor Jerry Shea and Professor Woodring Wright, cytobiology and nervous system scientists in the Southwest Medical Science Center of the Texas University, USA, made such an experiment. After inserting some gene into the collected foreskin cell (a by-product from circumcision), the gene caused the cell to generate telomerase. Usually, before getting old, the foreskin cell can split for 60 times or so. But after the aforesaid experiment, the cell has already split for more 300 times, and there is no sign of stopping at all, and no abnormal sign has appeared, either. "At the action of the telomerase, the cell is just like a bunny that has been injected with an excitant," said Shea, "it is just splitting and splitting without end." In the meantime, Shea and Wright's partners, US Geron Company's researchers did the same experiment by using the human body's retinal cells. As a result, these cells seem to get the ability to live forever, too.

(III) Control the "Life's Clock": Cell Telomere

The research on the telomerase has undoubtedly brought about a new great hope for realization of mankind's dream of "eternal life", so, there is no reason now to say man cannot live to the age of 1000, 10000, or even higher. Scientists call the telomere "life's clock" and believe that by finding the way to set back this clock, we may realize our dream of "living forever".

Scientists have also found that death has something to do with the "death hormone" produced by the person himself. Biologists' experiment on the octopus has proved so. Once the female octopus has "given birth to a child", it will die silently. Scientists have uncovered the secret: A pair of glands behind the eye socket of the octopus will secrete a chemical substance at a certain time, which leads to the death of the octopus. Biologists call such a chemical substance "death hormone". Is there such a "death gland" in mankind like the octopus? Research has found there is such a thing indeed. But mankind's "death gland" is not behind the eyes, but in the brain, and it is the pituitary gland. Scientists' research has proved that human pituitary gland also releases a "death hormone" at a certain time causing the person to go towards death.

(IV) Discovery of "Death Hormone"

Some scientists made a contrasting test with anile white rats. After cutting off the pituitary gland, in order not to affect the secretion of other hormones, they transplanted an artificial thyroxine into the rat. A miracle appeared: The anile white rat's immunologic function and cardiovascular system recovered their youth! The "death hormone" secreted by the pituitary gland is really a very crucial cause of man's death. Since the key has been discovered, we have a new hope for prolonging the lifespan. Of course, simply removing the pituitary gland is not a good policy. Scientists will go on to make clear what cells of the pituitary gland produce the "death hormone" and how it plays its role, then, use drugs or surgical operations or other modern techniques to prevent the generation of the "death hormone", or delay its appearance greatly so as to prolong people's life greatly.

Gene, stem cell, telomerase, death hormone... They are just incidental findings as the scientists have not intentionally made great efforts to tackle the key problems in eternal life. Although they have not made great efforts specially for the goal, scientists' progress in eternal life in the recent years has already been enough to strike us dumb, as those findings have uncovered a corner of the myth of eternal life. Along with the complete deciphering of the gene map, no one can imagine or estimate what power life sciences will release. But one point is certain: that power is extremely great, unthinkably great, and ultimately it will enable mankind to realize the dream of eternal life.

Someone may say these findings have only revealed a future prospect and after all, they have neither become reality nor have been put into practice yet. It is right, but those who say so, may not be aware of one point. All these mark the start of a new epoch and will change mankind's fundamental fate.

After Faraday discovered the electromagnetic law, he used a small electric generator to perform a program in public. When he was performing on the stage, a gentlewoman came to see it, holding a baby in her arms. She asked: "What use will your gizmo have in future?" At that time, Faraday did not know his finding would start a new epoch, so he found no words to answer clearly, but he gave an off-hand answer: "Madam, do you know what the baby you are holding will do in future? Since you do not know what your child will do in future, this gizmo is my child, too, so, I do not know what it will do in future, either." Faraday could not answer, but history can answer. It was just because of such a gizmo that we had the electric generator, electromotor, electric lamp, telephone, TV, broadcast and so on later, which has started a new era!

But now, it is much easier to predict the fine prospect the achievements in life sciences will bring about than Faraday could predict for his discovery, as the future is showing a much clearer picture before us. Although gene engineering, biotech and so on are still "small children", they will grow very fast, and finally will turn the ancient dream of eternal life into reality.

Do not pray to God, but only rely on science. If we say our forefathers' pursuit of eternal life could not succeed owing to the restriction of the times, now, with modern life sciences, we have the wisdom and power to succeed. So long as the whole world joins up, makes all-out efforts to tackle key problems, we are hopeful to let everyone enjoy longevity and eternal life, completely change the fatality of a short life and gain the greatest freedom and liberation in time and space.

The 21st century will be a century of the life. Life sciences and life technologies will surely open the way to eternal life sooner or later. **Bill Gates' slogan "Let each desk have a personal computer" has already begun to get old-fashioned, now, let's put forward a new slogan that is full of revolutionary enthusiasm: "Let everyone be able to live forever!"**

XIV. LEAPS IN HIGH NEW MEDICINE

No "incurable" disease will be "incurable" forever, and every disease will find a cure at last.

Since the ancient times, medical science has played a great role in mankind's conquest of diseases. Today, in the 21st century, along with continual advance in high and new technology and with continual progress in high and new medicine, the flower of life will bloom more splendidly and will keep flourishing, bringing about new wonders.

The high and new medical science of the 21st century will be another great power for realization of mankind's dream of eternal life.

(I) Overcoming Incurable Diseases: From "Incurable" to "Curable"

Hardly anyone dies naturally, but dies because of a disease or other kinds of unnatural death. When you see their death certificates, you can never find a single one in which the "cause of death" is written as "old age", but instead, it is always marked with some disease or accident.

Incurable disease is Azrael's greatest accomplice, being our arch enemy in prolonging our life. Mr. Sore Kent, an American expert in longevity research, said the following interesting words: *"In most cases, when we announce a person is 'dead', most parts of his or her organism are still working quite well. The reason for announcing death is only that the doctors have no means to revive the patient, not that the patient has already reached the absolute state of inability to live."*

Thus, it can be seen: **"Incurable disease" is a disease that doctors (we) cannot cure at a certain time, in a certain place and on a certain occasion, but never a disease that is "incurable" forever.** In the 18th century, smallpox was an incurable disease, running wildly though the whole world, putting people into a helpless position, but has it not died out on our earth already? In the 20th century, pulmonary disease was a terrible "incurable disease", but can we still call it "an incurable disease" now?

No "incurable" disease will be "incurable" forever, and any disease will find a cure at last. Quantum leaps in life sciences have caused modern medical science to rise into high and new medical science and it will go on advancing by leaps and bounds. Just take the gene for example: research has shown that all human diseases are related directly or indirectly to gene, and all human diseases can be regarded as "gene diseases" or "hereditary diseases", either being "single gene disease", or "polygene disease" or "acquired gene diseases". All the so-called "incurable diseases" now such as AIDS, cancer, cardiovascular diseases, congenital dementia, senile dementia, diabetes and so on, are gene diseases, and all of them will turn from "incurable" to "curable" one after another. Along with the fast development of gene technology, that day will come sooner or later when all those "terrible specters" will disappear from every corner of the world.

(II) Organ Transplantation: Freeing the Body from Damage

The only cause of many people's death is the damage of an organ which endangers the life. Nevertheless, now the day will come soon when all the organs of the human body can receive artificial transplantation and the function of the artificial organs may improve limitlessly. Organ transplantation is one of the disciplines with the fastest speed of development in the high and new medicine.

In the ancient book *Liezi*, there is a story about Bian Que (famous doctor in the Spring and Autumn Period(770-476 BC) changing hearts for two persons to cure a disease, and of course, it is a pure fantasy. In Pu Songling's *Liao Zhai Zhi Yi*, there is a description about Lu Pan changing the head for Zhu Sheng's wife: *"He took out the dagger from his boot, pressed the lady's neck with one hand, cut her head like cutting a piece of bean curd, and the head dropped beside the pillow. Then he hurriedly took the head of the belle from Zhu Sheng's arms, put it onto the neck, adjusted the position until it was proper, and then pressed it..."* He moved the head of a belle onto the neck of an ugly woman. All such fantasies as Bian Que's transplantation of the heart, Pu Songling's description of the head transplantation have become reality by the aid of science nowadays.

All the human organs can be replaced by high-function artificial organs, which may result in an entire high-function artificial "body". Everyone can change or renew any organ, so the body is just like a Rolls-Royce compact car: whenever a part breaks down, you may change and renew it and no part is unchangeable. Or like a computer, any part of it, such as the CPU, hard disk, motherboard, CD driver, memory and so on, can be changed at any time. In future, if any organ of the human body goes wrong, you may immediately replace it. It will be so easy, just like changing a part for the car or the computer. The reason why human needs the body is only for keeping the survival of the brain (mind), and, the true essence of a human just lies in the brain (mind). **By organ transplantation, we can keep the human body healthy forever, moreover, its function can improve limitlessly, thus, we shall have ideal "Jingang body" (free of damage) described in the ancient myth.**

In the past, organ transplantation met with two big problems: one is the too few sources of organs for transplantation; the other is the rejection reaction after the organ transplantation. The rejection reaction often leads to a high failure rate and besides this, there are too few sources of organs, so, the feasibility is very low. In transplantation of animals' organ for people, there is also rejection reaction and we have to take the risk of being infected by some viruses typically of the animal. Kang Youwei (famous scholar and reformer in the late Qing Dynasty) once received a pair of transplanted testis from the ape. At first, he walked lightly, became very vigorous, as if having been rejuvenated, but later, things went from bad to worse, he became listless, and collapsed at last. It just resulted from the rejection reaction.

Nowadays, great developments in clone technology and stem cell technology (especially the latter) may solve both these big problems completely. Clone technology can guarantee sufficient sources of human organs, and stem cell technology can help us in making new organs very easily, such as new hearts, livers, kidneys, nerves, limbs, bones, hairs, eyes, skins, blood vessels and so on and the rejection reaction can be avoided, too. What is more, if one of your organs goes wrong, just by transplanting the stem cell with self-repair function, you may re-make any new and good organ and it is the best solution for rejection reaction and shortage of organs for transplanting.

Along with the continual improvement of surgical operation technology, and with the fast development of pharmacology, immunology, genetics and gene engineering technology, organ transplantation will be improved all the time, its future development prospect is bound to be boundlessly splendid, which will offer a great hand to mankind's realization of the dream of eternal life.

(III) Gene Therapy: Permanent Cure Is Better than Temporary Relief of Symptoms

Since the 1970's, mankind has spent more money in research on diseases than ever before. The USA spent a huge amount of money in a tumor program but its effects were not good. Later, people found that many human diseases such as cancer and so on are directly or indirectly related to gene.

Soon afterwards, people realized another fact: as a matter of fact, all human diseases are gene diseases. Besides single gene diseases, there are polygene diseases such as; malignant tumor, cardiovascular and cerebrovascular diseases, mental diseases, diabetes, rheumatism, immunological disease and so on. Moreover, there are acquired gene diseases derived from the invasion of pathogenic microorganisms into the human body, such as AIDS, hepatitis B, tuberculosis and so on. Nearly all diseases are heavily under the yoke of genetic gene. So long as we master the secrets of the gene and make some simple modification in the gene, we may cure completely many pertinacious diseases or incurable diseases in modern medicine, such as cancer, diabetes, asthma, hypertension and so on. We have good reason to believe that by further reconstruction of gene, it is possible to enable people to live forever. Scientists have drawn such a picture for us: The doctor records the person's gene map on a CD. While diagnosing, the doctor first reads the CD, and just by comparing your gene with the normal gene map, he will find out your disease. Then he checks several possible "candidate genes", and clarifies the key areas and key locations. Then, he will decide on a proper solution. Owing to different "conditions" of the genome, some kind of drug is effective on some people, but is not effective on others, or even will endanger the life. In accordance with the individual variations, the doctor can suggest an individualized "wonder drug".

The above is a low-leveled gene treatment and the more veritable "gene therapy" is to send the disease curing gene directly into the human cell to give play to its curing effect, that is, to cure the disease directly according to the disease causing gene. There are many ways to send the gene into the human body. There is a "gene gun", which is like a gun, whose bullet is made of gold, very very small, so small as to be able to get into the cell and then be spat out. It can stick to the "curing cell" and get into the cell, and after being ejected out by the cell, the "curing carrier" will stay inside. In this way, by shooting a "gene bullet" into you to change your disease causing gene, your disease can be cured by the root.

If we say all the past therapies only "provide temporary relief of symptoms" then, what gene therapy does is "to offer a permanent cure". Genetics has shown as many as 6500 inherited diseases are caused by single gene defect, and by using the gene therapy to implant the related gene, we shall overcome many incurable diseases of mankind. The gene therapy has boosted medicine to a new prospect. Hence, gene therapy is called "the fourth revolution" in mankind's medical treatment history.

(IV) First Medicine: The "Bian Ya" Medicine of the 21ˢᵗ Century

The current medicines are chiefly curing medicines and can only be called "Second Medicines", because only preventive medicines are the "First Medicines". Developing from curing to prevention, from taking curing first to taking prevention first, is the mega trend of development of the present medicine in the world.

The great Chinese medical scientist Sun Simiao of the Tang Dynasty pointed out long ago: *"The upper-leveled doctor treats a disease that has not appeared, the middle-leveled doctor treats a disease that is going to appear, and the lower-leveled doctor treats a disease that already appeared."* He used the "the way of the upper-leveled doctor", so he lived to the high age of 101. Judging from the present medicine, most doctors deal with "diseases that have already appeared". But preventive medicine usually only deals with "diseases that are going to appear", and high and new preventive medicine chiefly deals with "diseases that have not appeared".

The "miraculous doctor" Bian Que in ancient China is widely known to everyone. As the story goes: Bian Que cured the intractable disease of an emperor. The emperor asked him: "Are you No. 1 doctor in the medical world of present time?" Bian Que said: "No, I am only No. 2." The emperor asked him: "Who is No. 1 then?" Bian Que said: "My elder brother Bian Ya." The emperor asked him: "How are Bian Ya's medical skills higher than yours?" Bian Que said: "When my father Bian Gong was dying, he passed two secret medical books to us. One was the *Way of Curing* passed to me, so I can cure a disease that has already appeared. The other was *Way of Prevention* passed to my elder brother Bian Ya, as

Bian Ya is cleverer than I am. Hence, he can cure a disease that has not appeared. Present people worship me so much just because I can cure diseases that have already appeared, but take no notice of Bian Ya, who is much higher than me as he can cure diseases that have not come. It is really unthinkable!"

Prevention of diseases is more important than curing of them, but people often only pay attention to treatment of diseases that have already appeared, and neglect prevention of diseases. Many people arrange their life this way: **In the first half of the life, they use their life to exchange for money and in the second half of their life, they use money to exchange for life.** Of course, it is very foolish. Chairman of the US Cardiology Society said, *"When the patient comes to see the doctor after getting the disease, what the doctor can do to help the patient is already very limited. Even if the disease is cured, the patient cannot recover to the state completely the same as before the disease."*

To realize the great ideal of eternal life, we need many Bian Ques all right, but we need more Bian Yas! We need to develop "Bian Que" medicine, but we need to develop "Bian Ya" medicine all the more.

Preventive medicine is the first medicine and is getting great development and progress with the drive from life sciences and biotechnology. Every person will have his own gene map, which clearly records the person's health condition, disease history, easily occurring diseases and other data, and from this, you may forecast; Will you gain flesh? Will you become bald, and when? Will you die of diabetes or cancer... You may not believe what the fortune-teller or prophesy teller says, but what the gene map says is true. Once you have got the disease causing gene, you can reconstruct or remove it before the disease comes.

A "gene chip" just as big as the nail can help us test our health condition by ourself. One minute is used for sampling, two minutes for measuring and the result can come out in three minutes. You only need to drop a blob of blood or saliva onto the gene chip, then, you can identify the disease gene, find out the disease, and even know accurately how many cigarettes you may smoke or how much food you may eat every day. The gene chip marks another profound revolution of technology, which will bring about marvelous changes to preventive medicine. Its super discrimination can identify diseases that have not appeared, discover all kinds of minute changes in the seemingly healthy human body. With such a gene chip, it will be just as if we carry a private advanced laboratory with us. We have an upper-leveled doctor Bian Ya working beside us all the time, monitoring our health condition at any moment and place, so that we may take all kinds of super preventive measures timely.

Preventive medicine also concerns people's food and drink, that is, nutrition. People's life time is closely related to nutrition. Bioscience will turn people's nutrition into nutrient substances processed with a special technology, which may help people live forever. For example, the "edible molecule food" that has already appeared. It is not the food we eat every-day, but a food in a form most convenient for absorption by the human body. Molecule food will melt together with medicines and foodstuff, bringing an illness-free state for mankind. In the case of traditional food, we eat a lot, but utilize very little. For example, in the case of a man living up to 70 and with a weight of 65kg, he will eat and digest more than 75 ton food in his life, but the nutriment he really utilizes is only about 5 ton, and all the other 70 ton is excreted out of the body as wastes. The molecule food is formulated reasonably according to the human nutrition need and the balance of chemical reaction, so as to promote or maintain the normal chemical reactions of all kinds in the human body, remove impediment in nutrient absorption, ensure the body's maintenance of the adolescent health status, and resist ageing. Scientists predict, by eating such a food a person who has lived to 120 still looks like one at 20, and a 1200-year-old person is still at the height of his(her) youth and vigor, and is still very pretty and charming.

According to experts' prediction, by the year 2020, even if a person is very healthy he\she may also need to take 3 pills every day in order to prevent all kinds of troublesome diseases. Some are for preventing all kinds of cancer, some are for preventing diabetes, some are for preventing hypertensive and so on. Taking such measures for prevention will be a healthful life style, just like sports exercise.

Because of the look-ahead predictive way of thinking, the first medicine of the 21st century, preventive medicine, will be able to do better in prevention of diseases, so as to prolong our healthy life greatly.

All in all the major means of health and longevity can not get away from medicine, especially the high and new medicine of the 21st century. The great leaps in the high and new medicine have brought about continual breakthrough in all aspects such as overcoming of incurable diseases, organ transplantation, gene therapy, preventive medicine and so on, which will certainly lead to the great ideal of eternal life.

XV. LONG LIFE SEEN FROM A THIRD EYE

In this world, so long as we have the "three sufficients", namely, sufficient intelligence, sufficient investment and sufficient time, any impossible wish may be turned into vivid reality. Eternal life is no exception, either.

If we say "the first eye" is life sciences "the second eye" is high and new medicine and "the third eye" is the vision of dialectical philosophy.

In the face of the twilight of life sciences and the great progress in high and new medicine, we cannot but feel a great hope for eternal life. However, it is hard to avoid the last doubt still existent in some people's mind about eternal life. Is eternal life really possible? Hence, it is very necessary for us to ponder over eternal life from the viewpoint of dialectical philosophy.

(I) Theory of Law and Theory of Condition

The motion of any matter in the world has its own objective laws and these motions are controlled by them. Whether we do things according to objective laws is the key to success or failure in all our activities for knowing or changing the world. Sun Bin (famous strategist in the Warring States Period) had a famous saying: *"Victory lies in knowing of the way." "Failure lies in ignorance of the way."* The "way" here refers to the objective law. When we esteem it and utilize it, the law will become our weapon and help us winning smoothly. Otherwise, if we contempt it or go against it, the law will become our opponent and make us running our heads against a wall and bite the ground.

It is generally agreed that for all things, since there is birth, there must be death. Births and deaths, one after another, constitute an inevitable and irreversible natural law. When people have taken it for granted that "Since there is birth, there must be death" is an unalterable law, whoever dares to seek eternal life will be regarded as talking nonsense, making the undermost and most ridiculous mistake, that is, disregarding objective laws.

However, we must know any law hangs on a certain condition. "Water runs to a lower place" is an objective law of nature. Nevertheless, under a certain condition, "Water runs to a lower place" can be changed into "Water runs to a higher place". When I was wandering in a park, watching the fountain rushing up like a jade pole, an idea suddenly occurred to me. So long as the conditions are changed, the natural law of "water flows downward" has been changed to the opposite artificially.

According to universal gravitation law, because of the gravitation, things heavier than air cannot fly, but under certain conditions, it is not so at all. They can not only get off the ground, but also fly with a heavy load.

All these have proved that law is changeable, but it does not mean you can change it arbitrarily. Only if the condition has changed can the law change accordingly. And in change of the condition, you must also follow a certain objective law. That is called "changing the law by the law".

On one hand, before the law everyone is equal. it "does not exist because of Yao (the good emperor), nor does it die because of Jie (the bad emperor)", that is, it is independent of people's will. On the other hand, the law changes with the condition: so long as the condition changes, the law will change correspondingly. People often only know the constancy of the law, but do not know its changeability, that is, they have not realized the truth that "we can change law by law".

In the birth, development and extinction of anything, there is an inherent law. Any law hangs on a certain condition, and all conditions are changeable. Hence, all laws can be changed.

In all ages, the life of a person resulted from the parents' joint work. Without the direct cooperation by both sexes and without the "ten months' pregnancy" in the female's body, there could never be "giving birth" to a new life. In the long river of history, uncountable persons' life came into being this way, without exception. It was regarded as a never-changeable objective law determined by nature. However, who could think, along with the development of high technology, we have been able to produce test-tube babies, to do asexual reproduction, to clone life... completely changing the law of birth. Since the law of birth could be changed so greatly, so the law of death can also be changed.

"Since there is birth, there must be death" is a "law" we have derived from a certain historical context. According to the truth of "changing law by law", so long as we create the condition and change the condition, we can also change "Since there is birth, there must be death" into "Since there is birth, there must be eternal life". Facts have proved that in organisms, birth is certain, but death is not necessarily certain. Scientists have discovered a germ that has lived for 250 million years! And they think, since it has already lived for 250 million years, there is no problem for it to live for another 250 million years.

In the past, scientists used the number of times of cell division and the time of division to deduce human lifespan. Human cells can only divide for 50 times or so and each division cycle is 2.4 years. So, the deduced limit of human lifespan is 120 years or so. Hence, scientists usually regard the age of 120 as the impassable limit of human lifespan. But, is it really insurmountable and unchangeable? Absolutely not! Today's scientific research has proved that: By the telomerase, we can increase the number of times of cell division or prolong the division cycle, so as to prolong human life span greatly. Thus, the lifetime limitation of 120 years has been declared to be wrong, facts have shown forcefully. So long as we change the condition, the law will change correspondingly for certain.

In a wide view of the life span history of mankind, along with mankind's advancement, people's life span has been advancing all the time, too. In the bronze ware period characterized by low productive forces 4000 year ago, people's average lifetime was only 18 years, and in the ancient Roman era, it was 23-25 years. In our country, the average lifetime of various periods has been like this: in the Xia and Shang Dynasties, it was 18 years, in the Qin and Han Dynasties, it was 20 years, in the Eastern Han Dynasty, it was 22 years, in the Tang Dynasty, it was 27 years, in the Song Dynasty, it was 30 years, in the Qing Dynasty, it was 33 years, in the period of Republic of China, it was 35 years, in 1985, it already reached 68.92 years, and in 2000, it climbed to 71 years... the Chinese people's lifespan inceased by more than one time in the short period of 50 years after liberation! In the past, "People seldom live to be seventy years old", but now "Seventy is no longer a high age", and in many countries, the average lifespan has already exceeded 70 years. The constant escalation of human lifespan results from the constant change of mankind's life conditions. Such a change is a gradual process. Now we are in the century of life sciences, so we shall usher in great leaps in this aspect. As far as the present is concerned, the breakthroughs in life sciences have already changed people's understanding of the lifespan greatly. Scientists are beginning to declare it is not a hard matter for human lifespan to reach 1200 years. It can be predicted with the continual advancement in life science and technology, 10 years, 50 years, or 100 years later, even greater breakthroughs will certainly appear in the limit of human lifespan. Thus it can be seen with the help of science and technology plus objective laws, we may change the existing condition of people's life, and the change of the existing condition of the life may in turn change the law of life. Hence, with the continual breakthrough in technology, there is no limit to human lifespan.

For a long time, we had only a smattering of knowledge about the secrets of life, simply thinking the law of life could not be changed and thus, we reached various incorrect conclusions arbitrarily. This has reminded me of the words by the financial wizard Sorrows who raised the financial storm of Southeast Asia: **"All our views about this world are defective and distorted"**. The same is true of our views on life. Perhaps, all our views on life have serious defects and distortions.

We have such an old saying: *"Nothing in the world is difficult for one who sets his mind on it"*. Any hard thing in the world will become easy to a willing mind. Such people are those who can understand and utilize the law comprehensively and profoundly, and those who are good at creating conditions to change the law. Before such people, the "hard thing" of eternal life will also become an "easy matter" ultimately.

(II) Checking by Practice and Correct Attribution

The saying "Practice is the only criteria for checking truth" raised an ideological revolution, becoming the basic turning point for China to walk up onto the broad road of reform and opening.

Truth must be checked by practice. What has succeeded in practice is reliable and correct. What has failed in practice is unreliable and wrong. Nevertheless, practice is a dynamic process and for all sorts of practical results we must do the correct attribution.

Attribution means making specific analysis of the causes of the successes and failures in practice. The attribution theory is a theory of analysis of the causal relationship in people's practical activities. By attribution, we explain, predict and control the results of success or failure in practice. The causes resulting from attribution may fall into three dimensions in terms of different characteristics; source of cause, stability and controllability. Whether we can attribute correctly plays a crucial role in our achievement. If we attribute the failure in practice to the unchangeable stable cause, we shall not make any further effort from now on, and the practice can be declared as a total failure. If we attribute the failure to the fact that we have not found the correct method, we shall make greater efforts to seek the method, and once we have found the correct method, we will be able to succeed at last.

For thousands of years, all the practices in people's pursuit of eternal life were failures, and we may even say they were complete fail. Hence, it is very easy for people to reach such a conclusion: Practice has given full proof that eternal life is unfeasible and impossible.

Let's take a brief retrospect to the practices of pursuit of eternal life. At first, First Emperor Qin and Emperor Wu of Hand Dynasty sent people overseas to look for the godly mountains and ended up in failure. If the alchemists had really found the "immortality drug" from the "godly mountains", they would have become the greatest savors of mankind, and no matter how high posts or how high salary they got, it would not have been regarded as improper. However, they failed. People attributed the failure to the fact that vegetative drugs could not prolong human life. So they changed their way of thinking, from vegetative drugs to mineral substances which are more stable; mica, quartz, jade, gold and so on, as they thought "those who ate gold would live as long as gold and those who ate jade would have a life as long as jade" (Inner Chapter of *Bao Pu Zi*: *"Godly Drugs"*). The idea that by eating gold or jade one could live forever prevailed for a time, so, the number of people preparing and eating pills of immortality made of gold or jade increased. Yet, what they got was a total collapse of their bodies. By the Wei and Jin Dynasties, people had changed from taking gold pills to taking the five-stone powder, as they thought the powder made of five kinds of stone could prolong life. But it had no effect, either. Later, people changed their ideas again, believing "The pill of immortality turns out to originate from the ego", thus, began to train the active substance—*qi*, in their own body. Yet, no one really succeeded. From begging gods, to begging the self, from vegetative drugs to metal or jade drugs, from taking gold or jade to preparing the cinnabar, from external training to internal training... All their practices failed, and together with the failures, they paid a huge price: from total financial loss to loss of their own lives and everything. How many monarchs, high officials, princesses and dukes, aristocrats, masterminds and personages have died in the efforts. Tragedies one after another and lessons one after another have forced people to hold their wrists in regret. Thus, no one dared to seek the so-called "eternal life" again, and this pursuit almost stopped completely.

Practice seems to have told people in the world very clearly: Eternal life can never be achieved. However, have we done the correct attribution for the failures in the practice?

When Nobel invented the explosive, he failed for hundreds of times within a period of three to four years. In one failure, the whole testing laboratory was bombed to nothing, and five testing workers lost their life, including Nobel's younger brother. Faced with such a heart-breaking fiasco, should Nobel declare the abandonment of the testing? Despite the fiascos one after another, Nobel still kept on his work. Was he a "madman"?

As it is well known, in the invention of the electric lamp, Edison failed more than ten thousand times. Faced with the thousands of failures, we could have two kinds of attribution: One is to attribute the failure to the fact that it is impossible to make an electric lamp. Then, when Edison had tested for 10 times, or 100 times, or 1000 times, he would have stopped his efforts to illuminate the world with the electric lamp. Then, we would not know for how many years mankind would go on living in the dim world of candles and oil lamps. **But luckily, when Edison had failed for ten thousand times, he did not attribute his failure to the fact that we could not make the electric lamp, but thought "I have not failed for ten thousand times, but only found ten thousand ways that won't do": he attributed his failure to the fact that he had not** found the correct way. Can you say that: The truth found out through the 9999th failure is that mankind can never invent the electric lamp?

Similarly, in the pursuit of eternal life, there were lots and lots of failures, but can we simply announce the death sentence of eternal life? In the trial production of the explosive, there were hundreds of failures, and in the test of the electric lamp there were ten thousand failures. In such a supreme undertaking as the pursuit of eternal life, especially in the very low level of science and technology in ancient times, even millions of failures are understandable. Can we declare its absolute impossibility just because it failed so many times?

So, it is obvious that we cannot look at the ups and downs in practice in a mechanistic way, should never consider the temporary failure(s) in practice simply as the unalterable ending, but should do correct attribution for the successes and failures in practice. If we just mechanistically understand "checking by practice", we will fall into such a mistake certainly and decide to give up all further efforts for all cases with repeated failures, without further analysis or discrimination, making it impossible to achieve the success that may be achieved finally.

We need to do realistic, comprehensive and correct attribution for practice, and use a positive attitude to make cool analysis of the subjective and objective causes of the failure. With regard to the ancients' failures in pursuit of eternal life, the greatest mistake we have made is that we have not done a correct attribution.

The ancients' failure in pursuit of eternal life resulted from the limitation of the conditions of the times. The level of science and technology of that time was very low, so they could not find out the correct ways and measures for eternal life. Now, we have possessed the high technology characteristic of continuous change and improvement incomparable to what the ancients had, especially the great leaps in life sciences have offered us many stunning findings, which have paved the golden way for eternal life. In the past, in our marching into the mother nature, almost all the laws we found were laws of the external substances. This time we are marching into the field of our own life, and we are bound to reveal the laws and secrets of our own life. The ancients' struggle for prolonging the life was a blind groping in the dark whereas life sciences have offered us the pearlescent lighthouse that can point to the correct direction. A horse drawn cart can never run as fast as a primitive train, no matter how perfectly it is made. **If we say people's exploration into the secrets of life in the past was even lower than a "horse drawn cart", then the exploration by means of modern life sciences is "a train", or even a "plane" or "rocket".** The genome project has just started, but it has brought us uncountable inspiring news. We have good reason to predict: Life sciences will finally decipher all the secret codes of human life and achieve an unprecedented revolution in the life of mankind.

Famous scientist Rutherford put a crocodile mark at the door of his laboratory. He thinks in scientific research we need a crocodile spirit, take our courage in both hands, and wobble everything ahead. In the pursuit of eternal life, mankind has experienced a journey full of storms, rains, blood and tears. Rains and storms, or blood and tears, should not be reasons for mankind's failure in this pursuit, and science and technology will help us win finally. Life sciences, gene technology and other high technologies will surely help us to realize the dream of eternal life.

We should never be misled by wrong attribution again, and never suffer loss again from wrong attribution. If we give up now, mankind will no longer have self-confidence, no longer make efforts, no longer aspire, then, the happiness of eternal life that might be gained by mankind in the present age will never be obtained.

(III) Absolute Impossibility and Three "Sufficients"

Someone says: Who has ever seen a person with eternal life?

None. From the ancient time till now, 80 billion individuals have lived in mankind and died, but not a single person with eternal life has appeared. Thus, it can be seen, eternal life is absolutely impossible.

However, to this answer "none", we have to add a modifier: "up to the present".

Not having seen, or not having appeared, does not equal "absolute impossible in future". Among the ancient people thousands of years ago, who had ever seen an automobile, a train, or a plane? Who had ever seen the telephone, TV, or computer? They had never seen and were totally unable to imagine them. If someone at that time said there would appear planes, TVs, computers, they must have thought that the person was talking about a dream. All those incredible and absolutely impossible things in the past have become ordinary things today.

We can never conclude that eternal life can never be realized just because it has not been achieved till now. We should never assert that something would never appear just because it has never appeared in the world till now. No one can use any reason to prove we can never make breakthrough to human lifetime limitation, hence, any judgment arbitrarily made about the impossibility of eternal life is a hooey uttered very irresponsibly to mankind.

Since there is birth, there must be death: this is a result of natural selection. Since there is birth, there must be eternal life: this is a choice made by mankind itself. We admit all laws are dependent on conditions, then, why can't mankind create conditions to change the present law and gain eternal life? From non-existence of the condition for eternal life to existence of the condition for eternal life, it is an earthshaking change, and we mankind have the power to bring about such an earthshaking change.

Eminent scholar Zhang Shaohua had a very wise saying in his *Spirit of the Universe*: "*Germs have a history several billion years longer than mankind, but they are still germs today, Ants have a history several hundred million years longer than mankind, but they are still ants today, Hundreds of millions of years have passed, but the turtle is still the turtle as before. On the contrary, look at the life of mankind of the present time, and then look at the life of mankind one hundred years ago, one thousand years ago, and ten thousand years ago... What great changes have taken place!*" People are neither ants, nor tortoises, people can change their own fate causing earthshaking and unprecedentedly great changes to their own fate. The development of modern science in the recent one hundred years can be described as "a thousand miles a day" and "changing with each passing day", especially, today mankind has already entered the century of life sciences, having made quantum jumps in the field of life research. We shall uncover all kinds of secrets of human life gradually, so, many "absolutely impossible" things will become reality in the measurable future.

In this world, so long as we have the "three sufficients", namely— sufficient intelligence, sufficient investment and sufficient time, any

impossible wish may be turned into vivid reality. Eternal life is no exception. So long as we have sufficient intelligence, sufficient money input, and sufficient time, eternal life will be realized for certain.

We have the sufficient intelligence, i.e. we can unite all the intelligence of all mankind, not merely utilizing the intelligence or wisdom of some people, but uniting all the mental power of mankind. By global computer networking, and by constant upgrading of intelligence, human intelligence will be amplified by zillions of times. By focusing such a matchlessly great intelligence on the technology of eternal life, the potency, efficiency and achievement of eternal life technology will be zillions of times more than now. By concentrating mankind's limitless intelligence on the undertaking of eternal life, we shall have the "sufficient intelligence", by the greatest input of financial resources, material resources and human resources all whole mankind in the undertaking of eternal life, we shall have the "sufficient investment". What we shall invest will not be several million dollars, or several billion dollars, but billions or zillions of dollars. The human power will not be a handful of men on the earth, but billions of people joining this great undertaking. To the sufficient intelligence and sufficient investment, we shall add the sufficient time. If 30 to 50 years is not enough, let's use 100 years, if 100 years is not enough, let's use 200 years, 300 years or longer. How can you say eternal life can not be realized?

Eternal life is mankind's eternal wish and in order to realize this wish, our ancestors have made unremitting explorations. In the present time characterized by great leaps and bounds in science and technology, and by unprecedented liberation of man's thinking, why can't we break through the old defective and superficial concept? Why should we restrict ourselves to a limited area like a prison and regard eternal life as a forbidden zone? All things in the world can be changed through mankind's efforts, then, why can't the duration of human life be changed through the joint efforts of all mankind? Science and technology have already turned lots and lots of uncanny fantasies into vivid realities, and what's more, now that the development of science and technology is being increasingly accelerated and their power is becoming more and more magical, why can't we turn the fantasy of eternal life into reality in future?

Any person with a good head and sufficient wisdom should believe firmly: So long as we have sufficient intelligence, sufficient money input, and sufficient time, mankind will realize eternal life for certain.

XVI. WISE PEOPLE' CONSIDERATIONS

Balzac engraved a motto in his walking stick: "I have broken obstacles one after another", but Kaffka said: "Obstacles one after another have broken me".

In the great undertaking for eternal life, it is very necessary for us to draw wisdom from Sunzi, the great strategist and sage of ancient China: *"A wise man's consideration must involve both benefit and harm. When it concerns benefit, it will lead to credible practice; when it focuses on harm, it will result in solution to problems."*

We should not just expect the benefits to be brought about by eternal life and underestimate the price for it.

We need to make clear what prices are, what we must pay and what prices are avoidable.

We should neither borrow trouble, nor be optimistic blindly. We need no shallow optimism, only profound optimism, that is, optimism that comes after seeing the misery.

We neither need flubdub or gag, nor do we need alarmism made deliberately for mystification, but we only need to think about problems in a rational, logical, objective and dialectical way.

On one hand, we should have a full imagination of the prospect of the great benefit to be brought about by eternal life and on the other hand, we should never evade or ignore the possible great worries and all kinds of problems to be brought about by eternal life.

When we take a deeper look at "eternal life" or "prolonging the lifespan", we will be sieged by various questions immediately.

Someone says, eternal life will result in population explosion and population explosion will overburden the earth...

Someone said, prolonging human life will aggravate ageing of population, and the 21st century will become an era of old people. In ancient times, there was once such a practice: people carried old people above 60 on their backs to the wild forest to feed them to the wolves. Who can guarantee such a tragedy will not happen again?

Someone said, population explosion and life-time dilation will cause an even greater unrest to the employment market, and the "food" problem will worry numerous pitiable people to death...

Someone also said, that along with food, energy sources will also be exhausted, so, the food crisis and energy crisis will follow in close succession...

All such problems have turned "eternal life", this most attractive dream, into a big headache for many people. Here, what we need most is **to shift from the infant's way of thinking of "black or white" to the wise man's way of thinking that considers both benefit and harm.** For the benefit, we should think to the extreme. For the harm, we should also think to the extreme. And in considering the harm, the purpose is to think up counter measures so as to prevent the harm.

Population Explosion

Of all the problems, the first is the population problem. The present-day population explosion has already become a great headache for mankind; if everyone lives without death, how terrible it will be!

In fact, the population problem is not insoluble. Above all things, we should be practical and realistic. The greatest problem being confronted by the developed countries now is not increasing population, but the declining population growth rate, and in some countries and regions, even zero growth or negative growth has appeared. According to the latest data published by the European Union's Statistics Department, the population of many countries in the European Union is showing a negative growth, so, population shortage has begun to be a big problem. The negative population growth problem is very serious in Germany, Italy and Sweden. The population growth of the other European countries is also declining increasingly, thus, United Nations suggests taking in immigrants to solve the problem. In most metropolises in the north of the USA, a negative growth has also appeared. Japan's 135 million people now will decrease to 50 million by the end of the 21st century.

The present population of Italy is 60 million, and according to the current rate of negative growth, it will be only 20 million at the end of the 21st century. Shanghai, in our country has also shown a negative population growth for 8 years running.

Population explosion or population shrinkage depends on people themselves. Production of people themselves is just like product production. It can be controlled and adjusted completely at our will. It is even so in the present time characterized by high level of civilization progress and high development of technology.

Gary S. Becker, winner of the Nobel for economics in 1992, has a very good saying: *"Education is the most effective contraceptive for a country to control its population."* Facts in the developed countries have shown, education, especially the education for women, may lower the fertility rate greatly, and even may result in many childless families. A low birth rate is an inevitable trend of civilization progress. The more the civilization progresses and the greater efforts we make in education, the easier it will become to solve the problem of population explosion.

We should never underrate people's ability to control their own childbearing. Developed education and advanced values may cause people to bear fewer or no children voluntarily. State laws and regulations or government policies may play a restricting role, causing people to bear fewer or no children. In developed regions and nations, the population growth is increasingly moving towards a standstill, and in developing countries, they have begun to take various measures in birth control. China is a successful example in birth control. In 1970, China's birth rate was as high as 33.59%, but at the end of the 1970's, the central government raised the call of "One Couple, One Child", so, in 1980, it dropped to 1.19%. From 33.59% to 1.19%: this is the power of the government policy! In a period of 20 years, a birth control policy reduced the birth by 300 million in China, which made a great contribution to mitigation of the population pressure of the world. Today, many demographers predict: If we continue executing the birth control policy well, 30 years later, a zero growth or negative growth of population will surely appear in China.

On one hand, mankind can control its own production, controlling the quantity of the population. On the other hand, mankind is able to utilize science and technology to exploit new fine worlds, building up a broader homestead for itself. As far as the present situation is concerned, the major ways out are the following: **March into the underground, march into the oceans, and march into the outer space... Underground cities, undersea worlds and outer space homesteads will become new living spaces for mankind.**

The most realistic and most promising one is, marching into the oceans. On our earth, the land area accounts for 29% of the total area whereas the sea area accounts for 71%. That is to say: 7/10 of the earth territory has not been exploited and utilized yet. No wonder many scientists say the 21st century is a century of the sea.

The sea is a matchlessly huge treasure house, can provide very rich food, energy source, minerals, water source, industrial chemicals and so on. The theoretical reserves of the regenerative energy such as the heat energy of seawater amount to 150 billion kilowatts or so and what can be exploited and utilized right now is more than 7 billion kilowatts; equivalent to more than a dozen times of the total amount of electricity generation of the present time. Seawater contains rich chemical elements and presently we can extract more than 80 kinds, including uranium, deuterium, tritium and so on. The fishery resources in the sea are very abundant: There are more than 200,000 marine organisms, and even the protein that can be provided now alone can account for about 1/3 of the edible protein of mankind. Judging from the most realistic aspect alone, by utilizing the sea, we can solve our energy crisis very easily. Moreover, we may also build floating towns, sea dormes, underwater cities, undersea worlds and so on to expand mankind's space of survival and development on a large scale. Almost all the great problems faced by the present society can be solved by utilization of the sea.

Today, constructing underground cities and undersea worlds and even outer space emigration are no longer pure imaginations, but are realistic. Taisei Kabushiki Kaisha of Tokyo is building an underground city, "Alice City". It may mitigate the problem of narrow land in Japan, and may also avoid the impact of earthquake. In some areas of the underground city, the residents can also take a glance of the sky and stars through the transparent vaulting. Japan has set forth its ambitious plan to build 25000 floating towns in the sea near Japan within the 21st century, each town will be attached to 10000 thick pillars fixed at the ocean floor, and people may reach the town by plane or hovercraft. So, it can be seen, while controlling the population, mankind can open up great new space and new worlds for holding its population. Hence, many scientists, such as the Nobel Prize winner in economics and US economist Gary S. Becker, think: *"The population problem of the third world, especially that of China, is being increasingly exaggerated. Even if the population may cause some troubles, all the problems can be solved."*

Old Age Is Still Best Age

In prolonging human lifespan, many people worry about not only population explosion, but also the resulting ageing problem, thinking that an ageing society may cause a great shortage of work force and a great drop in productivity, thus greatly lowering the living standards.

Is the ageing problem so terrible? In terms of the standard of International Population Organization, when the ratio of the ageing population is over 7%, the aging of population begins. By this criterion, the world falls into three types of countries: First, the least developed countries in Asia and Africa such as Rwanda, Chad and Kenya of Africa, and Maldives of Asia. Second are youth-type countries, with the lowest ratio of old people and the highest ratio of young people. Third are the adult-type countries mostly concentrate in the developing areas of Asia, Africa and Latin America. Almost all the old-age-type countries are developed countries, for example, the ageing population of the USA accounts for 12.6%, so it is a country with heavy ageing. So, you see, ageing cannot lower productivity, nor will it lead to shortage of work force necessarily.

We need work force all right, but what we need is an advanced work force. An advanced work force does neither refer to people toiling in the fields all day long, nor does it refer to laborers working with both hands and feet in the deafening noise of a factory. Only by possessing wisdom, information, science and technology can we possess the most powerful work force. According to a scientific estimation, the annual workload of the USA with only 200 million at the end of the 1980's is equivalent to the annual workload of 40 billion people in the world. What they rely on is not giving birth to many more young people, but several million computers only.

In the USA, just two persons, father and son, own 1800-acre farmland equivalent to the cultivated area of a middle-sized township in

China. The father and son can do the work of several thousand farmers of China, and besides this they raise 1000 pigs and scores of milk cows. They also use their domestic computer to inquire about the information on the grain market of the world, and sell their own grain through the futures market...

We have passed the agriculture age and the industrial age and entered the information and science age. Thus, our work force has changed from using physical strength mainly to using brainpower mainly. The disadvantages of the old age, such as physical strength and energy, are no longer the most important factors, and what are most valuable are knowledge, wisdom and experience, that grow together with the age. Premier Zhu Rongji, just like Jiang Ziya of ancient China, became a premier at seventy and he showed a brand-new image before the whole country and the whole world, winning admiration from all walks of life. Can you let him retire at 60? Does he contribute to the society less than young workmen? Can't you see that many state heads are senior citizens? Isn't their wisdom and experience still the most precious wealth of the society? Aren't they still dominating and determining the state's fate and the people's well being?

Ageing may increase burdens for the society, but will also accelerate social progress. Burden or progress, just depends on how we utilize it. In fact, the oldsters today are much better than the people of the same age in the past, both in health and intelligence. It is not only because of medical progress, but also because of the work they are doing.

World famous managerialist Peter F. Drucker suggested: **"Now the proper retirement age should be 79"**. According to the anticipation of the long life in future, people's retirement age may be put off greatly.

We should never keep the definition unchanged. If someone asked a farmer living in the 14th century whether he wished to live to be 70 years old, he would feel very laughable for certain, as their anile parents might have passed away at 40, and living to the age of 70 would mean being more decrepit than their parents. In fact, people at 70 may be younger and stronger than people at 40 of that time. In ancient novels, we can see such a description: *"The door was opened, and in came an old woman in her thirties."* Even now, the definition of old people is also different from the recent past. In the past, the definition of each age group was like this: 4 - 14 as the childhood period, 15 - 34 as the young period, 35 - 44 as the middle age period, 45 - 59 as the earlier old age period, 60 - 89 as the old age period and people above 90 were considered as longevous oldsters. But now, according to the definition of the World Health Organization, it is like this: Under 44 as young people, 45 - 60 as middle-aged people, 60 - 75 as youthful oldsters, 75 - 90 as oldsters and above 90 as longliving oldsters. After the revolutions in modern science and technology, biotech, medical treatment technology, food and drink structure and life style, it will not be rare to have 80-year-old boys or 100-year-old lads.

By that time, a 100-year-old person will be just in his (her) "flowery years" just like a boy or girl of 20, hot and strong... 100 or 1000 of "venerable age" will still be "youth". Then, there will be 600-year-old boys or 800-year-old girls everywhere in the society. The "getting old" of that time will be "reaching a high age only" in fact and to be precise it can be called "high age development of youth". How can that seem terrible?

Wise Ways to Create Jobs

Unemployment, this common pertinacious disease of all countries, is also a terrible headache for many people. If you lose the job, means lose the right to work, your right to exist will also be affected directly. But this hard problem is not insolvable, either.

Eminent economist Zhong Pengrong is famous for creating jobs for China, and he suggested four ways to create jobs: First, land exploitation. By opening up wasteland, utilize surplus labor to make forests, pastures, fishing banks and fertile land. Second, road construction. By expanding infrastructural construction, utilize surplus labor to build railways, highways and water resource facilities. Third, human resource development. Encourage social investment for great development of education. Fourth, environment construction. Let more people engage themselves in work of amelioration of the living environment of cities and towns, such as urban greening, city beautifying projects, decontamination and so on.

Of the four ways, I think the most important is human resource development. That is the fundamental solution to job creation. Make great efforts to develop education, not only lay stress on elementary education and higher education, but also pay close attention to all kinds of vocational education, support and encourage all kinds of skill training and quality training, especially encourage private investment in educational training. What is more important; we are in dire need of constructing a really sophisticated life-long education system, so that anybody may get the education he needs at any age, in any place and in any case.

A great characteristic of human resources is the "changeability"; resting on past achievements may devalue an "able person" into a "mediocre person" whereas by training, a mediocre person or idler may become an "able person". In the new economy era, there is a slogan: *"Any time is a new start."* Things are changing all the time. Our era is just characterized by abrupt changes. Unemployment often results from the inability to adapt to the new condition, and by education development we may help people adapt themselves to new environments, new conditions and new requirements at any time. For example, the USA takes technical training as the basic solution to employment. According to statistics, more than a half of American people have found jobs after participating in employment training.

Education development may contribute to the solution of the unemployment problem in three major ways: First, it adds many employment posts, enabling more people to engage themselves directly or indirectly in education training work. Meanwhile, as far the trainees are concerned, it may postpone their time for employment, which may mitigate the employment pressure. What is more important that it can raise personal quality, so, the laid-off workers may become competent for new occupations and new posts. This can solve the unemployment problem by the root.

In an era when science is the first productive force, it is easy to find that out of all trades one trade is special: here, the more the people, the better. No matter how many people are involved in it, it does not matter at all. What is the trade? It is the research industry. The more people participating (in basic research, application research, development research and so on), the better, and there is never a problem of "overcrowding". You may respond at once, how difficult it is! In fact, it is not so hard if you look at the matter from a different angle. The key lies in work from the source, education. So long as education is really improved, it will not be a hard thing for many people to go into research.

The present time has given the common people more and more conditions for scientific research. The major "means of production" for scientific research is knowledge. The fundamental difference of knowledge and other means of production is that any other means of production can only be monopolized for example when others have it, you will lose it, and in the period of use, it will be worn off and devalued all the time. But knowledge as a means of production can be shared and will not be devalued because both you and I know it, or because both you and I have it, or because both you and I use it. In the Internet era characterized by "all mankind is one family", it has become easy for anybody to get whatever knowledge and information they need. **Today, even an ordinary child can make the atomic bomb by using the atomic bomb design data provided by the Internet!** Our time has become a time for individuals. So long as you have a certain academic ability, by sharing wisdom with others through the Internet, you may be competent for scientific research. The megatrend of future development is characterized by more and more people doing scientific research.

Judging from the macro aspect, unemployment is often a structural unemployment. While people are complaining about saturation of posts, many things which need to be done, are not being done by anyone. According to statistics: in the seven big cities of China alone, the vacant posts for house upkeep and mending workers, hour-workers and home tutors are nearly 1 million. This fact alone may offer jobs to millions of people in the country. Of course, there are many other trades with vacant posts, especially those newly arising trades and high technology trades.

Moreover, **there might be a very great revolution in the future salary system, that is, people will not get a monthly salary or yearly salary, but get an "achievement salary".** Nobody will need to work for one institution forever, and anyone can sell his own "achievement" to anyone, any enterprise or any institution. Everyone's working place, working hours and working mode will be personalized, individualized, free and flexible, without any constraint, that is, so long as you can provide achievement, it is fine. This measure will also broaden employment channels greatly.

In solving of the unemployment problem, we may also give play to the power of policies and go all out to encourage people to create new jobs. Take the USA for example, in 1950, it set up 93,000 new enterprises, and by the 1980's, it could set up 600,000 new enterprise every year. Of all the new job opportunities, 2/3rd are offered by minor enterprises that employ no more than 20 persons. In Britain, the required registered capital for setting up a company is only 1 pound, and it is basically zero in the USA, Australia and some other countries. The condition of our country has much improved, in comparison to the complicated approval system characterized by numerous levels of examination and many barriers in the past. However, if we further reform to reduce the registered capital requirement to 1 Yuan RMB, it will make enterprise establishing much easier, which will certainly encourage many people to join the rank of self-employment. This policy alone may bring about a great solution to the unemployment problem. Thus, it is obvious that there are great potentials in the policy aspect for solution of the unemployment problem.

"The More People, The More Chickens"

Mankind's eating problem is serious of course. Living without death might cause population explosion and shrinkage of per capita cultivated area with each passing day, then, will the people have enough food to eat?

Mr. Lolin Clark, former dean of the Agricultural Economy Academy of Oxford University said: **"Suppose all farmers have adopted the best farming method, the food they produce may be enough to meet the need of 35.1 billion people at the standard similar to American dietary standard, and such a population is equivalent to 7 seven times of the present world population. If we use the comparatively economical dietary standard of Japanese style, we might use the food to supply 21 times of the current world population.** In the USA, farmers only account for 2.5% of the total population, but they can not only feed the people of their own country, but also export a lot of agricultural products to the world. Since they can use a little over 2 million farmers to produce food to feed 1/5 of the population of the world. why can't other countries use only 2.5% of the population to solve their own food problem?

US thinker and socialite Henry George used a metaphor to express the epistemological foundation on which many economists can reach consensus easily: *"Both eagles and people eat chickens, but the more the eagles are, the fewer the chickens become, whereas the more people, the more the chickens."* More and more scientists have come to realize that we have the ability to produce food at a speed higher than the speed of population growth. It reminds me of Malthus' saying in his Malthusia-nism: *"The growth of the population is in a geometric progression while the growth of food is an arithmetic linear growth, so, population growth will be faster than food growth"*, thus, he concluded that *"Mankind is bound to live in poverty and famine"*. Nevertheless, two hundred years have passed after his saying, our life is not only free of poverty and famine, but is much richer and happier than two hundred years ago.

What is more important: by adopting biotechnology, we may free our agriculture from the bondage of the natural world, solving the grain problem very easily. In 1997, China released its *No. 5 Report on National Condition*, which forecast the grain problem of China in the 21st century. The report's opinion is: *"The final way out for China's agriculture is the biological engineering."* Along with the development of high and new biotechnology, the future grain production will not depend on the land, but will depend on the factory: In the factory alone, we will be able to produce food enough to meet mankind's demand. Then, one pill, or one bean may be equivalent to the nutrition of three meals a day.

By using the factory to produce food directly, we can save a lot of land and hence we may optimize and beautify our earth according to the ecological law, so that every place will become a green homestead with a beautiful environment, agreeable climate, green mountains, clear waters and clean air. Thus turning all parts of the world into places more beautiful and healthful than the famous five major longevity zones of the world. Biotechnology is a good remedy for environmental protection. First, we may develop new species of crops without the need of pesticides, herbicides or fertilizers, so as to prevent environmental pollution. Second, we can do monitoring and rectification for the already polluted environment. For example; scientists have used the gene engineering to work out a weed able to absorb heavy metals, so that all the harmful substances may gather into the weed. The environment after being treated by this grass will become very clean. So we can see, all big problems caused by eternal life that have worried many people, such as the grain problem, the environmental problem and so on, may find the best solutions from science. In resources, we shall depend less and less on conventional resources such as ore, timber, metals and so on. All the high-tech products we need; such as the chip of the integrated circuit, rely on silicon as the raw material and silicon is extracted from sand, which can be found everywhere. And the biotechnology as our major technology in future will use biotic resources, which are inexhaustible in fact.

In the measurable future, we may have the "technology that can solve the energy resource problem of the world completely." It is, the "technology of changing seawater into gasoline". By using 1 liter seawater, the energy generated from the controlled thermonuclear fusion will be equivalent to 300 liter gasoline. Seawater can be considered as limitless; hence mankind will get an inexhaustible source of energy.

The future era will be a "new solar era", as the solar electric power generation has limitless potentials. According to scientists' estimation the energy from the sun illumination on the earth in 30 minutes is equivalent to all the electric power consumed by the world in a whole year. According to Qian Xuesen (famous Chinese scientist)'s calculation: on the 9.6 million square meters of land of China, the solar energy received in one year is equivalent to more than 1600 billion ton standard coal. If we install the devices for conversion of solar energy into electric energy on 1/10 of the land, calculated according to the energy conversion rate 10%, we may get the energy equivalent to 16.5 billion ton standard coal in one year.

World renowned scholar Symon said: *In talking of the future, it is unavoidable for people to worry about some "crises" such as resource depletion, environmental pollution and so on, but it is just these "crises" that are forcing mankind to look for new resources, new energy sources, and new "ways of survival".* If people had not met with the crisis of insufficient wild rabbits and root crops ten thousand years ago, maybe mankind would still be eating wild rabbits and root crops even today. He said: *"knowledge is the impetus to world progress and only a lack of well-educated people and a lack of imagination constitute the impeding forces to world progress."*

In the past, in the hunting and fishing civilization period; food crisis occurred for a time. In the agriculture civilization period, land crisis appeared for a time. In the industry civilization period energy crisis took place for a period. All those crises have been overcome by mankind. Then, in the present time with the power of high technology becoming more and more magical, why can't we overcome the crises to be brought about by our pursuit of eternal life?

Anything has both advantages and disadvantages, and you can never find anything that has only advantages without disadvantages. Automobile transport is very advantageous, but traffic accidents occur everyday, resulting in many deaths. The airplane is more advanced, but once an air crash happens, it will result in much more deaths. If we give up something that has a great value and benefit just because of its disadvantages, then it will be just like giving up eating for fear of choking, which is never a wise choice, of course. Hence, we should never give up our research and exploration into this great undertaking of eternal life just because of some possible problems, or some worries and doubt. We must be wise people, must make clear whether the advantages are greater than the disadvantages or the opposite, and make clear where the advantages lie and where the disadvantages lie. Any in-depth analyses and research on the possible disadvantages are for finding out the solutions, so as to make the disadvantages "removable" and the troubles "soluble".

Balzac engraved a motto in his walking stick: *"I have broken obstacles one after another"*. But Kaffka said: *"Obstacles one after another have broken me"*. While bringing about huge benefit, eternal life will also bring about obstacles one after another. The question only lies in: "Whether "we have broken the obstacles" or "the obstacles have broken us". It depends on us completely.

PART III - MEGATRENDS OF FUTURE HUMAN DEVELOPMENT

夢星

INTRODUCTION

Shao Xiaobo is a student of Class 8 in Grade 1 of Shanghai Yuanfang Senior Middle School. In the eyes of his teachers, classmates, and his neighbors, he has always been regarded as an incorrigible numskull.

In the English test this morning, Shao Xiaobo failed again, and was scolded by his English teacher Lao Cai thoroughly. The teacher cursed Shao Xiaobo as a fool, saying he had held back the whole class, saying he had a "granite head"... The teacher poured out a stream of abuses against him, with spittle flying everywhere.

Shao Xiaobo hung his head without saying anything, as he was used to the English teacher's cursing. Fat Zhang sitting behind him was speaking something in a whisper, but he turned a deaf ear to it. The class was over at last, and he just could not wait to dash out of the classroom to the playground. Playing basketball is the only fun he had in his dull campus life. While playing basketball, he chatted with his fellows.

"Don't you know? Hawking is coming to visit China." Shao Xiaobo spoke up.

Anybody knows Hawking is Shao Xiaobo's idol. He is the man who contracted the incurable disease of muscular dystrophy as early as the age of 21. He cannot speak, can hardly move (even raising his head will take him a great effort), is leaning in the electric wheelchair all day long, only relying on several fingers of his right hand that can be moved freely, to drive his wheelchair or control the computer... Such a man should be able to become a great star in the domain of science! Shao Xiaobo has always been full of great respect for Hawking's genius and perseverance. His greatest dream is to go to University of Cambridge and be able to chat with Hawking. But he can only bury this dream in the bottom of his heart, otherwise, others will laugh their heads off.

Just then, a basketball flying from one end of the playground hits his head. He felt darkness before his eyes suddenly and fainted down in spite of himself...

When he wakes up, he hears some strange voice of talking besides the singing of some birds.

"He has woken up!"

"Where has he come from?"

"I have just tested him. He has been pushed here by the airflow in the space-time tunnel. It is a possibility of part per trillion, but he has fallen across it!"

Shao Xiaobo opens his eyes with difficulty. A pretty girl with dark brown hair smiles to him, "Welcome to the world of 2050! You have 5 minutes to stay. What wish do you have? What do you want to know about the world of 2050?"

Shao Xiaobo opens his mouth in surprise. Luckily, he is an earnest reader of Wesley's novels from childhood, so he says: "Is what you said true?"

The pretty girl smiles without remark. "Your time is limited. Please visit the supermarket nearby. Today is the date for free dispatch of 'body optimization' products. All the products here are labeled with any of the three markers, 'birth optimization through technology', 'education optimization through technology' and 'body optimization through technology', so as to attract people to buy. Oh! You may not understand what a 'body optimization' product is? Look, now Mr. Seim is handing out 'I Am Tender and Kind' hamburgers and 'You Are Vinci' biscuits. As soon you eat the bread, you will have the painting talent of Da Vinci. Also, we have A-typed fruit juice that can change your eyes into phoenix-typed eyes, and we have C-typed almond milk that will change your eyes into the shape of a crescent..."

"It sounds so interesting. Do you have something that can make people clever? That is, something that can help me learn English easily."

"That is so simple." In talking, the pretty gal passes a piece of semi lunar chocolate to him smilingly, "Swallow it, and it can help you learn any language in the world easily." "Wow! So wonderful!" Shao Xiaobo picks it up half believing and half doubting, puts it into his mouth. Oh, very delicious!

"What other wishes do you have?"

"Hum, I...I..." Shao Xiaobo scratches his scalp, and says in a low voice: "I also wish... Like Hawking, go to Cambridge and become a physicist..." "Hawking?" The pretty gal frowns a little.

"Hawking is the greatest physicist of our world; is the present 'Einstein'!" Shao Xiaobo is protecting his idol whole-heartedly, without hesitation.

"I know all this. Hey!" The gal gives a deep sigh, saying: "Today, for that kind of disease Hawking contracted, just by eating three BOB specially-made pineapples continually, he could recover completely."

"That's too good! May I take some for him?"

"No way. The space-time tunnel does not allow anyone to take away anything belonging to this era. If people of your era make concerted efforts to develop the technology for 'birth optimizing', 'body optimizing' and 'education optimizing', then, the time of breakthrough may be moved up greatly, so, your idol, Hawking, may also stand and walk in his life time." After saying this, the gal passes a bottle of milk shake to him hurriedly, "Oh, your time will be over soon. Drink this bottle of milk shake, hurry, then you will become a genius in physics." Shao Xiaobo grabs the transparent sucker and drinks at a draught.

Just after drinking, Shao Xiaobo feels a golden glint before his eyes all of a sudden. His body is rolled up as if by a tornado, and then, he is sinking, sinking... "Xiaobo, Xiaobo, wake up, wake up!..."

"Xiaobao has woken!" A group of classmates were crowding around him.

"I..." Shao Xiaobo was looking at his classmates hesitantly, seeming both familiar and strange.

"Nothing wrong with you? You were hit and fainted down just now. However hard we shook you, you did not wake then." Seeing Shao Xiaobo was safe, they scattered.

In the final examination before long, Shao Xiaobo won No. 1 in English and Physics in the whole city. It was beyond belief for all the people, especially his English teacher Lao Cai who was shocked so much that his eyeglasses dropped from the bridge of his nose and the lenses broke into pieces! Soon after, Shao Xiaobo won the first place in the national Physics contest of middle school students, and then he participated in the world Physics contest of middle school students and won the Gold Cup. What surprised the world were not only his gift in Physics, but also his talent in language. Shao Xiaobo not only had mastered English, but also could speak scores of languages in the world, including the ancient language of a tribe in Africa! *Thames Morning Paper, New York Times* and other newspapers in the world poured praises one after another, saying Shao Xiaobo was "the youngest and most promising language genius, and the youngest and most promising physics wizard". Oxford University, University of Cambridge, Harvard, Stanford University, MIT and other prestige schools of the world were vying with each other in holding out the olive branch to him. Shao Xiaobo's former teachers, classmates, and his neighbors began to take delight in talking on everything about Shao Xiaobo, and all those foolish stories of his in the past became interesting anecdotes about his "Still water runs deep".

Just as Shao Xiaobo accepted the invitation from University of Cambridge and was going to study for the doctorate degree in physics, he heard a piece of news: UN Headquarters was going to have a global debate concerning the fate of mankind, and the attendants included top scientists and sociologists of the world, leaders of all countries, some representatives of private associations and so on. The subject of the debate was "Agree (Disagree) to Using High Technology for Development of Mankind Itself". At the meeting, the opposition group got the upper hand, and some scientists even thought: Faced with unknowable prospects, stop the research without further ado! The opposition group was using various means such as journalism, radio and television broadcasting, the Internet and so on to exaggerate the panic about mankind's future, so, more and more people began to fear the development of high technology, and more and more people agreed that the whole world should make a law to prohibit the research on the use of high technology by mankind itself.

Seeing this, Shao Xiaobo was burning with anxiety: What is to be done? Mankind's future is not so!" I must tell all people! We must have confidence in the future!" Shao Xiaobo wanted to speak out. But on a second thought, he found: If he told the world his five-minute short experience in the year 2050, then, from now on, everyone would no longer regard him as a genius and worship him, and hence there would be no eye-catching aureole, no happiness and peace in his life, people would take him as a monster, or even take him as a subject for research... "what should I do? Say or not? Xiaobo, are you just for your happiness, or for the happiness of the 6 billion people?" Thinking this, Shao Xiaobo made up his mind to stand out to tell his own experience to everyone. Tell the world to develop high technology to improve ourselves, and tell them our future will be wonderful beyond description. At once, Shao Xiaobo contacted the several hundred eminent mass media of the world and held a press conference at Shanghai, to open his secret to the world...

Soon after the meeting, Shao Xiaobo gave up resolutely the chance to go to Cambridge for study, and began to devote his own body and soul to the cause of propaganda for the bright future of mankind. He got the support from many people in the world. Among them, there were people who had been in the most radical opposition group. Some admired Xiaobo for his noble mind in giving up his own interest for mankind's happiness. Some donated all their properties to support Xiaobo's public welfare undertaking... At their strong support, in a very short time, Shao Xiaobo's web site and Shao Xiaobo's "Future Confidence Foundation" came into being in succession.

On the Head Page of Xiaobo's Web Site, he wrote impressively:

Mankind's fine future is the consummate overall development of everybody!

Mankind should take resolute moves to practice the undertaking of birth optimization through technology, body optimization through technology and education optimization through technology.

XVII. NEW DEVELOPMENT THEORY IN THE NEW CENTURY

All problems are chiefly ascribable to the development problem, whereas the soul of the development problem lies in the development theory.

The problem of becoming "gods" is actually the problem of human development.

Development is the theme of the world of the present age, and is also the theme of the present China. *"Only development is the hard truth"*, this famous saying by Deng Xiaoping has already become the tenet in the minds of all the Chinese. Development concerns success or failure and survival or death. Only development can save China and save the world. The whole China tries its best to seek development, and so does the whole world.

All problems are chiefly ascribable to the development problem whereas the soul of the development problem lies in the development theory. Regarding development what we must think first of all is: What kind of development do we need the most? Today, people tend to regard development just as social development and economic development. They even simply use economic growth to replace development in general, and economic growth is chiefly reflected by a few economic indicators such as the GDP. Such a traditional development theory may easily lead people towards the simplex market development, commercial development and material asset development. For this kind of development, they play hard, even if it may cause resource depletion or environmental deterioration. Thus, man has degenerated as "an economic animal", "a slave of technology". Although the appeal to sustainable development is getting more and more people's consent but hardly any governmental decision makers have really responded to it in action. The sticking point of all these problems just lies in the problem of development theory.

The development theory plays a guiding and decisive role for development. The old development theory taking economic growth as the major purpose may lead people to serious unavoidable mistakes. Unlike the old development theory, the new development theory may not only help us to come out of the mistakes, but also bring about the best development for us. The major meanings of the new development theory are like this:

Take human development as the core value, and let human development and social development interact with each other.

While valuing economic development, value human development all the more.

First of all, human being is the purpose not the means. We do not develop economy just for the sake of economic development, nor do we develop the society just for the sake of social development, but completely for the human beings. The purpose Marx stressed repeatedly from beginning to end is to realize *"the thorough liberation of mankind and the overall development of the humanity"*. In Marx's eyes, the humans have always been the purpose rather than the means. Nevertheless, in the eyes of many Chinese, such a notion as *"It is not easy to find a three-legged toad, but two-legged persons can be found everywhere"* is still very popular. Such a notion results in overlooking of people's life, people's survival and people's development. For example; the scholar of a local area doing AIDS investigation was dry-docked and deprived of his salary, because the local officials thought disclosing the AIDS condition of the place would affect the achievement of their post and damage the image of the government. Later, they showed a concern for AIDS, but it was not owing to their concern for the people, only because some scholar's estimation worried the leadership of the department concerned. Once AIDS got spread, would it affect the economic construction achievement since the reform and opening? Look, the standard of value of all things is the economic construction achievement not people's life and people's sufferings. Isn't it a saddening heterization of the purpose?

Secondly, human development is the foundation stone of all kinds

of development. All problems in the world are problems of people. There is such a story: While a man was busy with his job, his son dashed up to pester him for playing. He could not stand the harassment, so he tore a world map to pieces and threw them to his son: "Darling child, go and play yourself, and come to see me again after putting this map together." In a while, his son came to him: "Papa, I have done it!" The man was surprised, as he had thought it would take the kid several hours to piece it together. How could he finish it so soon? His son said, "The back of the world map is a person's portrait. After piecing the person together, the world is pieced together at the same time." These words are just like a direct sharp warning, containing a long-lasting aftertaste: So long as the people are upright, the world will be upright.

Look at this world full of splendors and crises at the same time. Does it result from God? No, it results from humans. As the old saying goes, *"What caused the six states to perish was not State Qin, but the six states themselves."* (In the Warring States period more than 2000 years ago, Qin, one of the states wiped out all the other states and unified China) Likewise, what can cause the world or mankind to perish is not others, but mankind itself. What can save the world or mankind is not any other thing, but mankind itself. Only when the people develope can the society and the world develop.

Development of the people is the foundation stone of social development. In general, the society's existence depends on the existence of the people as its precondition. Without people, there would be no society and without development of the people, the society would have no way to develop. Meanwhile, development of the people and development of the society are in a two-in-one organic system, supporting each other and acting on each other. Only when everyone is fully developed can it promote the full development of the society. And only when the society is fully developed can it provide the sufficient conditions for full development of the people.

As it is well known, some aspects of the present society have already entered the information society and knowledge economy society, yet, if the majority of people in this society are still in the agricultural age tilling the field all day long without knowing what a computer is, how can we build up a knowledge economy society? How will such a society be able to keep a high-speed of development, and how can it grow by leaps and bounds all the time? We have to face such data: A car that can be produced on the flow production line just in a couple of minutes in Japan equals the total income of a farmer working for 100 years in Shaanxi, China. The value created by an American professional working on the computer for one hour is equivalent to the total value of thousands of Chinese farmers working in one day... The social development of any nation or country depends on the brainpower of the people of that nation or country, and depends on the development of their intelligence. The final winner can only be the nation that has the most people mastering high technology and having high intelligence.

Not only human intelligence development concerns the development of the society directly, but human morality development is also directly related to social development. In a society composed of people with a very low moral standard and people intriguing against each other, there is a serious internal friction of the social powers. This mutual distrust causes the social transaction cost to rise day by day. Lies and flames fly everywhere, the meat is injected with water, the wine contains toxin, the doctorate or master degree diploma is bought from a street hawker, the speeches at the official meeting are not believable to anyone, those who curse corruption most loudly are just the most greedy on the sly, false contracts, false data and false accounts are everywhere, even to the extent that the school motto Premier Zhu Rongji gave to an accounting college was *"Do Not Make False Accounts"*... Nothing is without falsehood, and nothing can be trusted. A society composed of such people is full of a sense of insecurity even in the air, being beset with crisis everywhere. Today, the ecological crisis has alarmed the people. But who knows the moral crisis is even more terrible than the ecological crisis and the blustering sand devil of the mind is even more heart breaking? Someone said, *today's China is full of people having no belief, no faith, no confidence and no credit.* If we still overlook people's moral development, can this society still develop? Even if it is very prosperous and rich on the surface, it can only be a transitory "bubble development", very easy to go to the dead end of perdition.

According to a study done by world-renowned scholar Ingles on modernization, the key to the modernization of a country is modernization of the people. If a country only borrows the fine blueprints and measures from other countries and the people carrying out these blueprints and measures have not undergone a transformation to modernization in ideology, mentality, attitudes and pattern of behavior, then, failure and lopsided development will be unavoidable. The foundation stone of the society is the people; the kind of people determines the kind of society. When we compare human development and social development, it is easy to come to the conclusion that human development is more fundamental.

In the comparison between human development and economic development, human development is more fundamental. The greatest obstacle in removing poverty is the obstacle in human quality. "In curing poverty, first we must cure folly" and "in curing poverty, first we must cure laziness" are truths we have summed up from poignant facts. The chief cause of the long-standing "poverty state" of many regions of our country does not lie in the disadvantageous factors in the natural or geographic environment, but lies in the low quality of the people. When the economist Zhong Pengrong did investigation in an area in West China, he found, every day, around 9 - 10 o'clock in the morning or around 2 - 3 o'clock in the afternoon, there were scores of sturdy men in each village sitting at one end of the village taking a sunbath. They have a famous saying: *"Although I am starving, yet I am sitting."* On the contrary, in another place, the holy land of the new economy, Silicon Valley, we find people working night and day, holding the tenet of *"If I fall asleep, I will fail"*. Every night, they work before the display screen till 4 - 5 o'clock in the next morning, and sometimes even till 6 o'clock. On one hand, there are many benighted lazybones who are sitting although starving, and on the other hand, there are many hard working brilliant people who believe in *"Sleeping is failing"*: This is just where the basic cause of poverty or richness lies. If we think little of human development, do not try our best to raise human quality, but only seek the one-sided economic development, then, achieving a sustained economical development will be nothing else but seeking fish from the woods.

In curing poverty, we can not get away from human development, but after getting out of poverty, we will need human development all the more. Whatever high indicator the economy has reached, only when the people have really developed can the economy have a sustainable development. Otherwise, even a rising economy may also fall. We should pay attention to the warning by the director general of the

United Nations Educational Scientific and Cultural Organization: *"If we detach elementary education from knowledge economy, a fissure might appear between poor countries and rich countries".* Likewise, a fissure will appear between rich people and poor people in the country, so the disparity between rich and poor may grow sharper and sharper. US futurist Toffler has expressed his good opinion on China's economy: *Whether China can "jump from the first wave to the third wave" will depend on whether its people can receive a good education.* **"You may jump over technology, but you can never jump over education". Being unable to jump over education, in fact, means being unable to jump over human development.**

Human development is both the purpose and the foundation stone of all developments. Human development is above everything, and it is more important than everything. That is the new development theory, quite different from the traditional development theory.

Since human development is matchlessly important, it goes without saying we should put it at the most important position and seek it by all means. Why do I think what is most important in futurological research is not the study on future society or future economy, but the study on future mankind? Why do we think the undertaking of mankind's striving to become immortals is the greatest undertaking unparalleled in history? All these are based on this new development theory.

The "god" can go up to the extreme height and go down to the extreme depths, like a wind or shadow, enjoys limitless freedom and boundless happiness, and is beautiful beyond description.

The ancients said: *"Those who age but do not die are called 'gods'."* In the definition of "god", he is not "someone who is old but not dying", but "someone who enjoys eternal life", with youth as eternal as life.

Meanwhile, the "god" is also the wisest person: he is as omniscient and omnipotent as God. He is the most virtuous person: with the heart of the Buddha, pitying for the entire world. He is the most perfect person and beautiful in all aspects, such as; appearance, temperament, character, behavior, language and so on. He is the happiest person: enjoying daylong ecstasy, as happy as Heaven. He is also the freest person: He has been freed from the trouble of birth, ageing, illness and death, and from the spatial-temporal bondage, enjoying the greatest emancipation in the world.

The present people do not have the eternal life like the "god", nor do they have the ageless youth of the "god", the wisdom of the "god", the virtues of the "god", the beauty of the "god", the joy of the "god", and they do not have the freedom of the "god".

Becoming "gods" just means developing the present humans into the future's most ideal super humans, i.e immortals.

The problem of becoming "gods" is actually the problem of human development.

Becoming "gods" is the ultimate objective of human development, being the supreme prospect of human development. Since we value human development, we should value becoming "gods". Since we seek human development, we should seek becoming "gods". Since human development is above everything, the value of becoming "gods" should also be greater than everything.

The new development theory not only values human development, but also believes that, in human development we should pay equal attention to both overall development and sustainable development.

The road to being "gods" is just the road to sustained and overall development of the human.

Seeking the overall development of the human is Marx' basic idea. However, the idea of all-round development has met with distortions and misunderstandings of varying degrees. Many people only take "over-all development" as a vague and general concept, instead of a great theory. Many people think passing all the school subjects in tests means overall development, or mastering more skills means "overall develop-ment". Such a serious misunderstanding of "overall development" is a profanity of the theory of human overall development, being more terrible than openly opposing overall development. Overall human development is never the same as passing some tests in the school, but should include the following three levels: Overall development of body and mind, overall development of morality and overall development of knowledge and ability. Meanwhile, a high-leveled overall development of humans is a complicated and enormous system engineering and can never be achieved by mere education.

By expounding the four major attributes of "the god", namely; "eternal life", "omniscience and omnipotence", "supreme benevolence and supreme beauty", and "daylong ecstasy", we have revealed for the first time the full-sensed overall development of the humans in six aspects, namely; body, intelligence, ability, morality, beauty and pleasure. As a person, only when he is alive or in existence he can have all other things. Thus, "body" development is primary and the supreme goal of "body" development is "eternal life". What a person lives for is pleasure, and it is the basic reason for the existence of a life. The "daylong ecstasy" and "supreme happiness" characteristics of gods are the supreme goals that can meet the need for the highest level of development of the human race.

If we look at the matter according to the three major needs of the humans summed up by Engels, namely; need for survival, need for enjoyment and need for development, we will find, "eternal life" and "daylong ecstasy" just meet the "need for survival" and "need for enjoyment" whereas "omniscience and omnipotence" and "supreme benevolence and supreme beauty" just meet the supreme level of need, "need for development". Thus, it can be seen, becoming "gods" is just the highest level of overall development, meeting all the three major needs perfectly.

The highest ideal Marx sought is "mankind's thorough emancipation". The antonym of emancipation is bondage, oppression and suppression, and its synonym is freedom. Seeking "mankind's thorough emancipation" is just freeing mankind from all external and internal bondages, repressions and oppressions, so as to move up from the realm of necessity to the realm of freedom. Only by letting everyone become "a god" can mankind realize its fullest and the most thoroughgoing emancipation? Only by evolution from mankind to immortals can we transcend space-time limitations, free our body and mind completely, realize omniscience and omnipotence, reach the highest kindness and highest beauty, namely, reach the most perfect prospect.

To reach such a perfect prospect, we must ensure a sustainable development for mankind. The development theory of materialist dialectics holds the following point: *"There is no ultimate, absolute, or holy matter... Nothing exists except occurrence and elimination, and the unceasing and limitless process from a lower level to a higher level."*(p.213 of Volume 4 of *Selected Works of Marx and Engels*). Human development is no exception, also being an "unceasing and limitless process from a lower level to a higher level". "Immortal" is the greatest goal of the limitless human development, and such a great development goal can not be attained in one action, requiring people's long-term and sustainable development. Hence, the process of becoming immortal is a continual development process.

In short, becoming "immortal" is a great generalization of high level of sustainable development and high level of all-round development of mankind, being the most comprehensive development of humanity, and the final necessary result of limitless and constant development of the human race.

If Marx wakes up from his tomb and we give him 24 hours to see the present world: First, let him spend 3 hours rambling on the Internet, then let him see the miraculous nanometer technology, then, let him listen to the news about deciphering of mankind's gene map and then let him take a look at a page of *Yangcheng Evening News* that contains more information than the information amount of a 17th century farmer's whole life… He would certainly be amazed by this magical world. Mankind's fast development and the miraculous power of technology would certainly make him believe that the goal he had struggled for all his life, "the through emancipation of mankind and the overall development of the human race", would be realized in the near future, and make him believe that becoming immortal, as the sublimation of overall human development, will be the natural trend of human development.

XVIII. HIGHEST HAPPINESS AND HAPPINESS FORMULA

If a person had no pleasure before him, he would not exist in this world.
—Makarenko

"If there are things I don't like in Heaven, I won't go there. If there are things I don't like in Hell, I won't go there. If there are things I don't like in your future golden world, I won't go there." This saying was uttered by the writer Mr. Lu Xun. His meaning is nothing else than this: Whether it is "Heaven", "Hell" or the future "golden world", what

I am seeking must be something I like.

Why do we aspire to become "immortals"? The only answer is: for pleasure. The ultimate goal of human development is seeking pleasure.

This answer sounds too simple, too ordinary, but the most important and most difficult thing in the world is just to hold the ordinary truth.

People do not do things just for doing things, nor do they pursue just for pursuing, or do they suffer just for suffering. Whatever we do, in the last analysis, it is for getting pleasure. Aristotle said: *"It seems that man's greatest characteristic is seeking pleasure and avoiding suffering."* Man, judging by his nature, is neither good nor evil. At any moment of his life, he seeks happiness. *"All his abilities are used for seeking pleasure and evading suffering."*(Horbach)

Our common humanity is going towards happiness and evading suffering. No enduring of suffering in our life is just for the sake of suffering. Suffering is only the means, and only "pleasure" is the purpose: For tomorrow's pleasure. As the saying goes, *"Only if you can stand the hardest of hardships can you hope to rise to the top of the society"*, so, the purpose is to get the pleasure of being on the top of the society. Why does a mother endure all kinds of hardships to foster her children? Just for the future pleasure from their healthy growing. Why were the revolutionaries ready to bear the suffering of cruel torturing? Just for the pleasure in realizing the ideal and belief in their mind, just for the pleasure of the majority of people... Although everyone's pleasure is different, one point is the same: *"All our choices are out of the pursuit of pleasure. Our ultimate objective is to get pleasure."* (Epicurus)

From the epicurean "Hedonism" philosophy to the "Seven Rules" of happiness calculated by Jeremy Benthan, we can see Western philosophers' great enthusiasm for happiness. China's serious school of learning, Confucianism, also lays a great stress on happiness, as the whole book of *Analects of Confucius* is alive with the mood of "happiness". The first sentence of *Analects of Confucius* is about "happiness": *"When our friends come from afar how happy we are!"* When Confucius enjoyed music, he was so intoxicated in it that he did not know the taste of the meat for three months. When he studied, he "made such a determined effort that he forgot eating, and he was so happy as to forget any worry". Confucius was both a saint and a master in music, as he enjoyed pleasure from all aspects of life; Reading and learning, friend making and traveling, getting political posts and entering the society, listening to music and viewing dance, summoning disciples and teaching, and so on. The "three pleasures" Confucius liked most were "pleasure in courtesy, pleasure in praising people, and pleasure in friend making." Mencius said *a gentleman has three pleasures": "Both parents are alive, and brothers are living in peace, it is the first pleasure. Feeling no shame before both Heaven and fellow human beings: the second pleasure. Getting brilliant students from the world to teach: the third pleasure."* Confucianists put the learning of "pleasure" on a vital position, parallel to political science and ethnics. The pleasure mentioned by Confucianists refers to all kinds of arts having an amusing effect, which are represented by music, and even include hunting and other activities. They defined "music" as "the art of pleasure". ("On Music" of *Xunzi*). Thus, it can be seen Confucianists' emphasis on music is actually the emphasis on pleasure, as they value pleasure greatly.

One of the founders of Chinese Communist Party, Mr. Li Dazhao, also advocated clearly: **"Only the life philosophy of seeking joy is the natural and true life philosophy."** Mr. Yu Guangyuan, the eminent scholar of our time, believes in the "joy" philosophy, and has named himself as "a great player", saying: *"Isn't socialism just for making people happy?"*

It can be seen that, whether in the west or in the east, whether in ancient times or in modern times, pleasure has always been the supreme goal of mankind.

Life is priceless, and the interest of life is the supreme matter. However, when life and pleasure conflicts with each other, people's pursuit of pleasure often overwhelms their thirst for life. Makarenko said: *"If a person had no pleasure before him, he would not exist in this world."* Mankind's suicidal behavior is a good example. Suicide is not for seeking suffering, but for ending the suffering. When a normal and reasonable man fails to see the joy before him and believes there is no way to avoid the suffering in reality, he is bound to choose suicide. "Euthanasia", as a great invention of mankind, is also a proof that the purpose of human existence is pleasure. When people are suffering from a serious ailment or suffered to the utmost limit, and are sure the joy of tomorrow will never exist, they will crave for death and will be eager for a fast death...

We cry up in joy extremely, but the joy mentioned here is never the joy of the squanderer as is imagined by some people, or the temporary corporeal joy. The rational hedonism is totally different from the common hedonism. The rational hedonism is the natural enemy of the common hedonism.

Hedonists are those so-called "fresh new mankind" who are proud in saying "they prostrate themselves before all kinds of desires". For example; the "belle writer" Wei Hui's self-complacent declaration: *I believe in the impulses from my heart all the time, submit to the burning in the depth of my soul, resist no offhand madness, and worship all kinds of desire*.

Hedonism only seeks the corporal joy. The so-called "impulses from my heart" and "offhand madness" are nothing more than indulgence in the instinct and chasing the corporal pleasure. A philosopher said, *If happiness lies in the corporeal pleasure, then we should say: when the ox has found fodder to eat, he is happy.* If we only submit to the pure corporal joy, we will only reduce the humans into an animal like the ox, the pig or the dog.

Hedonism only cares about the joy under their very nose. For one moment's sensation of pleasure, one time of apolaustic drug taking might make someone fall into the abyss that could ruin all his life. The price of one-night "romance" might be a terrible burden of AIDS, and a boundless sea of woes. Gambling night after night can make someone become mad and can take his own life in his hands... The price of being addicted to such "small joys" at hand is often the loss of the lasting "great joys" of life. It seems to be seeking pleasure, but actually is losing pleasure totally.

Ordinary hedonism also only cares about the joy of one person, even to the extent of basing his own joy on the suffering of other people. "Believe in the impulses from the heart any time": when wanting to steal, just steal; when wanting to rob, just rob... So long as they are glad themselves, who cares about the monstrous flood in the world? So, the boss of a game playing hall should wave his knife to two young lives and kill them "justly", just for 1-2 Yuan, because doing so could "vent his anger", making him feel happy for a moment. Let's take another example in the environmental problem. Many people go to the mountain to cut trees for money, or dump garbage into the Yangtze River. They feel happy themselves, but others will suffer, and the whole next generation will also suffer.

For thousands of years, mankind's pursuit of happiness has resulted in unhappiness actually, which is the ill effect of hedonism. We are seeking happiness to the top of our bent, but the happiness we are seeking is—

Happy combination between corporal joy and spiritual joy.

Happy combination between the present joy and the long-term joy.

Happy combination between individuals' joy and the collective's joy.

While valuing corporal joy, value spiritual joy all the more.

While valuing the immediate joy, value the long-term joy all the more.

While valuing the individual's joy, value the collective's joy all the more.

Only by doing so, can we get the greatest joy, and only that is the joy the rational hedonists seek.

Can joy be calculated? Of all ages, many people have tried to find out the formula for calculating joy. We have got a joy calculation formula:

> **Total Amount of Happiness = Daily Amount of Happiness X Number of Days of Happiness**

The two major goals of becoming gods are "daylong ecstasy" and "eternal life". Daylong ecstasy is the maximization of "the daily amount of joy". Eternal life will maximize the "number of days of joy", so, the time effect of joy will have no bound or end, reaching eternity. Hence, according to the joy calculation formula, the joy to be brought about by becoming "gods" is like this:

Daylong ecstasy × eternal life = maximization of happiness making happiness limitless = highest happiness

In Part II "Subversion of Tradition and Fetal Movement of Future", I have expounded the great possibility of eternal life in detail. Likewise, the possibility of "daylong ecstasy" is also very great.

Whether a person is happy or not, is determined jointly by his external and internal worlds. Only by reforming both external and internal worlds can we make people happy all day long.

For reforming a person's internal world, there can be many ways:

Faced with a the same half cup of water, someone may cry joyously: *"I still have a half cup!"* whereas someone else may sigh sadly: *"Why is there only a half cup left?"* In our life, some people are joyful by nature, like Cheng Yaojin (famous general in the Tang Dynasty), and some people are sad by nature, like Lin Daiyu (heroine in *A Dream of Red Mansions*), as joy comes in part from natural instinct. Research has also found: the gene plays a decisive role in a person's optimistic character or pessimistic character. By the gene map we can find the gene that makes people feel happy and optimistic, thus, we can implant the optimism gene or joy gene into a pessimist.

Some scientists conducted an experiment on a rat: A very hungry rat (it had already learned to get a pleasant sensation by treading the lever) did not eat the food, but ran straight to the lever that could stimulate its "joy centrum" and did "self stimulation" madly, as its strong pursuit of joy overwhelmed its thirst for food, making it forget to eat and enjoy the pleasure without tiredness (See *Ability Panic* for details). The rat has a joy centrum and there should be a similar joy centrum in the human brain, too. Once we have found the joy centrum of man, by using some high and new technology to stimulate it effectively for all day, especially in time of study or work, our life may be full of joy every moment.

As our education has the function to cultivate the soul, we should stress "pleasure education", cultivate people's optimistic consciousness, optimistic thinking, optimistic attitude and optimistic character, so that everyone may be good at finding pleasure, tasting pleasure and making pleasure. The best education should give everyone a joyous soul.

In reforming the external world for "daylong ecstasy", there are even more methods:

By dint of the power of high technology, especially the virtual reality technology, we may help people realize any wish, so as to expand the channels of joy greatly.

If you are fond of playing in the bar but you do not have the economic strength for frequenting the bar every day, you may realize your wish at home. The home computer equipment has adopted the virtual reality technique, so, it can simulate the romantic mood in the bar.

If you are fond of tourism, you need not learn from San Mao (an adventurous writer) to take the pains in traveling everywhere. At home, you can visit Eiffel Tower, or move Palais de Versailles into your home.

If you are a girl fond of following fashion, there is no need to roam in the clothing market in hot sunlight. You may design your favorite fashion at your will on the computer, and then simulate putting on the fashionable dress you have designed.

If you are a dance fan, so long as you implant a tiny computer chip into your body, you may be able to dance skillfully all kinds of beautiful movements, whether it is classical, national or modern dancing. Moreover, in future, the protagonist in a teleplay may talk with you on TV, and invite you to join the story together.

If you wish to play the piano music of "Maiden's Prayer" but you have never learned to play musical instruments, and if you hope to compose your own opus but you cannot read music, it won't matter at all. In future, even if you are ignorant of music, you will also be able to play the lyre in a free flowing style like a musician, and able to compose moving musical works.

In the measurable future, many innovative ways of amusement will appear, and more and more high and new technologies will add joyous wings to us, so that mankind will get inexhaustible joy at last.

One point is specially worth noting: what we seek as daylong ecstasy is not only "ecstasy", but also "daylong ecstasy", which is to say, we must put people into a state of utmost joy for all the 24 hours of a day. On one hand, we may use technological means to reduce human sleeping time. A person uses one third of his lifetime to sleep, that is, in one third of the life, he is in a state of unconsciousness, just like a dead body. When we think of this, we cannot help feeling sorry. We may explore into the secret of human sleeping, so as to develop a high technology that enables people to sleep less or get rid of sleep (there have been some great breakthroughs in this respect), to ensure them to enjoy pleasure in all the 24 hours of the day. On the other hand, even if we have to have a certain amount of sleeping time, we may also optimize the sleeping time, for example; letting us have our favourite dreams all the time so as to enjoy the beautiful sleep. In short, we may keep people happy every hour, every minute and every second, whether awake or asleep.

Daylong ecstasy may maximize the daily amount of happiness. Eternal life may maximize the number of days of happiness. The result of the multiplication of above two will maximize the joy of the whole life.

Eternal life and daylong ecstasy constitute the overall goal of becoming immortals, and the two together determine the whole amount of happiness. The kernel objectives of becoming gods, "omniscience and omnipotence" and "supreme kindness and supreme beauty" also play a great role for happiness. Eternal life and daylong ecstasy will chiefly depend on the great development of science and technology, and on the other hand, the great development of science and technology also depends on human "omniscience and omnipotence". Meanwhile, only when everyone has reached "the supreme kindness and supreme beauty" can we form a harmonious and uniform great society, build up a globally integrated commonwealth and focus all the forces of the society and the world onto the most important aspect of human happiness, so as to ensure the realization of eternal life and daylong ecstasy.

At the same time, omniscience and omnipotence, supreme kindness and supreme beauty represent the highest prospect of mankind's age-old pursuit of "truth, benevolence and beauty". And they represent the highest level of happiness.

Omniscience and omnipotence means reaching the peak in wisdom and ability.

For a person, wisdom and ability are the roots of his joy. Sukhomlinskkie said quite well: *"One person's happiness is just the happiness of a society, and after all, it depends on the abilities shown in the person and depends on the outstanding abilities that make his life glitter."* Joy lies in happiness, happiness lies in success, and success lies in wisdom and ability. A success may bring about a joy directly, and omniscience and omnipotence are the guarantee of limitless success. Omniscience and omnipotence can enable a person to follow his inclinations, fulfill whatever he wishes, and get the greatest joy.

Supreme kindness and supreme beauty means reaching the peak in morality and beauty.

"Those who love others are always loved by others, and those who respect others are always respected by others". If you have the "supreme kindness" of the god, love and respect others to the greatest extent, you will win the greatest love and respect from others. Gaining unlimited love and respect means gaining unlimited pleasure. Just as Rousseau said: *"Only when you love others you can get the love from others. If you want to live a happy life, you must turn yourself into a person loved by others. If you want others to follow you, you must make yourself worthy for others' respect. Only when you treasure your own dignity can you get the praise from others."* If you take others as your enemies and are jealous of others' joy, others' joy will anguish you. When you are kind and friendly to others, and feel happy for others' happiness, you will share the joy of all people, and become the happiest person in the world.

Beauty is the source of happiness. Human beauty is shown in various aspects, such as beauty of the shape, beauty of manners, beauty of temperament, beauty of style and so on. Loving beauty is human nature. When a baby sees an ugly person, he will be frightened. People seek beauty in any aspect of their life: eating, dressing, dwelling and so on. Since ancient times, "It is hard for a hero to pass the test of a belle". The power of beauty is invincible. Even such a great saint as Confucius was intoxicated in talking about the beautiful looks of Nanzi. In order to contend for Belle Helen, Greece and Troy fought for a decade, causing one hundred thousand deaths. As soon as Helen appeared at the top of the city wall, the oldsters in the senatus were amazed by her beauty, and could not help gasping in admiration: "Really a goddess! It is totally worthwhile fighting for her for ten years!" Why was Helen's beauty so valuable? To put it bluntly, only because; Beauty can bring joy to people. The supreme beauty can bring the greatest joy and happiness to people.

Thus it can be seen: both the overall goals of becoming gods, "eternal life" and "daylong ecstasy" and its kernel objects, "omniscience and omnipotence" and "supreme kindness and supreme beauty" are meant for joy.

All developments of the world are for human development and all human developments are for joy. Only by becoming "gods" can we maximize people's joy, and lead to the Elysium characterized by happiness as great as Heaven.

XIX. MANKIND'S TWO DEITIES

In the past, Adam Smith called the market "an invisible hand", today, we call technology "an invisible head".

In the dictionary, the original meaning of "deity" is "a supernatural power's existence in religion, superstition or myth, with personality and human consciousness, being an illusory and distorted reflection of the external power in the human mind", and its extended meanings are: "extraordinary and unimaginable", "an existence with limitless power".

Ancient people entrusted their god-becoming dream to the deity. Along with the gradual flaking of the garment of the deity and the complete smashing of its idol, the dream of becoming "god" was getting more and more unreal and unreliable.

The deity in the eyes of the ancients is only illusory, not existent at all. No one has ever seen a deity coming to the secular world to help people become "gods". However, the divine power of today's "deities" may really amaze people and what's more, it can be controlled and used by mankind. Only by relying on the two "deities" of today can mankind have the hope to become "gods".

The two "deities" are: science and society.

When we say science is a deity, it is not a compliment to science.

Marx defined science as *"a revolutionary power in the highest sense"*. Such a revolutionary power is nothing less than a matchlessly huge supernatural power, as it can change everything. In fact, it has already shaken the whole world, turning uncountable dreams and myths into vivid realities. In the ancient times, in order to beg for rain, people spared no pain to beg gods to have pity, and sometimes they kneeled for three years in vain. Today, by artificial rainfall we only need a moment to summon wind and rain back and forth. Never in history have we met with such profound changes of the world as today. Never have we felt such a limitless possibility of mankind as today. *"So long as you can imagine, you can realize it."* The precondition for making this saying correct is: relying on the divine power of science.

"Qitian Dasheng (Mahatma as High as Heaven)" Sun Wukong had miraculously great power, but all his divine powers can be achieved through science today. Sun Wukong's one somersault was one hundred and eight thousand miles away, we can do so by taking the ionic space-ship flying in outer space. Sun Wukong used his fine hairs on the body to change into uncountable little Sun Wukongs, now the clone technology can duplicate everyone into uncountable selves. Sun Wukong once changed himself into a worm and got into the belly of Iron Fan Princess, now the "nanometer robot" or "nanometer soldier" produced by the nanometer technology is much smaller than the worm, and it can go in and out of any place inside our body freely, including blood capillaries... All those miraculous and uncanny fantasies have been realized by science. Then can't we say Sun Wukong's fantasy of eternal life will be realized by science too? All the beautiful myths: They either have been realized by science or are being realized, or will be realized for certain in future.

Science has won brilliant victories one after another, and science is speeding up mankind's progress increasingly, making all impossibilities into possibilities very soon. As early as the 17th century, Bacon said, **"Science has made us closer to God"**. In the present 21st century, the organism gene technology and life science and technology have already put the laurel of God onto the heads of scientists.

We cannot forget: the Industrial Revolution was started by the steam engine modified by Watt, and it enabled mankind to create a productive force in less than one century that was even greater than the total pro-ductive forces of all the past ages. Today, science is not only "the first productive force" but also the major productive force. According to economist Douglas' function, in the total amount of growth of social productive forces and economy of developed countries in a certain period, the factor of scientific-technical progress accounts for a greater and greater part. At the beginning of the last century, it was about 5%, in the 1950s it was about 20%, in the 1960s it was about 40%, in the 1970s it was about 50%, and in the 1980s it already mounted up to 80%.

In the past, Adam Smith called the market "an invisible hand", today, we call science and technology "an invisible head". Along with the fast high technology revolution, science is playing an increasingly greater role as the "head", and it is dominating all aspects of the world and the fate of all mankind in an unprecedented depth and to an unprecedented extent.

While seeing the great magical power of science and technology, we should also see the breathtaking potentials of science and technology.

First: let's look at the investment in science.

In the whole society and the whole world, the investment in science is greatly different from that in military affairs and war. The difference is so great as if between sky and sea. In 2004, the military expenditures all over the world exceeded one trillion US$, whereas the total investment in the genome project in one decade was only 3 billion US$ although this project is very important to mankind's bright future. The former is destructive investment to mankind. The latter is an investment that will greatly benefit mankind. The laughable thing is the investment in the former should be thousands of times of that in the latter. If we did general and complete disarmament so as to save about one trillion dollars each year and saved more money consumed in unreasonable high cost of living and high waste, used all such money in high technology, especially for research in life sciences, and united all the forces of the world to tackle key problems, then the breakthroughs in life sciences would surely have been thousands or millions of times of the present ones. Would we need to worry about the inability to realize the great dream of becoming "gods"?

Second: let's look at the condition of scientific personnel.

Of the 6 billion people of the world, those doing science work are less than 1%. Even in this less than 1%, more than 25% of them are doing research in the military aspect. In our world, on one hand, the rate of unemployment of the society is high, and keeps rising. On the other hand, scientific manpower and great able people are in great shortage. If we raise the quantity of scientific personnel from 1% to 10%, 20%, 50%, or even 80% in future, and greatly improve their level, our scientific research power will be thousands of times of the present level, and our research findings will also be thousands of times more. Great increase of the quantity of full-time scientific personnel is an inexorable trend of future. In the agricultural society, the peasant population accounts for more than 80%, but the peasant population of the USA now is only 2% and their industrial workers also account for a low rate of 5%. Such proportions will be further reduced sharply. Then, what will the rest of the people do? Now they are chiefly engaged in the service industry, but in future they will shift largely to the research industry and there will be millions or billions of people doing scientific research. Then, the potential magical power of science will certainly be exploited by many more times, thus, the progress of mankind's leap to immortals will also increase by many more times.

Third: let's take a look at the intelligence revolution.

High technology originates from people's high intelligence or high brain power. By the greatest and all-inclusive investment to exploit the intelligence and to trigger an "intelligence revolution" or "intelligence explosion", we may trigger a "science explosion". We shall utilize all forces, as mentioned later, to do the "three optimizations"; optimization of birth through technology, optimization of body through technology and optimization of education through technology. Meanwhile, turn the computer network into an intelligent network, so as to unite all the human intelligence together. Morover, we shall go all out to develop the AI (Artificial Intelligence), the IA (Intelligence Amplification), the man-machine integration system and global intelligence integration total system, so as to use the intelligence of uncountable great minds, the intelligence of uncountable magical machines, the compound intelligence of uncountable man-machine systems, the intelligence of uncountable expert systems and the network integrated intelligence to interact with each other, which may lead to a chained fusion effect, trigger an intelligence explosion, hereby resulting in a science explosion, That will be an amazingly fast development of science, like an explosion.

Thus it can be seen, how amazing the great power of science and its huge potentials are. By dint of the boundless power of science, we shall certainly be able to turn the dream of eternal life and daylong ecstasy into reality. So long as we rely on science, use science and develop science and our great dream of being immortals together will come true step by step for certain. Hence, science is our first deity.

The high wind of science blowing against our faces is causing earth-shaking changes in the whole world, so it is easy for people to sense the great power of science, but people often under-estimate the great power of social forces, and even have no awareness for it. Science is a deity and the society too, is a deity. The undertaking of mankind's becoming immortals together should depend on both these deities

The ancient sage Xunzi said: *"Human power is not as great as the ox, and a human beings cannot walk as fast as the horse, but we can use oxen and horses to serve us. Why? The answer is People can group, but the other two things cannot."* "People can group", i.e. people can form society.

The society is a collective body of many individuals. It may organize and take together the power of many people, which may not only result in a quantitative additive effect, but also lead to a multiplier effect of 1+1>2, 2+2>8 and so on. In an ideal society formed by 1 billion people, its power is never merely the 1 billion times, or 10 billion times, or 100 billion times, or 1000 billion times of the individual. In comparison with the isolated individual power of 1 billion people, it is different in nature: as different as sky and sea. If we take the society as a person, or an organism, then the individual is just one cell of the organism. If the cell leaves its human body, it will separate from the nutrition, so what awaits him will only be death. When a person leaves the society, he (she) will have no means to survive or develop. Bill Gates possesses the most wealth in the world, and has a gifted brain. He is so powerful. But, if you throw him onto a detached island, let him be Robinson, his power will disappear immediately. Suppose every one of the 6 billion people of the world stays on a detached island, never in contact with each other, the whole world will not be like the world now, human civilization will never appear, and no science or technology will appear.

The great social power includes its cultural power, economic power, political power, system power, power of laws and regulations, policy power and so on. Now we only take the system power and policy power for example to consider the matter.

On the gate of the National Patent Bureau, USA is engraved a saying by Lincoln: *"Pour the oil of interest onto the fire of mankind's wisdom"*. The patent system embodies a high respect and reward of man's wisdom, thus, it has greatly helped the fire of man's wisdom to burn better, more strongly and faster. Why did the new economy originate from the USA instead of Britain or Japan? One important cause is that American capitalism has adopted the stock equity system. Only because of such a system did the sleepless people of Silicon Valley appear and did the world-famous Silicon Valley miracle appear. And, in order to follow the American new economy, the European nations and Japan also began to abandon the traditional "banker capitalism". In China, because of a wrong criticism to Ma Yinchu (scholar on population theory), we had an increase of 0.3 billion people. Later, because of a change in policy, we reduced the birth by 0.3 billion people. In the same nation, with the same people, the change from a terrible explosion to an effective control in population was just because of a policy. Earthshaking changes have been taking place for the past 20 years, since the economic reforms started and with the opening up of China. They have also chiefly resulted from the reform of the society. Then, in reflective thinking on the reason why the great China became so backward in recent times, we may also find the answer from the social system. We should not ask ourselves: why Chinese did not invent the steam engine? But we should ask why we did not make the Patent Law? Why we did not progress from the enterprise partnership system to the joint-stock company, why we did not develop the private money shop into the modern bank... The system power alone has such a great effect on the nation's fate and mankind's fate that we should never underrate the magical power of the society.

Many problems we are facing today, actually resulted from social causes. Poverty, crime, corruption, environmental pollution, terrorism, war... all these can find their starting point in the society. In a sense, every criminal is guiltless, because it was chiefly the social causes that have pushed him or her into the abyss of disaster. Even those corrupt elements are not completely wrong themselves in committing the monstrous crimes, as the society lacks a sound supervision mechanism. If the supervision mechanism was sound, they might have behaved differently and have had a different fate. On the contrary, if the society does not establish and improve its inspection system, corruption may also "follow the normal rail", going from bad to worse.

While praising the magical power of science, we have to admit that this deity should not get out of the control of the society. That is to say, the great power of science must be managed and controlled by the society. In modern times, science deals a fatal blow to God, "God has died" and however, the devil has come accordingly, too. For science is a double-edge sword. At first, the invention of the powder was only out of the need in production, but later it became a high-energy weapon for mankind's fratricidal fighting, and thousands upon thousands of people lost their life under its power. Atomic energy was also a "masterpiece" dedicated to mankind by scientists, since August 6 and 9, 1945 when two atomic bombs were dropped over Japan, mankind has always have had the nightmare of nuclear terror. Of course, it is not the fault of science for all those tragedies. We should never have all kinds of prejudice against science just because of those phenomena, like some people who attributed environmental pollution, ecological damage, war terror and all such bad things to science. What caused all those? Who should be blamed to let the Devil open the pandora's box? It is the society! Just because the society, this deity is "dozing off" and failed to give play to its adjusting and controlling role.

In the writing by Zhuangzi, a technique called "Buguishou drug" could be used to raise the efficiency of clothes washing in winter, and could also be used to decide the victory or defeat of a war or the survival or death of a nation. In a time characterized by numerous breakthroughs appearing with each passing day in science and technology, whether all kinds of technology can be used for great matters or small matters, for good purposes or evil purposes, just depends on the society's control. The society is shouldering a much greater responsibility now. Only by making the best use of the deity of society, only by greatly enhancing, opening up and utilizing the great power of the society, can we prevent science from becoming a Devil. This can boost the development of science more effectively, and can maximize the huge power of science so as to maximize the benefit brought about by science to mankind.

However, we have done little in calling forth the social power. The divine power of the society is still very incommensurate with the divine power of science, and the potentials in the society now are much greater than that in science. We still have a long way to go before building up a perfect and ideal society.

In my opinion, social power is decided jointly by the following eight major factors: the quality of the members of the society, the level of the leadership of the society, the size of the society, the vitality of the society, the degree of unity of the society, the degree of harmony of the society, the degree of rationality of the society, and the level of civilization of the society. Judging from the eight major factors, there are limitlessly huge potentialities in our society.

Quality of the Members of the Society

The society is composed of individuals, and the great system of society is based on the numerous subsystems of the individuals. The social members' quality determines essentially the magnitude of social forces and the speed of social development. Only by strengthening the people we can enrich the nation, and only by making the people wise can we make the nation strong. Only by raising the quality of all the social members and by exploiting the potentialities of all the social members so as to enable all of them to achieve all-round and fast development can we strengthen the social power tremendously. China has an illiterate or semiliterate population of 0.2 to 0.3 billion, its proportion of college graduates is only 2% of its total population, being a twentieth of that in western countries. Thus, China's social potentialities are far from being fully exploited. Once we raise the quality of the people, our society will advance by leaps and bounds.

Level of the Leadership of the Society

The level of the society's leadership is decisive to the social power. The chief reason is that the leaders are the decision makers of the society. Once the decision-making is wrong, the whole society may get into a calamity. Once the decision-making is wise, it may lead the society into an ideal prospect. During the Cultural Revolution, the absurdity of the leaders in decision-making brought the society to the verge of collapse. After the reforms, the correct decision making of the leadership has brought about an amazing progress of the society. If the level of the leadership of the society is high and any decision is reasonable and wise, the huge energy of the society will be fully opened up. What is more, the leadership has an example-setting function to the society, just as the saying goes, *"the below follow the behavior of the above"*. The king of State Qi (in the Warring States Period of China) liked wearing a purple robe, so all the people of the state wore purple clothes, making the prices of purple cloth and purple silk soar up. The king of State Cu liked females with a wasp waist, so all the women of the state began to do slimming even at the cost of hunger. Likewise, when the leaders are corrupt, the whole society will be corrupt. When the leaders are virtuous, the whole society will tend to be virtuous.

Size of the Society

The size of the society can be big or small. In ancient times, there was a clan society, and later, a state society appeared. The state society also falls into different sizes. Today, there are states composed of scores of people only, and the scores of people form a little society on an island as big as a basket ball court. Needless to say, such a small society has only a small energy. Although our country ranks behind in the world in per capita resources and per capita wealth, even behind some small countries, yet, ours is after all a big country with 1.3 billion people. It is just because of such an enormous social size that many countries think high of this broadest market and that the only superpower, the USA, regards it as a potential adversary. After introducing the concept of "size of the society", we should not only talk about the state society, but also talk about the international society or the global society. The global society is the society with the largest size. The concept of "global village" was put forward long ago, and the "globalization" tide is just coming to us vigorously. **When we have really established a borderless, democratic and sound global society, it will bring about the greatest "scale effect"**, and the energy of the whole human society will be released to the best effect.

Vitality of the society

Whether a society is rigid and conservative like a pond of dead water or full of vitality and up-going forces, will also bring about a striking difference. A vigorous society is a society that has the greatest power to kindle every member's enthusiasm, initiative and creativity, and is a society that has the greatest capacity to trigger every enterprise' vitality, creativity and competitive power. If each member of the society is passive and lazy, like the insensible viewer described by Mr. Lu Xun in his writing, and if the enterprises in the society are also "corpse-like" organizations full of inertness, being conservative and rigescent, then the energy of this society will surely be greatly weakened.

Degree of Social Unity

If a society is in a state of disunity, falling apart, its energy will certainly be greatly damaged. On the contrary, the more the society is integrated, the more possible it will be for it to produce a fusion effect, thus, its energy will be amplified by uncountable times. History has proved repeatedly, unity or splitting may mean a difference between heaven and hell. Only unity can gather all forces into one rope, and focus all these forces for one goal, so as to work great miracles. A state needs unity and so does the whole world. When the society of the whole world is integrated highly, and a real global integration is achieved from economy to technology, from culture to politics, earth-shaking changes will take place in the power of the human society.

Degree of Social Harmony

A society not only need unity, but also needs harmony. If there is only a uniform appearance but the inside is in great disharmony and people intrigue against each other, the power will be wasted seriously from the inside. Just as Chinese sayings go, *"If the family lives in harmony, all affairs will prosper"*, *"The time isn't as important as the terrain, but the terrain isn't as important as unity with the people"*. Unity with the people is of paramount importance. Only when the society is harmonious inside can it have a great cohesive strength and centripetal fore, and can it generate a huge resultant force.

Degree of Rationality of the Society

All aspects of a society should be rationalized, such as the guideline, policy, route, system, law, system, mechanism, structure and so on. For example; just one unreasonable system may impede social progress greatly. The domiciliary register system of China is unreasonable: It is a regional discrimination and birth discrimination similar to the racial discrimination. Once a farmer's child is born, his "residence booklet" will be branded with "rural population", just because of "being born in a wrong place", they will not be able to get equal treatment and development chances, which has greatly restricted the development of human resources. The birth control system is not so reasonable, either: For people with very good educational conditions for their children, the birth control is very strict, but for people with very poor educational conditions for their children, the birth control is quite loose, which will certainly lead to the "reverse elimination" in the population quality, being very harmful to the society.

Level of Civilization of the Society

In measuring the level of civilization of a society, we should check five aspects, namely; human-centered civilization, spiritual civilization, system civilization, material civilization and environment civilization. The five aspects of civilization promote each other, interact with each other and problems in any of them may affect the whole civilization of the society seriously. We have great potentialities in all the five aspects of civilization. The huge potentialities of the five major civilizations are just where the great potentialities of the society lie.

If we make efforts in the above eight major aspects, the great magical power of the society will show up before us.

The two deities, science and society, are not contradictory and isolated to each other, but are complementary to each other. If we only use the power of science or only start up the power of the society, we will not be able to realize the great dream of eternal life. Only by the joint work of the two deities can we work matchlessly great wonders. **If we say science is the father to become "gods", the society will be the mother for it.**

In the alliance of science and society, we should realize socialization of science and "scientization" of society, and establish a science society.

The future society should be a science society. In the science society, not only all people in the society will uphold science and devote themselves to science, but also all aspects of the society will be "scientized". In the conceptions about the society we have heard so far, there are "postindustrial society", "information society", "knowledge economy society" and so on, but none of them is comparable to the "science society". The science society will exploit the huge energy of both deities of science and society to the best effect, and combine them together, so as to boost the fast development of mankind itself. Only in the science society can we make both powers more magical. Only then shall we be able to realize what Hugo said: *"Only mankind is the demiurge whereas God is only a pitiful creature."* Only then can we enter the era of becoming "gods" together.

So long as all mankind reaches consensus, so long as the whole mankind is fully aware of these "two deities", fully rely on these "two deities", and make the best of the boundless power and limitless composite force of the "two deities", then mankind will certainly have the most ideal and most brilliant future and will certainly evolve into immortals at a fast rate.

XX. FIRST WAY TO BECOME IMMORTAL: BIRTH IMPROVEMENT THROUGH TECHNOLOGY

Not long from now, we shall execute both "compulsory birth improvement" and "compulsory education", and only then can we really turn the great slogan declared as early as two centuries ago, "All men are created equal", into a reality, instead of a mere wish or dream.

In mankind's evolution, we should not wait passively for a favor from nature. *"Although mankind is on the top level of natural evolution, it is not perfect yet, moreover, the natural evolution is very slow. In the 5 million years after humans separated from apes, the developmental change of human DNA was less than 2%."*(Tong Tianxiang) Mankind's natural evolution is so slow. How can we only rely on the natural evolution process? Now, it is high time for us to control our own evolution direction and speed up. Birth improvement through technology should be taken into the important agenda of mankind now.

In 1883, the British biologist Galton put forward the concept of "eugenics" and more than 100 years have passed since then. According to Galton's explanation of his concept of eugenics, it is "using better breeding methods to improve the race".

Birth improvement through technology is different from the ordinary eugenics.

The eugenics in the ordinary sense is nothing else but the following: extending the mating sphere and avoiding intermarriage, avoiding genetic diseases, doing premarital and prenatal examination, noticing the best age for pregnancy, noticing nutrition in the gestational period, creating a good environment and doing foetus education and so on.

To be precise, birth improvement through technology is eugenics through high technology, that is, using the great breakthrough in today's genetics, the amazing gene engineering, to reform human genes. By all-round reform of the genes, we can optimize human natural predisposition in all aspects, so as to provide the optimal material basis (material source) for human's overall development, provide the most optimal natural predisposition (basic quality) for the quality education. This will create "superhumans" who have been optimized in all aspects such as physique, intelligence, ability, virtue, beauty, pleasure and so on. Such a superhuman is an all-round transcendence of the present human, possessing the inherent qualities of the "god" basically.

Factors determining a person are nothing more than the two major aspects: Congenital factors and acquired factors. Since the ancient times, people have been arguing endlessly about nature and nurture. Some educationists assert *"one ounce of inheritance may outmatch one ton of education"* and there is a folk saying spread from mouth to mouth: *"A dragon bears a dragon, a phoenix bears a phoenix, and a rat's son is able to burrow naturally"*. On the other hand, some scholars claim: *Give me a dozen infants, I can cultivate them into pickpockets, robbers or lawyers, professors, according to your requirements.*

Congenital and postnatal, which is more important?

To my way of thinking: human being is a result of the joint function of both congenital and postnatal factors. Human is decided congenitally, and also postnatally. That is to say, human is decided both by the gene and by the environment. We are gene determinists, but also education determinists. The human gene has decided by the root that the human is a human, instead of an animal. However, a human is both a biologic human and a social human; the congenital and postnatal factors determine the human together and determine the person's fate together.

If it is required to say the order, then rationally speaking, the congenital is first and the postnatal is second.

Matter is always the first factor. Judging from the great difference among different species groups, we can see the great role the innate factor plays. All the properties a species group shows are closely related to its congenital gene. *"Plant melons and get melons, sow beans and get beans"*. The seed of the melon can never grow into a bean and the seed of the bean can never grow into a melon. A person with congenital gene defect can never be cultivated into a healthy person. It would be as hard as ascending heaven to nurture a child of innate dementia into a world famous genius.

We venerate life, but we should not deify it. Life is no mystery and the essence of life is the gene, namely DNA. Scientists' research has increasingly proved the decisive effect of the gene on all aspects of human life.

Research materials have shown, "**The degree of genetic effect on mankind's brainpower is 70 - 80% or so and certain specific intelligent behaviors (such as the linguistic competence, the mathematical skill) may be specially dependent on hereditary predisposition**" (See p.412 of *Encyclopedia on Dialectics of Nature*). Although the congenital factor is the primary factor, yet in the past, we had no means to control it, to say nothing of changing it, and we could only do something in a very superficial scope. Now, we can use gene design to optimize human brainpower to a great extent.

Scholars of Princeton University headed by the Chinese-American scientist Qian Zhuo added a gene into the embryo of a rat successfully, and the newly born rat showed a strong memory and a high learning capacity. This clever rat called "Doogie" could recognize the label plate it had met, could learn to find out the hidden underwater platform, could sense the signal that might mean a coming slight electric shock and then dodged away. All these abilities are beyond the ability of other rats. Just by a simple gene reform, it became a clever rat with "a high intelligence quotient" among the rats. The magazine, Nature, announced this research result, which aroused a sensation in the world. This achievement is really inspiring. The appearance of the "clever rat" signifies that mankind has taken a key step forward in optimization of human brainpower. If we make good use of and develop such an achievement, it is hopeful for us to optimize human intelligence, and work out babies with "a high intelligence quotient" or a super "high intelligence quotient".

The gene not only determine intelligence, but also determines physical performance, appearance, character, morality and so on.

Does the appearance of a dissolute "rogue" result from a poor moral education or just from an innate character? The neurologists in the Emery College, USA, made an experiment with mice. In their gene molecular structure, the grassland mice practicing "monogamy" and liking a social life are quite different from the mountain mice practicing polygamy and loving a solitary life. They transplanted the grassland mice's gene into the bodies of the mountain mice, and a miracle appeared: Those fierce, pugnacious and promiscuous mountain male mice changed abruptly in behavior, becoming much milder, and their relation with their mates seemed to have become sweeter and warmer, too. Research on humans has also proved personality trait is related to the gene. Is greediness only a personal penchant? Australian scientists found a sort of gene related to appetite lately, which could produce a protein called "directing sign" that could induce people's desire to eat. In addition, there are such reports: After scientists did tracking investigation on some twins, they have found people's eating habit is also determined by the genes...

After removing the "dissolute gene", we might change a dissolute and promiscuous animal into a loyal, gentle and faithful animal. Likewise, by changing the gene of a dissolute man, we may turn him into a model husband. After Kennedy Jr. of America flew a plane into a wreck, some scientists thought: it was the "adventure gene" that resulted in the "frequent accidents" of the Kennedy family. The latest research findings have proved, some people have a special love for adventures and always seek novelty, at least partly because the genetic gene in their body is different. Lately, some scientists have discovered a "suicide gene". When we allege there is a "suicide gene", it does not mean all those with such a gene will certainly commit suicide, but those committing suicide must have that "suicide gene"... Whether dissolute or not, adventurous or not, suicidal or not ... we can find the root of all those from the genes.

Thus, it can be seen, by reforming the gene we can not only optimize or improve intelligence, but also optimize and beautify the character, morality and so on.

Gene also plays a decisive role in deciding whether we are merry or not, whether we are happy or not. Recently, some researchers found: whether a person is happy or not is closely related to his sense of happiness, and the sense of happiness is determined by the genes. Such a "happy gene" is located in the inboard section of the forehead. If this part is damaged, a person will have no sense of happiness. Moreover, earlier research has already found: the more the serotonin in the human brain is, the happier the person will feel, and contrarily, the person will have little sense of happiness. All this has nothing to do with the economic situation or social class.

All the four major attributes of the "immortal", namely; eternal life, omniscience and omnipotence, supreme kindness and supreme beauty, and daylong ecstasy, or the six major aspects corresponding to them, namely; physique, intelligence, ability, virtue, beauty and pleasure, can be achieved by way of birth improvement through technology. That is, we shall be able to optimize the human being in all aspects. As early as 1994, US scholar C. S. Lewis wrote: *"The ultimate moment is approaching when mankind gains the complete control over himself by eugenics, by means of setup of the prenatal conditions..."*

Gene technology will not only be able to help the doctor find out what genetic disease the embryo has, but also be able to help people cure diseases caused by defective genes. Moreover, we can design the future infant's body constitution, intelligence quotient, individuality, skin color, eye color and so on. Then, as early as the embryo stage, we can modify and optimize the gene into; the gene of eternal life, the gene of omniscience and omnipotence, the gene of supreme kindness and supreme beauty, the gene of daylong ecstasy... that is, provide all the DNA conditions necessary for the "god".

Owing to the continual progress of gene technology, in future, we shall be able to do overall optimization or super optimization of human gene, realize the gene leap in mankind's evolution to immortals, thus, mankind will basically be able to evolve into immortals independently and quickly. Through the high-leveled gene eugenics and gene optimization, we shall realize eternal life and eternal youth, enjoy a healthy physique, enchanting looks, a fine and smooth skin. Moreover, our memory, thinking capacity and creativity will be surprisingly high. We will have a kind and gentle character, showing concern for the country and the people all the time. Meanwhile, in life, we shall keep an optimistic spirit and happy mood every hour and every minute.

Someone may think: It is too good, too good to be credible, but after all it is still faraway from us, the present level of birth improvement through technology has not reached such a stage. Indeed, such a super eugenics through technology has only shown some initial inkling now. However, the endless sustainable development of birth improvement through technology will turn all those into realities, or even work more unimaginable wonders.

In general, the production in the world falls into human production and material production. Human production, namely; production of mankind itself, refers to the population problem in the usual sense. The population problem is never only a problem of quantity of population, but what is more important is the population quality problem. Now we always focus on population quantity, rather than population quality, and such thinking and practice is very shallow and very harmful. An over-large population will have a negative effect on economic development, but the more serious problem is the population quality problem. The population quantity problem is primarily the population density problem, that is, the population quantity is chiefly relative to the land area. Japan's population density is much greater than China, but its economy is much more developed than that of China, the population density in East China is much greater than that of West China, but its economy is much more developed than the latter on the contrary. So it can be seen that the effect of population on economy is clearly chiefly in its quality, instead of its quantity. Moreover, the improvement of population quality can restrain the growth of population quantity most effectively. That has already been proved by the history of developed western countries, and has also been proved by the facts in China. Hence, the best policy of population control lies in improvement of population quality.

Birth improvement through technology has offered the most powerful weapon for improvement of population quality. In China, about 8 million infants with congenital anomalies are born every year; in Sichuan Province alone, the annual rate of birth of infants with congenital anomalies is as high as 4%. In one small place alone, 200 dements were found, they gave birth to 5 children averagely, and each of the children was a defective infant. When such children grow up, they will be heavy burdens of the society, the family, and even of themselves. It will be also the greatest life tragedy and a lifelong hurt. If we carry out birth improvement through technology, we may weed out congenital anomalies before birth. State Committee of Family Planning has already launched the "Congenital Anomalies Intervention Project", aiming to eliminate malformed poor embryos, so as to prevent heavy society burdens, family tragedies and personal tragedies.

In birth improvement through technology, we should not only carry out the project of intervention of congenital anomalies, but also go all out to carry out the project of birth quality optimization. In all ages, a genius prodigy appears only in one hundred years or even one thousand years, like the century-old submarine pearl or the millennium-old glossy ganoderma on the snow berg. Nowadays, we have artificial pearls, artificial glossy ganoderma, but why can't we produce genius prodigies or super genius prodigies artificially? **By birth improvement through technology we may usher in an era in which groups of genius prodigies or saints appear.** In our descendants, we will not only be able to prevent defects, but may also cause large groups of Lei Feng, Lu Xun or Mozart to come forth, and there will be more and more superhumans coming forth who are much greater than the great people in the past. The kind of people decides the kind of society. How can a society composed of very high-quality geniuses and saints not develop at a high speed or super speed? How can such a society not lead to the ideal world commonwealth very soon?

Of course, any technological progress is accompanied by violent controversies. Birth improvement through technology is no exception. As this field is still lacking in rules, some investigators are hesitating to move forward, but others have began to "act recklessly", which has intensified the argument. Argument is a good thing. Mr. Kaplan, head of Bio-ethics Center of the Pennsylvania University, USA, cried out: *"We should argue faster, more clearly and more loudly."*

Quite a few people show worry about birth improvement through technology, and about gene improvement. Hawking, reputed as Einstein's successor, once worried about such a prospect that mankind can work out "superhumans" who are much higher in intelligence than other people and can rule the world. But later, he modified his view gradually, and became an advocate for gene improvement and acceleration of human evolution. Gene eugenics not only means optimization of physical performance and intelligence, but has offered a possibility for multi-dimension optimization. **The "superhumans", we are going to create will not only be mental superhumans, but also be moral superhumans.** By gene reform, we may get rid of the selfishness, greed, cruelty and bestiality in human genes. We can transplant the gene of noble-minded and kind-hearted people extensively, thus, the fine wish of *"Let the world be filled with love"* can be realized. Mr. Lu Xun once pointed out from a superhuman foresight: *"Judging from the present human races in the world, we can believe, in the future, a nobler and nearly perfect mankind will appear."* The fast development of today's science and technology has brought the day predicted by Mr. Lu Xun closer and closer to us. The future Lu Xun believed will come for certain, and a noble and perfect mankind will appear soon. Then, this world will become extremely fine.

But some people turn pale at the mere mention of "eugenics", as it reminds them of the "eugenics" of Hitler. Mr. Qiu Renzong, member of International Human Genome Ethic Committee, said: *"If we use the gene research findings for improvement of the human race, it will be nothing else than Hitler's 'eugenics' of racial annihilation."* As soon as such a saying came out, many judges made a big red cross on "improvement of the human race". Is such a saying really reasonable? Hitler's eugenics was totally different from today's birth improvement through technology. Hitler used "eugenism" as a pretence and waved the whittle madly to the Jews and other weak nations, practicing a brutal and inhuman massacre. But using gene technology to produce babies with "a high intelligence quotient" and "a high moral quotient" is meant to meet the human need on a profound level, so as to promote human development deeply and bring mankind into the finest future. It is just where the greatest interest of mankind lies. Professor Rao of Religion Institute of Virginia University of India said: *"The knowledge we have acquired should be used to benefit all mankind. If a charming appearance, a strong physique and a bright wisdom can be worked out by design, we should let more and more people have them."* Only such a saying is a wise one, and really stands to reason.

Many experts worry that if we use gene technology to "produce" infants of high intelligence quotients, some rich parents may use it to make their children's intelligence quotient grow above normal, resulting in a social inequality.

The pursuit of equality may be passive or active. In China, those who are able to go to collage and receive higher education are only a small fraction of the population, whereas many people can not even finish their primary school. Then, can we, for the sake of pursuit of "equality", close up all the colleges to ensure everyone may have the equality of being unable to go to collage? On the contrary, what we should do is not to close up colleges, but go all out to develop colleges, including building more colleges, and should also encourage private education, trying every means to let more and more people go to college, and finally enable all people to get the best education. Likewise, in birth improvement through technology, if we, for the sake of pursuit of a temporary and passive "equality", give up its research, exploitation and utilization, it will not only be a loss of many individuals, but also be a great loss of the society and mankind. On the contrary, just for the sake of eliminating inequity, we should make great efforts to develop the technology for birth improvement, so as to reduce the cost continually until everyone can afford it. Since letting some people "become rich first" is reasonable, letting some people "use it first" is reasonable, too. It is reasonable and beneficial for the development of all mankind.

What is more important: birth improvement through technology is a development trend and no force can prevent it anyway. Maybe most people are not so interested in duplicating themselves, but, in this world of fierce competition, which parents do not hope to give birth to the most perfect children? Which parents do not wish their own children will have a high intelligence quotient, a strong constitution and a pretty look? Now in Beijing, in order to give birth to a clever and healthy baby, a man wants to give up smoking, give up drinking, give up tea, give up the computer, and even give up the mobile phone and the TV... *"I have decided to have a child, so, from today, I will not drink a blob of wine, not drink a mouthful of tea"* When hearing a man in his 30s saying so, no one will feel surprised. Such men can be found everywhere. In order to have a healthy and clever child, they can do any hard things, and pay any high price. What's in their mind is just one point: Either have no child, or have a best one. In order to give birth to an excellent child, they can endure any great hardships. Hence, the future trend must be like what the geneticist John Campbell said: *"Every generation of parents will wish to give the newest and best reformed quality to their children, and will not submit to the will of Heaven by passively accepting the inherited chromosome."* Children's natural predisposition determines everything in their future by the root, so, all parents will try every means to give the best to their children. "Owing to such a human nature, the embryo gene therapy will be worked out sooner or later" said Anderson: *"So long as we have new ways to think up, no one is willing to give the fatal gene to the child. That is just the impetus to development of the embryo gene therapy."* The same is true of the development of the whole undertaking of birth improvement through technology.

As far as any parents or any family is concerned, birth improvement through technology is their greatest need. The same is true of any nation or country. Making the nation prosper through education is a development strategy of many countries. Education is a postnatal optimization, and is not the primary determinative factor for personal quality, but birth improvement through technology is an innate optimization, being the primary factor. In the past ages, we never had the super technology and conditions for birth improvement, but now the conditions are already available. We have the capacity and method to highly optimize people's natural predisposition. Such a eugenics project will have a matchlessly far-reaching and decisive effect on the quality of the whole nation and the future of the whole country. In so much that it is quite common now for many countries to practice "compulsory education", taking "compulsory education" as one of the fundamental national policies, **they will, in the measurable future, also practice "compulsory eugenics" and regard "compulsory eugenics" as a fundamental national policy.** Not long from now, we shall execute both "compulsory birth improvement" and "compulsory education", and only then can we really turn the great slogan declared as early as two centuries ago, *"All men are created equal"*, into a reality, rather than a mere wish or dream. Only by doing so can everyone get the best congenital quality equally, get the best postnatal education equally, and possess equally the key conditions for overall development and optimal development.

Compulsory birth improvement through technology may lay the best material basis for becoming immortals, so that everyone, as soon as he (she) is born, may be destined to become the greatest and perfect superhuman equally. We can predict that compulsory birth improvement through technology will surely become the most important development strategy of a state or nation.

Mr. Sun Zhongshan (Sun Yatsen) said: *"The tide of history is vast and mighty, those who are with it will thrive and those who are against it will perish."* Birth improvement through technology will certainly become a historic tide and no one can stop it for any reason. Whosoever, for some prejudice or wrong thinking, snuffs out this great undertaking that is so important to everyone, every family, every nation and all mankind, will be condemned as a sinner through the ages.

XXI. SECOND WAY TO BECOME IMMORTAL THROUGH TECHNOLOGY: BODY IMPROVEMENT THROUGH TECHNOLOGY

With the aid of the man-machine integration technology, we may acquire any knowledge or ability without learning, and reach the level of the grand-master directly.

Birth improvement through technology has built the bridge for our descendants to become "gods". Those infants who have gone through the gene optimization process will take a smooth road to becoming "gods". Yet, for those who are unable to return to their mothers' wombs, what is to be done? Will they only sigh before the fine dream? In the coming competition, will they be under the yoke of "those" who have gone through the birth improvement process? What will be our way out?

As far as we are concerned, the way out lies in "body improvement through technology".

Just as its name implies, this project means using high and new technology such as biotechnology, information technology, nanometer technology and so on, to optimize our own body. Body improvement through technology has opened up an approach to amend and reform the innate condition.

There may be three ways for body improvement through technology: Gene reconstruction, man-machine integration, and multi-dimension upgrading.

I. Gene Reform

In the previous section, I talked about the great effect of the gene on many aspects of a person such as body, intelligence, ability, virtue, beauty, pleasure and so on. So there is no need to detail it here. For the embryo, we can repair and reform gene defects; for people already born, we may also do gene therapy, gene repair and gene optimization.

A paper published in the authoritative journal Nature describes the attractive prospect: *"Experimental results have shown, it is feasible to use the biological means of gene reform to improve mammalians' intelligence."* That means, maybe not long from now, if you feel you are not clever enough, you may use the gene therapy as a remedy. Scientists predict, just within the first quarter of this century, it is very likely that people may produce an injection or tablet that can turn a person cleverer. That will not be "a tonic to the brain" in the usual sense, but "a change of the brain". Popularization of gene drugs may turn a fool into a bright person and help a bright person become even brighter, and even create super intelligence. Gene reform will also become the leading force in beauty treatment, being widely used in beauty treatment projects. We may add some "beauty gene", "fresh-keeping gene", or "whitening gene", etc. to the gene organization of a normal person. What is more important; we may modify or implant an eternal life gene to cause the most fundamental gene reform to people.

Of course, in the in-depth research on gene technology, we have just made a start, and we need further in-depth study to fulfill all kinds of gene reform tasks needed for becoming "gods". In gene technology, we have only spent very little time and money, but have brought about such great breakthroughs. In future, we shall certainly increase our efforts and investment in this area, hence, we shall certainly make more and greater breakthroughs. Any secret that is not revealed now will be revealed, any technique that is not mastered now will be mastered, then, we shall be able to do the best postnatal reform and optimization for human beings. The ephemeral may become long-lived, those easy to become anile will keep young forever, the ugly may become pretty, the foolish may become bright, the bad-tempered may become good-tempered, woebegone people may become people beaming with smiles...

The great leaps of gene technology in future will change mankind's material basis by the root, providing the fundamental conditions for men to become "gods". All aspects of the natural predisposition endued with the "god" may be acquired through gene reform.

Man-machine integration falls into two types: One is the combination with the human brain, and the other is the combination with the human body.

Combination of machinery with the human brain means using artificial intelligence (AI) and life science and technology to join and combine human intelligence with machine intelligence, so as to cause abrupt changes and great leaps to human intelligence, enabling people to reach the "omniscience and omnipotence" prospect of the "god" with ease.

In Europe of the middle ages, there was a theological question: How many angels could stand on one needlepoint? In terms of the past common thinking, this question is unthinkable. But now we can record all the contents of *Encyclopedia Britannica* onto a biochip as big as a needlepoint. Along with the development of nanometer technology, information technology and biotechnology, such a biochip may become smaller and smaller, with higher and higher functions, and can be connected directly to the human brain cell. **A person may master the knowledge of the whole set of Encyclopedia Britannica in an instant through the chip implanted in his brain, so, in an instant, he can become a great scholar others could not become even by scores of years of efforts.** Not only implanting the whole set of *Encyclopedia Britannica* into the human brain, even implanting all the books in all the libraries of the world, and all the cultural wealth of all mankind gained for thousands of years, is also possible! Thus, won't mankind be like "gods" in having super knowledge, super wisdom or even omniscience and omnipotence?

By implanting the knowledge chip and intelligence chip directly connected to the brain cells, we may not only expand our memorizing ability, computing ability, inferential ability and problem-solving ability greatly, but also enhance our capacity for appreciation of music, poetry and other arts. It seems hard to imagine. According to common sense, music and poetry belong to the emotional life field typical of human, and leave no room for machinery to join. However, David Cope has made a music intelligent program, it can filter Mozart's music DNA from his works, so, the new music composed in that way just sounds like the style of "Mozart". Some musicians were amazed by the naturalness and quality of the new work. The EMI (Experiments in Musical

Intelligence) made some traditional artists feel a great disillusionment, and a German musicologist rushed roaring to Cope, wanting to beat him. If we put this music intelligent program made by Cope into a nanometer chip and then implant the chip into the brain of someone who knows nothing about music, he might have the music gift of Mozart in an instant. So, we can see, with the aid of the man-machine integration, we may acquire any knowledge or ability without learning, and reach the level of the grandmaster directly. Hence, we are actually not far away from the omniscience and omnipotence of the god!

The key component of a human is the brain. Mr. Zhang Xiangtong, famous brain scientist of China, thinks: *"The brain is the material basis of all intelligent behaviors. The rise and fall of a state or nation is closely related to the mental power of its people."* Mr. Li Lanqing (vice premier of China) pointed out clearly: *"The significance of full development of the human brain is even greater than the exploitation and use of matter. The key to gaining the upper hand in the 21ʰ century is manpower. The coming decade is a decade for brain science. Any country that can make breakthrough in this area will have the greatest vitality."* In the new era, full development of the human brain should include two aspects: On one hand, it refers to digging up the greatest potentiality of the human brain by using all sorts of achievements in the brain science, including using the ever expending gene technology to reform the gene of human brain thoroughly; on the other hand, it refers to man-machine integration, that is, the high combination between the human brain and the computer, thus, the computer may become an organic component of the human brain, so that the human brain will be upgraded constantly, until reaching the level of the omniscient and omnipotent "godly brain". The computer's development speed is amazing. In the multimillion years of development, mankind's brain structure did not change very much, but in only half a century after its birth, the computer (the electronic brain) developed to such a degree that it has defeated the top-rate chess grandmaster. The fastest speed of human brain's response is a millisecond, but the computer's lowest speed is a microsecond. The human brain's nerve perception conduction speed is only 10 m per second while the transmission speed of computerized information is 300,000 km per second. Moreover, the developing speed of the computer is extremely striking: in the near future, the computing speed of the photon computer will be

1000 times or even 10000 times of that of the present computer. Development of the computer from now on will surely be accelerated. The current 10 years have exceeded the past 50 years, and in future, one day may outmatch 100 years!

Man-machine integration also includes the combination of the machinery with the human body. British Professor Warwick implanted a chip into his own arm, and achieved a success. He could use this chip to open the gate of his home, accompanied by a greeting, then, the lamp would be turned on automatically, the bath tub would be filled with water automatically, and even the wine keg could be warmed up automatically. Hence, Warwick declared proudly that he has become the first "cyborg" in the world. Meanwhile, the Jacobus family (including three members) of the US became the first group of people who volunteered to accept the implantation of a new computer chip called "VeriChip", thus, it was declared that the first "cyborg" family in the world appeared in Florida. The chip implanted by the doctor could help the doctor inspect the internal condition of the human body via the scanner, so as to make a fast evaluation of the patient's condition.

In 2004, a research group composed of 9 top-leveled scientists over the world made a great breakthrough in bioengineering. After they implanted the "organic USB interface" into the hindbrain of a man called "Gali Carrel", the man's brain could operate a series of computer peripheral equipment directly, including the CD driver, scanner, printer and so on. According to the scientific personnel, after the operation, Gali Carrel's intelligence quotient was as high as 265, naturally becoming "the most clever man" in the world.

Bill Gates said recently at a seminar held in Singapore: In future, the computer will be implanted into the human body. He said: *"Many Microsoft employees often say to me: 'I am ready, please turn me into a computer'"*. These techniques are already helping mankind repair their physiological defects. In Germany, in more than seven hundred deaf-mutes' heads, the artificial ear cochlea and word processor have been implanted, thus, they have restored their hearing and language function. French scientists are working hard to develop a microcomputer called "artificial retina", and are going to implant it into the eyes of blind people, connecting it to the nerve of the human brain, so as to enable many blind people to see things again. By man-machine integration, the deaf can hear, the blind can see, and normal people may hence acquire superhuman functions, such as "clairvoyance", "ears that can hear sounds a thousand miles away", and they even can walk a thousand miles a day or fly ten thousand miles a night and so on. Then, all such things will no longer be imaginations, but realities.

In information storage and computing skills, the computer has already surpassed mankind by many many times, and the future computer will have the abilities to perceive, think and judge, and have a learning capacity. Meanwhile, its form will be smaller and smaller. Dejus Koessler, an advocate of ultramicro technology, says, **the future computer may be "as small as a millionth of the cell,** then, it can guide the molecule machinery to sense the structure inside the cell, and repair the damaged section of the cell". This means, future computers will not only be able to suffuse the human body like germs, but also be able to permeate human cells. And, the relationship of all kinds of ultramicro intelligence machines and ultramicro computers with human brains and human bodies will become more and more friendly and harmonious. Finally, the computer will reach such a prospect as is described by John J Anderson in the journal *Mac User*: *"The ultimate objective of computer technology can be said to let the computer disappear, let the user feel a 'transparent' technology, as a matter of fact, the computer will not exist..."* **The computer or machinery will become "our own parts" increasingly, become the organic members in our own body**, melt with people into one increasingly, without boundary or distance to us.

By dint of man-machine integration, mankind's evolutional steps will be quickened at an unprecedented rate, accompanied by many leaps and bounds; so, we shall reach the "godly realm" directly.

Super things in many aspects: "super food", "super medicines", "super reshaping", "super dress" and so on.

I. Super Food

Someone asserts: All geniuses come out of eating. The nutrients contained in food are energy sources for meeting the human body's needs and keeping the human body's functions. By developing super foods through high technology, we may greatly optimize people's nutrients in every aspect. In future, we may use biotechnology to work out the most reasonable, most ideal "longevity dietary pattern", "ever-young dietary pattern", "genius dietary pattern", "beautifying dietary pattern" and so on as suitable for each individual, to provide the optimal material conditions for enhancing people's natural quality.

Nutrients in food have a great effect on people. Once a certain nutrient is lacking, the person may become a fool. There is such a story: A group of distinguished persons went to an out-of-the-way village for investigation. They found everyone in the village was a fool. They could only count the coins of five cents and one dime. If you want them to turn the mill, they would go on without stoping... Only the headman and the children were still normal persons. When these visiting writers and painters held those very lovely children in their arms, seeing the 8- or-9-year-old children looking so innocent and clever, they could not help feeling sad in their hearts, as they knew, two or three years later, these children would also become as foolish as the adults in the village. In fact, such a disease is easy to cure: So long as they eat a little salt and food containing iodine, they can prevent it. Mr. Henry Labouisse, former Executive Chairman of the United Nations Children's Fund said: *"The disease of iodine deficiency is very easy to prevent. Even if there is only one child showing mental deficiency because of lack of iodine, it will be our sin."* Just because of a lack of a little iodine, a person may become a fool. Contrarily, if we eat some kind of nutrient substance more, maybe we will become clever or even become a genius. The Jews only account for 0.3% of the world population, but in the 72 years from 1901 to 1973, the winners of the Nobel Prize from the Jews accounted for 16% of the total number. Why? Through the study by the Medical School of Yale University, USA in succession, they have found the secret: The reason why Jews have a high intelligence quotient is that they often add an excited-state substance called "EGB" in their food, which makes their memory surprisingly good. If we add such a substance to the food of all people, everyone may become a genius easily. Another example: Modern nutritional research has found a biochemical substance called "acetylcholine" that may improve human memory. Thus, an American scientist added some sinkaline to the flour and made noodle. When people eat such noodle, they may improve their memory. In future, we may not only eat the noodle that can improve our memory, but also eat the twisted dough-strips that can improve our creativity, the bread that can improve our mental quality, the milk that may raise our morality...

II. Super Medicines

A little medicine may put someone to death, or raise someone from the dead.

In future, we may develop super medicines for optimization of the human body. According to research findings, nearly all diseases are gene diseases. If we work out "customized medicines" according to everyone's gene condition and suit the remedy to the case, we shall certainly get twice the result with half the effort. One of the trends of the present gene medicine research is to realize "individualized treatment" in future, namely according to different patients' disease causing genes, work out the therapeutic drug most suitable for them. "It may not only cure pertinacious diseases most effectively, but also optimize our bodies to the best effect. An American corporation has recently invented a medicine that can slow down an organism's ageing process, and now it is in the process of application for clinical trial on the human body. Although the aim of the experiment is to test the effect of the new drug on patients of apoplexy and skin heat injury, some scientists think this drug is hopeful to break mankind's fatality of birth, ageing, illness and death. We can hope confidently: The "immortality drug" First Emperor of Qin Dynasty sought with tremendous efforts may be obtained and used by everyone in future.

Super medicines can not only cure diseases and prolong life, but also improve intelligence, enhance beauty and add to pleasure. Worfson Research Center of Medical School of University of London of the UK improved the mice' learning capacity by reforming their gene. We may also make great efforts to develop the "memory drug" for improving mankind's memory, the "rejuvenating drug" that can help people to return to youth, the "happy drug" that can lead to boundless happiness... Such "elixirs" will be incomparable to the ancient cinnabar, as they will greatly help us to become "gods" in deed.

III. Super Reshaping

In future, the beauty treatment through technology will also be very magical. Would you like to change your appearance? Your face can be fabricated in the beauty treatment laboratory at your will. High and new science and technology may enable people to make their appearance or figure as desired. In other words, you can have whatever looks you wish to have.

Nowadays, science and technology have already made surprising progress in artificial skin, artificial bones and artificial organs. Our country has already mastered the high technology of artificial skin, can make very perfect artificial skin, which is very tight, very elastic, and as bright as jade. In the past, in the breast raising operation, people used silica gel that is very unsafe, but now scientific personnel are studying how to utilize human cell tissue to cultivate a breast tissue, which is not only safe but also is the person's own. Any organ of the body may be changed at will, just like changing a part in a machine. If a leg or a hand is broken, just go to the hospital to have it changed. Moreover, the new leg or hand or any organ will be stronger and prettier than the original one.

The female writer Bi Shumin made a statistic and found that 100% of females are dissatisfied with their looks. The future super reshaping technique may make all the females and all the males satisfied with their looks by 100%. Any dissatisfactory part can be made satisfactory. Any person who worries about his (her) appearance may become very beautiful or handsome, and full of self-confidence. According to the investigation by a college in London, the money earned by tall salesmen is 1/4 times more than that of earned by their short colleagues. Investigators have tracked 17733 persons born after March 1958, and their results show, the money earned by tall men is more than the money earned by short men by 1000 pounds to 10,000 pound, and the short females' average income is less than females of average stature by 5%. In future, people's height and shape can be changed at will, hence, those who worry about their shape may get a higher salary, and have greater self-confidence.

The super reshaping techniques will certainly develop all the time, so, the day is not far when mankind can turn out millions of Xishi, Hellen or Pan An. By the top high technology and along with the constant breakthrough in life science and technology, everyone may have godly looks like flowers or the moon.

IV. Super Dress

Super dress belongs to the category of body improvement in the broad sense. A super dress is not like the ordinary dress, as it has turned from "it" into "he", becoming intelligent.

In his *Digitized Life*, Negroponte talks about "wearing the computer on the body". He says: *"The material of the future digitized clothes may be corduroy having a computing power, or tabby crocus cloth with a memory capacity or silk of solar energy, so, I will not need to take my laptop, but just wear it on my body."* As early as the Tokyo Exhibition of 1998, IBM displayed its wearable computer.

All kinds of intelligent clothes will appear in the future market. Scientists of the Massachusetts Institute of Technology, USA are just attempting to embed some electronic devices such as the telephone and computer into clothes. The future digital clothes' screen will be as slippery and bright as silk, with little keystrokes like luxurious buttons. Just by pressing the button, the waistline may be widened or narrowed at your will. A "Star Lab" in Brussels of Belgium has already worked out a group of models of novel intelligent clothes. The "Prompting Dress": When you have forgotten to take your key or lost the key, it will sense it timely and remind you; when you are in a danger zone, it will remind you timely to take more care. The "Sports Information Dress": It may measure the athlete's heartthrob, and use E-mail to transmit the training data about the athlete to the club where the athlete belongs. The "Record Storing Dress": It can memorize the condition of the place you pass, such as the air freshness, the background sound and so on. The "Relaxing and Amusing Dress": When you feel afraid, worried or painful, it will play an appropriate piece of music through the mini mike, to relax your mood... In short, such super clothes can help or replace people to handle all kinds of daily affairs, adding to your abilities of all kinds. It is estimated that such high-tech clothes will be put onto the market soon.

In future, **all we wear may be intelligentized, ,becoming something "with a brain"**, Our wrist watch will have a brain, our handbag will have a brain, even the buttons on our clothes may have a brain, the socks and shoes on our feet may have a brain, each having some "godly power"... They will become able helpmates of all kinds in our life, so our abilities will become stronger and more magical.

In conclusion, the connotation of body improvement through technology is very rich. By gene reform, man-machine integration, multi-dimension upgrading and other means, all of us may become "gods" gradually. Hence, body improvement through technology is another major approach for becoming "gods".

XXII. THIRD WAY TO BECOME IMMORTAL THROUGH TECHNOLOGY: EDUCATION IMPROVEMENT THROUGH TECHNOLOGY

By dint of the magical power of high technology, we shall do quality education, individualized education, automatic education and lifetime education, so as to realize a revolution in education in the real sense.

As the old saying goes: *"The greatest way to benefit other people is education."* The eastern saint Mencius said: *"A good politics is not as good as a good education."* The western sage Kant said: *"Only through education can a man grow up. People are purely results of education."* Whether a person has received an education or not, and what kind of education he receives, determine his qualities and fate.

Why is education so important? For education is the reproduction and re-creation of mankind itself. The kind of education decides the kind of people produced or created. Becoming "gods" is mankind's highest goal in its own development and evolution. In the process of mankind's evolution to immortals, in no way can we do without education, as it shoulders the important task of "human reproduction".

A person's fate is determined jointed by nature and nurture. "Birth improvement through technology" starts from the congenital aspect, to optimize the congenital elements of a person, so as to lay the optimal congenital material basis for becoming "immortals"; "Body improvement through technology" starts from the postnatal aspect, to optimize the postnatal elements of a person, so as to provide the optimal realistic conditions of all kinds for becoming "immortals". On the one hand, "body improvement through technology" lays stress on changing of a person's physical structure and function; on the other hand, "education improvement through technology" places emphasis on changing human mental structure and function, that is, through the chain of "knowledge–thought–behavior", to optimize all the aspects of human quality, such as virtue, intelligence, physique, beauty, pleasure and so on, and improve the person continually, until he turns into an "immortal".

The evolution process depends on education, but it is not the present education. It should be an education infused with the philosophy of mankind's evolution into immortals, and infused with the culture about mankind's evolution into immortals, and an education optimized through high technology.

Today, nearly all fields and behaviors have undergone earthshaking changes owing to the super power of high technology. Nevertheless, while all the other fields are vying with each other for modernization, the educational field has stopped almost on the level 100 years ago. Gordon Deliton pointed out: *"If we revived a person who died 100 years ago, he would marvel at the social transformation of the U.S. The typewriter has been replaced by the computer, the oxcart or horse cart has been replaced by the automobile, the hand-powered telephone has been replaced by satellite communication and fiber cable communication... But in only one area, this visitor would find it is basically the same, that is, the schools and classrooms of America."* Even American schools and classrooms are in such a bad state, to say nothing of other countries! Although we know education is fundamental, education should come first, yet, the revolution and development in education has lagged behind. In developed countries, all the other fields have shifted from the labor-intensive type to the capital-and-technology-intensive type, but only the field of education still remains to be the labor-intensive type, like a handwork mill. For thousands of years, teachers have been using a book, a piece of chalk and a blackboard as their tools. The serious backwardness in education also signifies that we have huge potentialities in education development.

Education improvement through technology means using high technology to realize the development of education from the labor-intensive type to the capital-technology-and-intelligence-intensive type, especially, using the fast-developing information technology to help education realize four transformations, namely; quality-oriented transformation, individualization-oriented transformation, automation-oriented transformation and lifetime-oriented transformation. Transformation here means a thorough change. By the four transformations, it is hopeful to optimize education to the best extent and help people become immortals most effectively.

Quality-oriented Transformation

The qualities of an "immortal" include six major aspects, namely; physique, intelligence, ability, virtue, beauty and pleasure. Optimization of education should be targeted at each of the six aspects. Take "physique" for example. Physical education is never a small-scaled physical training of several sports items, but is an all-inclusive revolution from the concept and the contents to the method. We shall shift from teaching sports skills, to enabling people to cultivate a health preserving consciousness, develop good habits for health care, and master all kinds of knowledge, methods and skills about health care, so as to enhance their health care ability and health preserving ability greatly, and open up their greatest potentiality in body building, until they reach the prospect of eternal life.

Quality is not knowledge. Quality education should be as Einstein said: *"Education is what has been left when the student has forgotten all the knowledge after leaving school."* Why is it so hard to carry out quality education now? The major reason is that we can not extricate ourselves from the bondage of only emphasizing knowledge and testing knowledge. That way of thinking can only lead to mad infusion of knowledge and mechanic memorizing of knowledge. With the aid of high and new technology, we will have no need to spend the time cost and energy cost of ten years or even a few decades in memorization of knowledge. We may record all the knowledge of the world into a mini chip and implant it into the human brain. With the help of nanometer technology, the tiny space smaller than a hair can store all the books of scores of Library of Congress, USA, and everyone may get a huge amount of knowledge with ease. Hence, knowledge education will quit the stage of history, and will be replaced by the true quality education. Only then can we focus on "things beyond knowledge", and direct our major energy to the cultivation of innovation ability, and the cultivation of all the attributes of the "god".

"Quality cannot be told"

That is what have I repeated many times in my *Ability Panic*. The basic mode of quality cultivation is the three-in-one of "knowing—understanding—acting", chiefly dependent upon experiencing, tasting, acting, or practicing. When a noted British biologist recalled a story of his childhood, he was very excited: *"When I was 5 years old, I got the most important scientific discovery in my life. I shut a caterpillar in a bottle, it began to spin a cocoon, and several days later, to my surprise, I found a butterfly came out. I observed the whole process carefully with surprise."* He regarded his finding of the change of the caterpillar into a butterfly as "the most important scientific discovery in his life". Although this discovery had been made by others long before, this finding lit up the scientist's spirit. If a teacher or his parents had told him this knowledge "A caterpillar may change into a butterfly", and told him to take down this conclusion in his notebook and recite ten times, what he gained might be only the boredom of memorization of knowledge, he would never have got such a great effect as through his own experience: An effect that impressed him for all his life. Such a process of learning through experience is hard to practice in ordinary education. However, with the aid of information technology, especially the multi-media technology, we may digitize images, sound and movies, so our education may enter a "virtual reality", we may simulate all kinds of reality, set up all kinds of situation, then, it will be very feasible and easy to feel and taste learning.

"A picture can outmatch a thousand sentences". A dynamic virtual learning situation is more powerful than thousands of pictures. The future school will set up many virtual reality situations, for example: the excavation spot of Colombia for study of ancient relics, the "Peace Team" village of Africa, the grand scene of enthroning of First Emperor of Qin Dynasty, the inspiration Newton got under an apple tree... By participating in all kinds of time-and-space-crossing "limitless virtual world", the student may cultivate his quality much better. He may even play the part of some historical figure to taste and cultivate the comprehensive quality that he could not learn from books.

In the last century, Mr. Lu Xun said: *"Using movies to teach students must be better than using the teacher's lecture sheets, and perhaps it will be so in the future."* Before he finished his words, the whole room burst into laughter. Lu Xun's prophesy will become true in the fast development of today's multi-media technology. The "movies" made by multi-media technology have already appeared, and they will grow further for certain. The lively, colorful and changeful virtual situations may impress people most profoundly, offering great enlightenment to them, and greatly helping them to improve their quality.

Individualization-oriented Transformation

"Immortals" never follow the same pattern, and are never made out of one mold. In becoming "immortals", what we pursue is to highlight the individuality, reaching the summit of personal wisdom and creativity.

Esteeming individuality is the same as esteeming the congenital, and by combining the postnatal with the congenital organically, we may make "the gifted flower bloom". If we want to train Chen Jingrun into an outstanding diplomat, or train Jourdan into a gifted mathematician, it will be as hard as climbing upto Heaven. The 20-year-old Zhouzhou was a mentally retarded child. He could not read and write, could not do addition and subtraction within 10, and could not distinguish between a square and a round, only had the intelligence of a 3-or-4-year-old child. But he could work as a conductor for Dworschak's famous music "From the New World", and his understanding of music shocked all people. According to the "multiple intelligence theory" any person has his own advantageous intelligence and disadvantageous intelligence. Everyone is both a fool and a genius. **Without individualized education, we may turn many genius prodigies into mediocre persons. With individualized education, we may turn everyone into a genius prodigy.**

The development trend of future education has a greater and greater emphasis on individualized learning and the slogan is: *"The best course is to let each student have a course of his own."* Not only everyone will have his own course (learning contents) but also everyone will have his own rate of progress, and his own learning style. Thus, there will be different knowledge structures, power structures and quality structures established by different methods and different rates of progress. It is basically the same as the theory "Teach according to the student's ability" put forward by the Chinese saint Confucius more than two thousand years ago. Such a perfect education ideal could not be realized in the feudal society, and nor could it be realized in the industrial society, but in the information society, by dint of information technology, we shall be able to realize it.

Although "Teach according to the student's ability" was already set as the general principle of quality education by Mr. Liu Bin, former deputy chief of State Education Commission, yet, without the participation by high technology, it could only be a dream. In a talk show program, a professor reputed for cannoning the college entrance examination system, argued with me heatedly. He regarded "Let each student have a course of his own" as a daydream only. Indeed, if we do not have information technology and internet technology, it would hardly be possible to "let each student have a course of his own". But, with high technology, such impossibility can be turned into a possibility and the individualized learning and individualized education will become an inexorable trend in future.

The Internet-based teaching has broken the spatial-temporal limitation. In the Internet age, on one hand, one person's knowledge and intelligence can be shared by many people; on the other hand, the knowledge and intelligence of the whole society can be used by everyone. An American youngster John Hartford invented the supermarket in 1951, which put all kinds of living resources into one place, for you to choose freely. In a grand feast, the kinds of dishes are no more than 20. In a buffet dinner, you can have 50 kinds to choose. By dint of the Internet technology, all the educational resources of the world can be shared by all people, just like the greatest supermarket or the greatest buffet dinner. You can choose yourself, organize your study yourself, arrange your study yourself, and control the rate of progress of your own abilities; so as to do the best individualized learning and individualized education. You can select the most needed from the matchlessly rich course resources of the world network, can design your own learning completely according to your own conditions, in short, you can achieve the greatest individualization in all aspects.

Automation-oriented Transformation

The "god" is someone who has the highest self-awareness and greatest freedom, and is most capable of automatic development.

What is the highest prospect of production? Automation! What is the highest prospect of education? Also automation! "Teaching is meant to stop repeated teaching": that is the true meaning of education. Only such an education is the best education: "No need to whip the horse as it will gallop as fast as he can at his own will". In Western education research, a fundamental shift has occurred: "A shift from research on teaching to research on learning". The quality or efficiency of teaching is shown finally in the quality or efficiency of learning. If learning can be automated, that is, if students can teach themselves self-consciously and automatically, not only educational costs can be reduced greatly, but educational benefit can be improved greatly as well. Information technology has provided the best condition for education automation and learning automation.

In the past, the learning resources were simplex, we could only rely on several teachers or several textbooks, but the open-ended world network can provide diversified, multi-channeled, multi-leveled and multi-formed resources. Any person, so long as he possesses a computer, no matter where he is on the globe, just by connecting to the Internet, can choose easily the course most necessary for him and most interesting to him, can choose any teacher he likes, can begin his learning at any time he likes, and he can also ask all kinds of hard questions to the most excellent or most authoritative expert in the field, or have a dialogue with the expert. The student will not be restricted by any rigidly specified grade or course, and will not be restricted by any time or place. He can choose freely what to learn, and choose the way and time for the learning. The teachers will be a combination of the artificial intelligence system with the greatest experts in the world. Nowadays, scientists have already worked out the "virtual talk" system, so, in future; it can simulate grand-masters such as Einstein to offer you great advice. In teaching and explaining, such "grandmaster systems" or "expert systems" will be more vivid, interesting and profound even than the best teacher, and moreover, they are much more patient and selfless.

Information technology has offered the optimum condition for learning automation and teaching automation. After having such a condition, the key will lie in whether you have the quality for self-study and auto-learning. If you lack such a self-study quality, no matter how advanced the techniques are, they can only be put on the shelf. Thus, the key point in future education is not teaching knowledge, but cultivating the consciousness, habit, interest and ability for auto-learning.

By automated education via information technology, any one in the world, whether he is in a noisy city or in a poor mountain village, can do auto-learning, can utilize the educational resources of the whole world at will, and improve his quality in the best way. Everyone will be able to share the best education equally, and such an education will maximize the scale effect and efficiency of global education as well. We seek both fairness and efficiency. The automatic optimization of education through technology is just the most fair and most efficient highway for people to grow up and become "gods".

Lifetime-oriented Transformation

The goal of becoming "gods" is extremely great and high, so, naturally we need sustainable development, continual learning and incessant education. Hence, we absolutely need a "lifetime" education. The lifetime here is neither a few decades, nor hundreds of years, but thousands of years or even an eternal "lifetime".

Lifetime means "from cradle to tomb", and we need education at every moment of our life. Usually, people only pay attention to the dozen or more years of education from primary school to college, but neglect other periods, regardless of whether it is more important or not.

Now, more and more scientists and educationists agree: Early education is the most important. Scientific research has discovered: About 50% of the intellectual development of a person is realized from the fetal period to the age of four, 30% is accomplished from four to eight, 20% is from eight to seventeen. The development of human intelligence in the first 4 years equals that of the later 13 years. Not only for intelligence development, but the same is true of other aspects, too. Great educationist Tao Xingzhi thought: *"All important habits, tendencies and attitudes needed for a person are basically cultivated successfully before six. In other words, the period before the age of six is the most important period for nurturing the personality."* If we neglect or miss this key period, the re-education afterwards will get half the result with twice the effort. Great educationist Ushinsky also thought: *"A person's character is mostly formed within the first few years, and what is built up in the person's character in these few years is very firm, becoming the person's second nature."* Great educationist Makarenko pointed out more clearly: *"The foundation of education is laid chiefly before the age of five. It accounts for 90% of the whole education process."* The facts about "wolf children" and "bear children" can serve as full proof that what these educationists and scientists say is correct. The period before 6 or 5 is the most crucial period of life, and the education in this period stands head and shoulders above the education at any other time! However, at present, we do nothing for this period, excluding it out of compulsory education completely. What a great mistake we have made!!

The whole year's work depends on a good start in Spring, an hour in the morning is worth two in the evening and the most important education happens in the early stage of a lifetime. Pavlov said: *"If we start education from the third day of the baby, it is already one day late."* From the birth of a baby, there should be a series of advanced education programs for it, for example; what kind of music is to be listened to, what kind of films to be seen, what kind of games to be played, what kind of enlightenment to be accepted, and so on. A country's compulsory education must start from infant education, and we should make a policy for it. If so, in one or two decades, the world will become a brand-new world. In the case of the noted lexicographer Webster, as soon as he was born, his father made an education program for him. His father only spoke English, his mother only spoke French, his grandfather only spoke German, and the nurse only spoke a northern European dialect. In this way, from childhood, Webster mastered four languages very easily and naturally. An ordinary family cannot have such a condition like Webster, so they can only rely on social forces. The viewpoint of the great educationist Tao Xingzhi is very profound: *"Without using social forces, education will be powerless."* Especially infant education, it must utilize social forces, or else it will be "powerless". The United Nations or governments of all countries should gather a lot of education experts and science experts and use modern information technology to work out a most rational program for infant education (software), in which the most advanced educational thoughts and methods should be applied, so as to get the best infrastructural construction and provide the optimal technical platform. In any kindergarten at any place in the world, so long as the teachers can master the easy operational method of this technology, all the children may get the most effective education. It is a paramount step for "lifetime" education.

The more important "lifetime-oriented transformation" is that, anybody may get the most needed development conditions and learning opportunities at any age and in any case. In the pursuit of equality, the key is to seek such an equality and only this is the greatest equality. The whole society must build up a life-long education system for exploiting all the potentialities of everyone completely. But the present educational system has seriously smothered people's intelligence, wasting their most precious life. The ten-year hard school learning is often just for a diploma. What's worse, many people stop learning as soon as they go out of school. The future information technology and Internet technology will customize the most necessary course for people of any age, any occupation and any interest according to their demand, so that they can get the best development in any case. By virtue of the ever growing information technology, especially the multi-media technology and Internet technology, we shall establish the most advanced and ideal life-long education system to be shared by everyone, thus, in the whole life, everyone can choose to work or study at every moment, their work can be a studying process as well, and their study can happen at any time, in any place and on any occasion.

By dint of the magical power of high technology, we shall do quality education, individualized education, automatic education and lifetime education, so as to realize a revolution in education in the real sense. Hence, education will not only be able to mould people into people in the real sense, but also be able to turn people into "gods" and it will always be playing the crucial part in the whole process of mankind's evolution into "gods".

In becoming "gods", we must rely on technology, whereas the source of technology lies in education. In development of high technology, the key lies in highly intelligent people. If we say technology is the first productive force, education is the mother of the first productive force. Mr. Tian Changlin, member of the US National Science Committee thinks: *"The correct saying is 'Education and science make a nation prosper', instead of 'Science and education make a nation prosper'. The reason is so simple: Education comes first, then come science and technology."* Without education, there would be no capable people; without capable people, there would be no science and technology; without science and technology, we would not be able to become "gods". Hence, judging from this aspect, for becoming "gods", we must optimize and develop ourselves all the time. Optimization of education through technology is as important as optimization of birth through technology and optimization of physique through technology.

XXIII. FIRST CORE TECHNOLOGY FOR LONGEVITY IN A TORTUOUS WAY: BODY PRESERVING TECHNOLOGY

The freezing-preservation technology has solved the most difficult problem of mankind, "death!"

The ways to eternal life may fall into "the straight way to eternal life" and "the tortuous way to eternal life".

Those taking the straight way may be the present young people, children, babies and their offspring. The current middle-aged people and some oldsters may also take the straight way to eternal life, as scientists believe "before 2030, it is hopeful to realize the dream to live to the age of 1000".

The "tortuous way to eternal life" is the guarantee for becoming "gods" provided for people who are already in their eighties or nineties, or patients of "incurable diseases".

The key to the tortuous way is the technology meant for eternal life in a tortuous way. What inspires us is that we have already possessed the core technology for becoming gods in a tortuous way: the "body preserving technology". The magical body-preserving technology is the most valuable technology in mankind's challenge to death.

What is a "body-preserving technology"? Just as its name implies: it is the science and technology for keeping the body. There are many kinds of body-preserving technologies, and the most realistic is human body refrigeration technology, or human body low temperature storing technology.

In ancient Egypt, people worked out the decay preventing mummy, just in order to raise someone from the dead someday. Some present-day Americans have set up corporations of human body refrigeration or corporations of life prolonging. Unlike the ancient Egyptians' idea to place the hope of revival on gods, **the body-preserving technology is to place the hope of "reviving from the dead" on the omnipotent science and technology, so it stakes the hope on the fast developing high technology.**

Today, we are often amazed at the fast progress of science and the rapid development of medicine, as many "incurable diseases" of the past have been cured with ease. If the patients of "incurable diseases" of those days lived in the present time, they would not have died. So, it is easy to draw such a conclusion: When someone has contracted an incurable disease and is dying, the greatest and only hope lies in the development of future medicine; any present-day incurable disease may be cured as easily as turning over the hand and what's more, the person may even be rejuvenated. Even if the disease is caused by ageing, it may also be curable and the youth may be restored.

Along this way of thinking, people have thought of human body refrigeration. That is, by quick freezing, preserve the patient suffering from an incurable disease or on the verge of death, wait until medical science develops to the level of being able to cure the disease, then unfreeze him and treat him, so as to give a new life to him. Science and technology are advancing at a continually accelerated speed, the ten years now may outmatch the past 100,000 years, and the coming one-day alone may outmatch the past 100,000 years! With the accelerated and amazingly fast development of high technology, maybe it won't take too long for people to reach eternal life, omniscience and omnipotence, supreme kindness and supreme beauty, and daylong ecstasy. Then, after unfreezing, we shall not only be able to cure "incurable diseases" and rejuvenate the person, but also be able let him become a "god" directly, and getting all sorts of bliss brought about by it.

Can we succeed in freezing-preservation of the human body?

Scientists discovered long ago, in Mother Nature, many organisms may keep alive when they are frozen at a low temperature or even ultra low temperature. A frog of 1 million years ago was dug out of the ice, after the unfreezing treatment, it revived! A breed variety of lizard already went extinct from the earth 5000 years ago, but now, after that kind of lizard was dug out from the ice, it has also been unfreezed and revived. Japanese scientists unfreezed the heart of a mouse, and it began to throb again. They also put a goldfish in the liquid nitrogen with a temperature of -210 for a period, then, after unfreezing, it also came to life again, and many functions of the body were not damaged at all.

Animals are so, and mankind is no exception.

A young man from Switzerland called "Wehl" was buried by heavy snow when he climbed the perilous peak of Alps in 1962 and met with an avalanche. In 1987, his dead body was discovered by chance, then, it was carried to Niss Medical College of France. After unfreezing, this person was brought back to life after refrigeration for 25 years. After reviving, Wehl was already 52, but he was still as young as 26.

In 1990, in a nivation cirque of Siberia of the former USSR, people found a miner who was frozen to death in 1921, and his heart had already stopped throbbing completely. The scientists took him back to the laboratory and "asked him questions" using the computer. This human head after being frozen for half a century should be able to answer questions such as "Pain or not?", "Do you want to go home?".

In the winter of 1980, a 19-year-old American girl Joan Hilia was frozen to death at -26, in the open air. A dozen or more hours later, she was unfreezed and revived, living healthily. Lately, a 13-month-old baby girl of Canada was frozen stiff, her heart stopped throbbing for at least 2 hours, but with the doctors' all-out effort, she was revived!

And we have examples of cells keeping alive in a frozen condition. In the former USSR, scientists took out the sperm from an ancient soldier who was frozen to death a thousand years ago, and planted it into the womb of a female scientist. Then, she gave birth to a baby successfully. In the USA, people also took the sperm of a man living 5000 years ago and "coupled it" with a female volunteer. She gave birth to a baby girl in 1995.

In a low temperature and freezing condition, why weren't some animals or people frozen to death but came to life again after unfreezing? Scientists have found: A low temperature environment may greatly lower the metabolic activity of living tissues. When the outside temperature drops to a certain level, the body's cells will neither get old nor degenerate, but be in a state of "standstill of vitality". Freezing may stop the splitting, ageing and death of cells, hence, their life may be "sealed up for keeping" forever. It is generally agreed that the temperature range for life stopping is -15 - 50. Once it passes this critical temperature range smoothly, the life will be kept safely no matter how long the freezing time is.

Can we use the cooling method to freeze the human body for a certain period and then unfreeze and revive it? As early as 1773, the most versatile genius in the world, Benjamin Franklin already considered a possibility similar to that. In the 1960s or so, even a special subject called "human cryogenics" appeared, which is a science about freezing the human body before its tissues begin to decay and reviving it when the remedy is found in future. The term "human body freezing" came from the book *Immortal Prospect* written by Robert Ettinger in 1965.

Now, we are already able to preserve all kinds of human organs and tissues, such as heart, kidney, skin, sperm, ova, embryo and so on. Many countries have already set up frozen sperm banks, frozen blood banks, frozen organ banks and so on. A heart, liver or kidney can be unfreezed and used after being frozen for some days, and its function is not damaged at all. Frozen sperm and frozen blood can be stored for years without problem and some can be stored even longer, or even be exported overseas.

Since human organs can be freezing-preserved, the whole human body can be treated this way, too. The revival of the frozen sperm and frozen organs has proved undoubtedly that there is a great hope in revival of frozen human bodies. In future, if people like to live in some different age, they may use the body-preserving technology to make a choice.

At present, we have already had the technology for deep-freezing of the human body, so, we can freeze someone until the unfreezing and reviving technology becomes mature. Now, in the US alone, there are 4 such "Human Body Freezing Companies" or "Life Extending Companies" they have already frozen several hundred bodies and brains and those ready to accept the freezing are getting more and more, 70% of which are eggheads, mostly being computer scientists, software development engineers and other high technology experts. "I like the feeling of being alive", "I wish to see what the future will be like", "I really hope to wake up someday several centuries later, and live healthily"... Those are the best reasons for their choosing the freezing technology. The author of the said book points out **People in the Silicon Valley are very confident of the advances in science and technology. They believe, so long as you put plentiful money and technology, you will be able to solve any problem. Why can't we live forever?"**

The method adopted in the present freezing technology is like this: First, an artificial heart force pump, to draw out the blood of the human body, is used, and then the freezing treatment is started. Firstly, a protective substance is infused into the body. Then, the artificial heart force pump is used to draw out this substance, and at the same time a solidification-preventing glycerol substance that cannot crystallize is injected. Then, two weeks are spent to use the liquid nitrogen to freeze the body to a low temperature below -190. It will then be stored in an elliptical vessel full of liquid nitrogen. Now, the freezing operation is finished. It is only the current freezing method. In future, the freezing technology will certainly be innovated continually, so as to keep the human body more simply, safely and effectively, waiting for the future revival technology to activate it.

The freezing technology now chiefly falls into partial (head) freezing and whole-body freezing, and the two are different in price. At present, the charge for freezing-preservation of the whole body is about 120,000 US$, that of partial preservation (only a head) is 50,000 US$. In order to be economical and practical, we can just preserve the head, for we may use clone to make a body or use an artificial body. A human needs the body, chiefly for maintaining the function of the brain. Those amnesiac patients, senile dementia patients, or human vegetables are people with healthy bodies and it is just because their brains have gone wrong that they cannot have the life state of normal people. Thus, so long as we can revive their heads, we will be able to revive their memory, thought and emotion, and that is the same as reviving their life.

Big star Elizabeth Taylor dislikes her body that is getting more and more obese, so, she only wishes to keep her head, that is, after her death. Her head can be cut off for quick freezing, so that when she is revived in future, she may be fitted with a slim and charming body much better than the original one. Thus, it can be seen, in a sense, there are special advantages in only preserving the head and then fitting an artificial body that is like the human body but more beautiful than the human body.

What is more encouraging is, the continually emerging new technologies have brought about a great hope for unfreezing and reviving. The nanometer technology and stem cell technology at the front edge of technology will take the spotlight and show their invincible might. Take nanometer technology for example: we can make a nanometer robot even much smaller than a hair, which can go into the human body to dredge the blood vessel. In future, it is very likely that we might utilize the nanometer robot to help to unfreeze the body, rectify the brain tissue cells and the cells of each part of the body. Nobel Prize holder Smalley predicts, the nanometer robot may come into being in 2010. Another famous expert in this field, Ferrettas thinks, the first case of revival of frozen bodies may happen between 2040 and 2050. So far, we have not put much investment and manpower in the research on cryosurgery or body preserving technology. If we get more scientists' participation, more support from the common people, spend more manpower, material resources and financial resources, regard it as a "Manhattan Project", we shall certainly win the thorough victory in human body freezing and reviving!

Just as Chairman of All-America Refrigeration Technology Association has declared, **"The freezing-preservation technology has solved the most difficult problem of mankind, death!"** The body keeping technology is the most realistic and most feasible way to fight against death, being the paramount core technology for becoming "gods" in a tortuous way.

Of course, the application of any technology will bring about some annoying problems unavoidably. Body preserving technology is no exception. What people worry about chiefly centers around the following aspects.

First, "the expenses". The present charge for body preserving is still quite high: for keeping the whole body, it is 120,000 US$; for preserving the head, it is also as high as 50,000 US$. With such a high cost, can it be popularized?

In our age, what is only unchangeable is change, so we should never look at the expense of body preserving technology statically. The expense needed by any technology depends on its cost, and the cost of high and new technology is always going down rapidly. Just a few decades ago, the computer was the most costly equipment in a large laboratory, but now it has entered the home of an ordinary citizen. So long as we try our best to tackle key problems in the technology and go all out to develop it, its cost can be lowered very soon and continually, and we may popularize it widely in a short time.

Second, "the burden". Will the extensive popularization of the body preserving technology increase the social burden so as to bring about a great impact on the whole social structure? On one hand, the expense of the body preserving technology will go down continuously, so the burden will become lighter and lighter and on the other hand, the social productive forces are going up all the time, so the bearing capacity of the society will increase continually. The flying progress of technology will lead our social productive forces into an era of accelerated development and amazingly fast development. As early as the last century, in their *Communist Manifesto*, Marx and Engels cried up the fast development of productive forces in their age, saying the productive force in one century surpassed the total productive forces of all the past ages. But today the growing rate of productive forces is already incomparable to the age of Marx and Engels, and in future ages, along with the knowledge explosion, we will certainly have a great explosion of productive forces. Then, the expense of the body preserving technology will be out of question.

Third, "the influence on the later generations". Someone imagines: If everyone adopts the body preserving technology, in future, people of several different times will live together in the same time and space. Will that occupy the living space of our offspring and lead to over-crowding?

Those who use the body preserving technology will not become burdens after revival, as they can work for the society and make contributions to mankind, too. What's more, the multiplication of our offspring is controllable, so we may adjust it according to the actual condition. Such an adjustment will not affect the interest of our offspring. Owing to the family planning policy, our country's population was reduced by preventing 0.3 billion births in 20 years. Can we say we have deprived the 0.3 billion offspring of their right to exist? No, lives before birth are inexistent in fact, and cannot be regarded as lives. Only real lives can be called lives, and we cannot mention the two in the same breath. The so-called "affecting our offspring" is purely groundless.

Just think, if we could raise Marx, Einstein and Lu Xun from the dead now, let them open their eyes and use their great brains to observe and think about our age, what great benefit we will get! if it is so for our age, and it will be so for future ages, too.

Maybe someone will regard the body preserving technology as an absurdity: Is it necessary to take so much trouble to preserve human bodies? We must know: judging something as "necessary" or "unnecessary" depends on different values. When we agree to the values that "life is absolutely fundamental, of absolute one-time nature, and is absolutely precious", we will make the right judgment about the body preserving technology.

Marx raised the requirement "**start from the realistic and living individual**" in a very clear manner, which is the starting point of his entire theory, and is also the starting point of our entire "god" becoming theory.

"**Start from the individual**" is an individual-oriented viewpoint. *"The overall and free development of the individual is the foundation of all people's overall and free development"* (Marx). The individual's becoming "immortal" is the foundation of all people's becoming "immortal". The body preserving technology has opened the door for individuals' becoming "immortal", so, it has also opened the door for the collective process of becoming "immortal".

"Start from the living individual" is a life-centered philosophy. Existence of the life is the living individual's fundamental interest and greatest interest. The famous writer Ba Jin said: *"Clinging to 'life' is not a sin. Every organism has the desire to live. When a grasshopper is hungry, it even eats off its own legs for maintaining his life. Such a 'foolish action' is not ridiculous, for it contains a solemnity."* The body preserving technology is to ensure the living individual may continue its life in future, that is, to ensure the person's fundamental interest and it is legitimate, definite, and serious.

"Start from the realistic living individual" is a reality-centered philosophy. An individual is neither what has disappeared, nor is what will be in the visionary world, but is a realistic existence.

Starting from the realistic and living individual, is just starting from the realistic life of the 6 billion people. The body preserving technology is to guarantee all realistic people, even those old ones who have the least possibility to realize eternal life in time may insure their interest of life and walk onto the road to become "immortal" in a tortuous way. The concept of death has already been changed from "stop of the heart throb" into "irreversible stupor". In the case of those who are treated by the body preserving technology, their "stupor" is reversible, so they can not be regarded as dead, they are still humans, still have human rights, thus, should still hold the existent right of real lives.

In conclusion, the body preserving technology is a great technology that may insure the realization of the god-becoming dream in a tortuous way. It may offer all of us a chance to work wonders of life, and enter the beautiful wonderland finally.

XXIV. SECOND CORE TECHNOLOGY FOR LONGEVITY IN A TORTUOUS WAY: MIND-PRESERVING TECHNOLOGY

Information can be retrieved, can be caught, can be kept, can be diverted and can be duplicated. If we treat the soul as information, we will be able to catch, divert, duplicate and preserve it, thus, we can realize the imperishability of the real soul.

The body preserving technology is meant for keeping the human body whereas the mind-preserving technology is targeted as a way to keep the soul. It is another core technology for becoming immortal in a tortuous way.

The human body can be preserved, but can "the inner man" be preserved, too?

All the major religions of the world have the theory of "imperishability of the soul". Since the ancient times, people have been fancying about immortalization of the soul in their sub-consciousness. First Emperor of Qin Dynasty duplicated a troop of soldiers made of pottery exactly like humans and of the same size as humans, as he believed these terra-cotta figures would follow him to show his great power in the nether world. When I was lecturing in Jinan, I took my free time to visit the Han Tomb at Luozhuang Village, and I saw a chafing dish for cooking lamb shashlik several thousand years ago. People think they would also live an ordinary life in the nether world. So, husband and wife should be buried together, the living should often send money or clothes to the dead, and render religious services for the dead. Today, when people look back at the dead, they often can't help praying in their mind and murmuring: *"Wish your soul in Heaven will...", "If Mother knows in the nether world..."* All such sayings express people's eternal hope—hoping the soul is immortal and longing for the eternal living of the soul.

Anyone, so long as his (her) spirit or soul goes wrong, will no longer be that "person" although his (her) body is in good shape in the biologic sense. In the case of a person who has eyes that can see, ears that can hear, mouth that can speak, legs that can walk, in short, has a sound body, if he gets a mental illness, he (she) will lose his (her) self-awareness, lose the memory, will not know who he (she) is, will not know his (her) family, will not know his (her) friends, will not know the whole world, that is, he (she) will no longer be the original person. Or to put it bluntly, the person can no longer be considered as a human. Therefore, it is not hard to draw such a conclusion. The soul (the inner man) is the essence of life.

By the mind-preserving technology, in the near future, we may realize the dream of immortalization of the soul. Revolutions of science and technology like rising winds and scudding clouds may lead to realization of preserving of the mind, which has offered a bright road for becoming immortal in a tortuous way.

Undoubtedly, the soul is illusory and its existence must rely on a certain material medium as its storing place. That is to say, only by storing the soul on a certain material medium can we keep it. In terms of modern technology, the soul is information, namely the bit. The river of souls may be turned into a digital river. Information, as one of the "three major elements of the universe" parallel to matter, can stand the wearing of the time. Information belongs to the world and will never die away along with the flesh. **Information can be retrieved, can be caught, can be kept, can be diverted and can be duplicated. If we process the soul in the way of information, we shall be able to catch, divert, duplicate and save it. The idea is like this: Copy and store the soul information, namely; the life information, wait for the time when the revolutionary**

development of high technology can result in the production of the emulational robot exactly like the human body, then copy the life information into the brain of the robot. Now, just like God creating Adam: Blow a puff of "godly air", and the new life will come into being.

Using technological power to preserve the soul is neither a whimsy, nor a fantasy, but a bold predication by many scientists struggling at the fore front of the revolution of science and technology.

A material scientist in Pennsylvania State University, Robert Newnham predicts: Silicon (computer machinery) may generate many kinds of new life forms. The major advantages of "silicon" life are that it will not die and it has an unthinkable intelligence, which will enlighten scientists to produce the man-silicon compound life form that "has a general consciousness that transcends all organisms".

A professor in the Massachusetts Institute of Technology, USA, Marvin Minski, says in his great science book *Mental Society*, "*It is possible to transplant consciousness into machinery*". His theory has triggered a research wave, which takes Massachusetts Institute of Technology as the center and has reached up to Japan. If consciousness can continue to exist after being diverted out of the human body, then before a man's body dies, he may store his own mind into a super computer, so as to reach the aim of immortality of the "soul". It is an inference derived from Marvin Minski's theory.

The famous grandmaster in cybernetics, Wiener said: "*If we transfer the human body as an entire mode, what will happen? .... Will there be an imaginary receiving tool that could restore such information properly to the original body and mind...?*" Wiener even asserts: "*We may transfer the human mode like the telegraph from one place to another. But the question lies in the difficulty in technology.*" But he adds at once: "*Such an idea has a high feasibility.*"

Mr. Sussman, a professor of electronic engineering and computer science of the Massachusetts Institute of Technology, uttered the wish and expectation of his and many of his colleagues. "*If you can make a machine that can hold your mind, the machine will be you...Now, the machine can last forever. Even if it cannot last forever, you can also transfer its mind into a magnetic tape as a backup at any time. If the first machine breaks down, you may load the information in the tape onto another machine... Everyone wishes to be immortal. Unluckily, I'm afraid I will belong to the last generation that has to face death."*

Professor Sussman does not need to worry that he will "belong to the last generation that has to face death". Today's technological development is at an amazing speed, and his assumption has already been partly realized or is being completely realized. For a technology called "Soul Catcher" is appearing. "Soul Catcher" is the name given by the researchers of British Telecom's Lab for this innovation.

Now, scientists may store a person's whole life memory into a tiny integrated circuit of silicon. In no distant future, it will be able to record every sober-minded instant of your life. Whatever you see or hear and whatever you say or write, can be recorded completely, saved, and can be analyzed and catalogued automatically, to be kept in your individual life file. The hardware unit needed for "immortality" of the soul is also very handy. Now, they use a mini video camera embedded in the rim of your eyeglass, to take down your daily life. These video cameras will be connected to IBM's most advanced hard disk. The hard disk is only as big as a 25-pence coin, can be installed in the ornament hanger, and can record 300 megabytes of data, enough for taking down 30 day's life. In future, memorizers with a much greater capacity will smooth the way for British Telecom's project of "Soul Catcher". It will be in the form of a super microcomputer that can be worn on the body. You cannot see it with your naked eyes. It can be set in your wristwatch, and connected to the micro-sensor able to transmit 5 kinds of signals, built under your scalp and in your nerve. In this way, there will be no need to fix a video camera on the rim of your glasses, and it can work without your consciousness of it.

We have already had the "soul catcher" and will turn out the matching system, the "soul reviver". Even if now we are still unable to transfer some signals of human thought and emotion easily, we may store such signals for the time being, until we can transform it into a silicon integrated circuit and put it into the mechanical "human body". Then, with the aid of the mind-preserving technology, you will be able to keep all your soul of this life, and then put it into the robot. When you have revived, you will feel, just like the princess in *Sleeping Beauty*, you have just taken a long sleep of one century!

The mind-preserving technology is very similar to preserving a man's memory. Memory duplicating and transplanting are closely related to the mind-preserving technology. Recently, many scientists have made great breakthroughs in memory transplanting. According to research findings, human memory process and the computer's computational process are in the same principle and scientists hope to work out a biochip to realize memory transplanting. If we can copy all the memory information stored in a person's brain into a biochip and then fix it into the brain of another person, memory transplanting will become a reality. In February 1992, Psycho Technology Research Center of Alabama University of America made such an attempt:

A boy called "Senior", Champion of the Undergraduates Gymnastics Contest USA, and amateur gymnast, was famous not only for a very great balancing capacity, but also for a good memory capacity. He was able to memorize a lot of gymnastics movements. These abilities of his were input into a chip, and the chip was implanted into the brain of a middle school student, Kelley, who could not stand and walk stably because the balancing function of his brain was damaged in a traffic accident. The operation was a great success. When Kelley was able to get down the bed and walk, people found he walked very stably, looking totally different from his past! Some experts took him to a lawn and asked him to do a set of gymnastics movements. Kelley stretched his waist and legs several times gracefully, then, ran a few steps, and made a jump: a smart somersault in midair... A great success! However, several days later, Kelley's sports memory declined. One week later, he felt he could no longer do any gym movements. Yet, his movement coordinating ability was still better than before. As the chip was powered by a battery, the experts were afraid its electric energy would be exhausted, so they had to take out the chip ahead of schedule. After taking out the chip, Kelley became the same as before again. Although this memory transplanting experiment did not get the final success, it has brought a bright prospect of memory transplanting for us.

Many scientists have also realized that **the significance of memory transplantation not only lie in fast learning of knowledge, what is more important, it may also play a key role in eternal life.** They think, "In future, mankind's death will not be described in the present term, and judgment of human death will be based on the memory or consciousness." That is to say, so long as a person's memory or consciousness is immortal, it will mean this person is immortal, living without death. Lately, Ien Pearson, director of Futurology Research Center of British telecom magnate "BT" Company, predicted: By 2050, it is hopeful that we shall have been able to "download" your thought into the computer's hard disk for keeping forever, thus, you will enjoy an "eternal life" in a sense. Hence, one efficient way to become "immortal" in a tortuous way is to copy the "soul" via the soul catcher or memory transplanting technology before your life comes to the end, then, transplant the information about your memory, thinking, character, etc. into the robot. Thus, just like changing a dress, we may get a new body while keeping the original memory, thinking, character and so on. That is a key "mind-preserving way" in our struggle for becoming immortal in a tortuous way.

Can the soul merge with machinery harmoniously? Someone says: Someday in future, after my "soul" is revived, I find myself should be living in a machine; such a feeling must be very bad. Obviously, people's impression of a machine still remains to be a black and shining iron guy full of metallic luster. If every human body becomes so, of course, we cannot accept it in emotion. However, we should never underrate the amazingly fast "evolution" of machinery.

Recently, an Australian scientist specializing in robot research made a robot model very similar to a film star, and after covering it with the emulational skin, it was almost the same as a real person, and the robot's limbs and organs could receive control information. Sichuan University has just developed a new generation of dummy man as a substitute for mankind. It is in the same size as a real person, its skin and internal organs are very similar to a real person in both tactile sensation and bone hardness, and its internal organs and skeletal composition are also close to a real person. Look, **just in the present age, we are already able to make a robot that is exactly like a real person and is as pretty as a star! Then, in future, when your soul is revived on a robot, you will not have an ugly mechanical body, but have a body more beautiful than the original you.** The robot body may have a fine and smooth skin of high resilience, have a charming appearance, and can smile enchantingly and glance bewitchingly. The past bucket-like waist is gone; the old big belly has disappeared, too. The shape of a world famous model is just your shape.

We can not only make a robot exactly like a human body, but also let the robot have emotion. In the past, people often said, *"A machine is emotionless."* but now the present robot has got emotion and expression. The robot made by the Massachusetts Institute of Technology, USA, has all kinds of expression such as curiosity, anger, fear, joy and so on. Of course, their expressions don't seem to be very obvious. But, along with technological development, they may have richer expressions of emotion. In the coming 20 - 50 years, a robot may marry a human beings, and fertile robots may also appear. Such a robot is exactly like a human form and "skin", inside it is installed an artificial womb, and once it is married to a man, it may give birth to children.

Meanwhile, any part of the mechanical "body" can be changed, just like the whole body, so, actually it is never damageable, that is, being the "Jingang unbreakable body". If we load the soul into the mechanical body that is "exactly like the real", "unbreakable" and "with supreme virtues and supreme beauty", a human being will be able to revive ideally.

The mind-preserving technology has turned the "immortalization of the soul" and "Jingang unbreakable body" of religions and myths into reality. Of course, the keeping of the soul via the mind-preserving technology is essentially different from the religious "soul immortalization". One is materialistic, scientific and realistic; the other is Platonic, anti-scientific and illusory.

In comparison between the body preserving technology and the mind-preserving technology, one is for keeping the body, and the other is for keeping the soul. The expense for keeping the body is much higher, whereas the advantages of the mind-preserving technology lie in its easiness in preserving, low expenses, small space occupation, and it can be popularized extensively: Everyone can use it. Treasuring and managing the souls of several billion people will not take much space: everyone only needs a thin multimedia compact disk, and moreover the CD can be kept forever. If the old CD goes wrong, you may copy and divert the information very easily. Preserving the soul is not only the privilege of a distinguished personage, but should be the innate right of all mankind. This may just take a part per hundred million of mankind's wealth. That little money will be enough for providing a shelter for all the souls of mankind. Moreover, there is no time limit; you may keep them until the time when mankind's technological forces and productive forces have grown to the level to be able to provide the perfect Jingang unbreakable body for each person.

The mind-preserving technology like the magic wand that can turn a stone into gold just by a touch, has opened up a limitlessly fine approach for mankind. It is a most effective and most magical shortcut leading to "immortals" in a tortuous way.

XXV. TRIAL IN THIS LIFE AND LIFE ENCOURAGEMENT

In the past, the dark old society reposed people's hope for fairness and justice in the religious "Last Judgment", but today, we may use the social technology of "Bank of Merits and Virtues" to turn the illusory "Last Judgment" into a real and fair "Trial in This Life" that can occur openly and at any time.

The Occidental World rests on two major pillars: one is science and the other is religion. Many scholars believe that the progress of Europe is closely related to the social basis laid by Christianity, and without Christianity there would be no social progress of Europe. Above 95% of Americans believe in religion although they live in the modern civilization. Along with the development of modern science and technology, people believing in God have not decreased, but increased more and more. Religion's social effect is not getting smaller, but becoming greater and greater. Although I am a thoroughgoing atheist, not believing in any religion, but I cannot face the great social function of religion in moralization.

The elementary theory framework of nearly all religions can not get away from the four interdependent basic tenets, namely; creationism, soul immortalization theory, heaven and hell theory, and retribution theory. The social function of religion is ultimately displayed in "retribution". God, soul immortalization, Heaven and Hell and so on, all such concepts center around "retribution", and all of them are meant for plowing the soil and creating the condition for "retribution".

When we make a comprehensive view of the three major religions in the world, Christianity, Islam and Buddhism, we find, that in all of them, there is a mechanism of punishing the evil and upholding the good through their own way of "Last Judgment". In Christianity, the "Last Judgment" means God "rewards or punishes each one according to their behavior". And all things people have done would be settled on the day when Christ revives. Righteous people will enjoy eternal happiness in the paradise where the ground is covered with gold, the roofs are made of jewels and they can see beautiful scenes, listen to melodious music and taste all kinds of yummy... It is an Elysium. Those who did many evil things will be dropped into the bottomless abyss, suffering pain in the fire and water of the Hell. In Islam, all human behaviors are recorded in detail in the "accounts" which are divided into "Book of Benevolence" and "Book of Malevolence". On the day of Last Judgment, Allah will award or punish all people according to the "accounts", all things will be settled fairly just like trading, some will go to "Hell", some will go to "Paradise", "without any wrong". In Buddhism, humans and animals can convert from each other, depending on their behavior in the world. If you do evil things, you will become an ox or horse or even meaner things in the great beyond , or from a noble to a pauper. If you do good things, you may change from an animal to a human, or from a pauper to a noble.

So we can see, **the religious retribution and Last Judgment let people fear something, wish for award for virtues and strengthen their willpower in doing good things and this is just the social function of the religion.**

Can we move the religious "Last Judgment" to reality and establish a bar of trial in the real world?

In real life, we have such trial institutions as courts, prisons and so on. However, such a trial is only to punish people for their crimes, especially the most heinous guilt, but the unreligious "Last Judgment" being a two-way return, will give each person (soul) both award and punishment. If there is only penalty without award, we can only avoid certain extreme maleficences passively. Both giving award and punishment to all conducts may help people to avoid the evil and encourage them to seek the good. The reins of power mentioned by Hanfeizi (an ancient statesman in the Warring States period) included both "award" and "punishment". Only by working in both ways can we get the best effect.

Encouragement means kindling one's motivation, generating a great impetus inside the person, so he will advance towards the expected goal. Encouragement is the key to action, being the key-press of power. Encouragement may kindle people's fire of enthusiasm, open up their door to wisdom, may turn laziness into diligence and turn improbity into justice, making people's impetus of benevolence grow by ten times, hundred times or more.

The power of encouragement is surprisingly great, and of all kinds of encouragement, the most effective one should be life encouragement. Life interest is a person's greatest interest; without life, nothing will exist. If we compare life encouragement with money encouragement, we will find, the former is greater and more suitable for people's natural propensity. A farmer in Huaxi Village, Wu Renbao, has a saying highly praised by Chairman Jiang Zemin: *"Even if you have tons of gold at home, you will have three meals a day only; Even if you have many luxurious houses, one person can occupy one bed space only."* Many billionaires have come to such a conclusion: when they have possessed a vast wealth, money will serve as a symbol only. People's pursuit of money is limited, but their pursuit of life is endless.

Life encouragement not only include a pursuit of the duration of life, but also includes a pursuit of quality of life. In the basic attributes of the "god", "longevity" means a very long lifetime; "ageless", "omniscience and omnipotence", "supreme kindness and supreme beauty" and "daylong ecstasy" refer to a very high quality of life. Hence, the comprehensive and maximal life encouragement is encouragement for becoming "gods".

If we utilize mankind's inherent nature to seek life interest limitlessly, and utilize mankind's limitless pursuit of life's time and life's quality, to mobilize human's subjective initiative, exploit human's maximum potentialities, encourage everyone to make contributions to the society and to the common cause of mankind, so as to build up a mechanism, namely a life encouragement mechanism, then, we will certainly be able to boost the great undertaking of mankind's becoming gods to the best effect. Just as the American stock equity system has triggered the new economy, with this life encouragement system, we will usher in the era of becoming "gods" at a faster rate.

Psychologist V. H. Frum has a famous encouragement formula:

$$MV = E$$

(M – encouragement effect or level, V – valence, namely the realized magnitude of value of the encouragement objective seen from the subjective perspective, E – expected value, namely expectation of the possibility of realization of such an encouragement objective)

The encouragement formula has shown, the two parameters, valence and expected value, i.e. "importance" and "possibility", decide the encouragement effect jointly. Judging from the encouragement formula, the valence of life encouragement is great beyond comparison, as we have discussed the great value of eternal life in Chapter II. The expected value of life encouragement is also extremely great, not only because the flying progress of life technology has brought the great possibility to break through the life limit in a short time and realize eternal life, but also because the practicing of the life encouragement mechanism has a tangible guarantee both in theory and in technology, thus, it has become very feasible now. Hence, the effect of life encouragement on people will surely be matchlessly great.

In exercising the life encouragement system, we shall need the implementation of the "Bank of Merits and Virtues" technology. The field of natural science falls into science and technology. Owing to the fast development of modern technology, it has become the idol of our age. However, our concept on technology is often limited to the natural science field, not including the social field. In fact, **in the social field, we do not only need social science, but also need social technology. The technology of Bank of Merits and Virtues is actually a social technology resulting from high technology.**

The Bank of Merits and Virtues is an information bank that records completely and truly the merits and virtues of each member of the whole society. Any person's dribs and drabs of merits and virtues will be recorded, especially their behavior in the undertaking of mankind's becoming "gods". And, any one may refer to the records freely. We can build up a great evaluation system for an individuals' merits and virtues worldwide, so as to improve the whole working process, from collecting and recording of information on merits and virtues to the evaluation, certification, reporting and inquiry systems. Meanwhile, the Immortal-becoming Undertaking Association that controls the key resources will evaluate each person's conduct, and give just award or punishment to them according to their conduct.

The merits and virtues evaluation system will examine both "merits" (contributions) and "virtues" (morality), each accounting for 50%. Finally, according to general grading of the merits and virtues, we shall set the grades for enjoying the god-becoming resources, deciding which people should get priority. That is to say, in the evaluation, we shall review both, the person's contribution (especially to the undertaking of mankind's becoming immortal) and his (her) credit standing, virtues and personality. Bank of Merits and Virtues system abides by such a rule: allocation according to merits and virtues, allocation according to distributions, using merits and virtues to change for life, and using contributions to change for life. Allocation according to "merits" is just like distribution according to work. Allocation according to "virtues" means distribution according to the degree of kindness, that is, good for good and evil for evil. We shall pay equal attention to both aspects, evaluate the overall contributions and morality including all kinds of performance, all kinds of awards won, all kinds of public benefit contributions and so on. Contributions to public benefit undertakings can be considered both as merits and as virtues. Especially the generous donation to the undertaking of mankind's becoming immortal should be regarded as an alms deed with boundless beneficence. The reason is: A donation to the undertaking of becoming "gods" together is both a great contribution and a great virtue, as it can accelerate the process of becoming "gods" together, and let more people enjoy the great life interest brought about by the undertaking. It is contributing to other people's benefit and happiness, being a display of great merits and virtues to mankind and the world.

Merits and virtues are in a dialectical unity. People usually set morality against utility. In fact, the foundation of morality is utility. An eminent economist said: *"Only when others can gain a benefit will you think it is worthwhile sacrificing your interest, and the benefit others get is usually greater than the interest you lose, which is just the normal basis of morality. Thus, performing a moral obligation is not for an illusory purpose, not for the sake of asceticism. If a moral code damages the interest of all people, it will no longer be considered as moral, but in fact, it will be just against morality. If we fail to see the utilitarianism of morality in essence, just see the anti-utility character of it on the surface, we shall never master the pith of morality, and shall get confused in commenting about moral questions."* The merits and virtues system is just a system stressing the harmony and unity of merits and virtues. The merits and virtues system has brought about the greatest impetus for realization of the fine ideal of mankind's becoming "gods" together, and this system itself is also the greatest act of morality.

Technically speaking, the merits and virtues system has also got highly mature conditions for operation. In the past age characterized by inefficient passing of information, adopting the merits and virtues system for life encouragement could only be a daydream. But in the present information age, by the Internet technology and information technology, we can organize all social members of the world, can do the fastest and best work in collecting, counting and examining the merits and virtues of each social member, and besides, we can publish such information to the whole society and the whole world, and can give award or punishment to all the members timely. Using the network intelligence system to manage everyone's information on merits and virtues can not only avoid any man-made interference, but also ensure a high efficiency and a high quality.

The merits and virtues system must be fair, just and open. Both fairness and justice should be ensured by openness. Only by resolutely stopping dark operation and putting everything in the sunlight can we ensure fairness and justice. Utilizing information technology to set up a democratic Internet-based system may open everything to the public, not like the old-time archives secrecy system in which the person concerned could hardly see his(her) own archives throughout his life, so the restricting or encouraging effect was very little. The merits and virtues system stresses the open principle. Anybody, including your business rivals, your friend you just met for the first time, or even a stranger who hears your name on the way for the first time, may know your condition of merits and virtues, and of course, you yourself can also know it completely. When you find false records about your merits and virtues, you may request for correction or even resort to legal proceedings. All in all, a merits and virtues management system established by utilizing the information technology can be the most direct, most effective and most fair rule for evaluation of people's behavior.

In the past, the dark old society reposed the hope for fairness and justice on the religion: "Before God, everyone is equal", when the last day comes, "God" will decide fairly whether you should "go down to Hell" or "go up to Heaven" according to your conduct in this life. Today, by using the "Bank of Merits and Virtues" technology, we can turn the illusory "Last Judgment" into a real "trial in this life" that is fair and just, and can happen any time.

The high feasibility of the merits-and-virtues system can be proved by the credit system practiced in the West for more than 150 years.

Development of capitalism can not get away from the credit system. In western countries, everyone has a life-long credit ID card and a credit report made and filed by the credit rating company. Any bank, company or business entity may inquire about this report by paying some charges. For instance, if you have violated the traffic rules but have not paid the fine after three fine notices, your record will be sent to the credit standing bureau, and the bureau will file this record, thus, your credit standing has got a black mark. And, the computer-recorded data of the credit standing bureau is open to all walks of life of the society. Hence, if you want to get the preferential treatment of installment in buying an expensive commodity, the shop will refuse you after they consult the record in the credit standing bureau. If you want to find a good job, the employer will find you unreliable when they refer to the record. If you want to apply for a loan from the bank, the bank may also give you a rebuff when they consult the record. In short, it will be impossible for you to live a life with dignity in this society. In this way, anyone who has a good conscience and wishes to live a life with dignity will dare not do illegal things, and will follow the rule naturally. Hence, the credit system has cultivated the credit consciousness of western citizens, and everyone values his(her) own credit very much.

In China, while practicing the market economy, we lack a corresponding credit system, so, the whole society has got into such a state described as "You do not trust me, and I do not trust you either". A businessman has a lot of money at hand but is not willing to invest, and his reason is like this: *"Even sitting idle and living off my past gains is better than being cheated of all the money in case of incaution."* Ordinary people fear being deceived when going out: When they drink, they fear being poisoned; when they go to the hospital, they dare not get injection for fear that the "one-shot needle tube" is a fake and if they get AIDS from it, it will be too late... As early as several thousand years ago, our saint Confucius already put credit above military affairs and the grain problem, and ours is an ancient country with a very good tradition of keeping faith, with numerous stories about that, such as Weisheng's "keeping faith by holding the pillar until death", Shang Yang's "erecting a piece of wood to offer a high award to people moving it just for showing credit", Ji Bu's "great reputation as a credible person is more valuable than a thousand ounce of gold"... Why are we not as honest and reliable as people in the capitalist society now? Just because we lack a credit system. In China, a person's fraud records, bad faith records or records of violation of traffic rules... are never filed. Even if they are on record, there are hardly any people who can see them. If a man sets up a small workshop to make and sell false wine or false cigarettes, he may just get a fine when caught, and he may change a place to re-register a small factory and change a brand, then his old "business" will continue. If he practices fraud in Sichuan, he may do so again tomorrow in Henan or Shandong. He may change a place whenever he swindles, and get away easily. If an employee lives on his employer while helping others secretly, and is found out, he may just change an employer and no one will punish him... If there is a credit system, all your delinquent conducts will be taken down, can never be erased, and others can look them up, hence, once they have found so, they will never do business with you, never be cheated by you.

If we say the credit system is the protecting god of market economy, the merits and virtues system will be the protecting god of the immortal-becoming undertaking. The feasibility of the credit card has provided a strong proof for the feasibility of the Bank of Merits and Virtues. The merits and virtues system is an incentive system based on the life interest, which is more powerful and more comprehensive than the credit system in encouraging effect. It is an overall stimulation for people's merits and virtues, not only including the restricting and encouraging contents of the credit system but also greatly enriching and transcending them. Thus, the merits and virtues system is more perfect and more effective, playing a greater role in promotion of civilization.

The merits and virtues system may greatly help moral evolution by making the society become better and better. Hugo said: *"A good-natured moral is the foundation of a society."* *Zuo Zhuan*, the ancient historic book of China, happened to have the same view as Hugo: *"Moral is the base of a state."* Both mankind's survival and society's development must take morality as the precondition. The key to the construction of a moral society lies in the construction of a moral mechanism. A moral mechanism takes such a system as the lever: in it, people have to do good things and dare not do bad things. Deng Xiao-ping said long ago: *"When the system is good, even bad people cannot do bad things. If the system is not good, good people may also do bad things."* Nietzschean also believed that kindness is out of interest. **Once our society has built up a sound moral mechanism characterized by "one good turn deserves another, and sow the wind and reap the whirlwind", the whole society will no longer be in a moral crisis.** Mr. Nan Huaijin (a famous scholar) thinks, all the cultural ideology of the five thousand years of China is based on karma (punitive justice). However, this "karma" is merely a religious belief. But now we can realize the full and real "punitive justice" in society by way of the high and new social technology. Those who have moral virtues will certainly be returned with good results. Those who are immoral and evil will certainly get evil returns. Only in this way can we cause everyone to keep improving moral virtues, and cause the whole society to progress into a moral society and highly civilized society.

Marx pointed out clearly long ago: *"All things people are struggling for are related to their interest"*. Seeking interest is human's natural propensity. Helvetius' opinion is: *"The individual's interest is the only and universal criterion for judging the value of human behavior."* Horbach also says: *"Interest is the only impetus to human conduct."* The life interest of a person is the greatest interest. So long as you are a human beings, you will have the demand for eternal life and daylong ecstasy. So long as we use a certain social technology, such as the Bank of Merits and Virtues technology and make the best use of the circumstances, we may make everyone vie with each other for making greater contributions and becoming more virtuous. Hence, once we carry out the merits and virtues system on a large scale in the society, we may turn the society into a moral society, a civil society and an ideal society very soon.

Moreover, in execution of the merits and virtues system, the reward offered by the system, prolonging one's life, may also promote morality. If the reward is money or a beautiful woman, the reward itself may not lead to moral progress. But the reward is life, and the prolonging of life is closely related to moral progress. The economist, Doctor Sheng Hong, made a penetrating analysis from the perspective of game theory. When the game is only once, both parties may take a treacherous policy. If the number of times of gaming has increased to infinite times, both parties will tend to mutual cooperation instead of mutual treachery gradually. Because any treachery may lead to the retaliation by the other party next time. Life is just like gaming, and the number of times of gaming is never only one. However, a human being must die, and death just means the end of the gaming. When someone knows he will quit the gaming stage soon, he may stop fearing any retaliation by others after his death, so, he might do any bad things to others. Thus, we have the saying: *"After my death, even the flood dashes to skies, I won't care!"* In the soul theory of many religions in the world, the prospect of "the great Beyond" may prolong one's life in people's imagination, so, anyone living in this life has to consider the retribution in the other world. If there is really an eternal life, the probability of "one good turn deserves another, or sow the wind and reap the whirlwind" will increase greatly. The life encouragement mechanism helps to prolong human life until eternal life, then, people's considerations about their own behavior will naturally become longer and farther, so it will be easier for them to cherish a "great aspiration", and seek a "greater morality". Hence, the life encouragement mechanism can encourage people to abide by moral rules and raise their moral quality, and offer a good remedy to those who have lost their moral sense and lack a firm belief. Meanwhile, the life encouragement mechanism may help the society to give just judgment to each one, that is, using the real merits and virtues to change for real life, which is more convincing and powerful than the illusory religious preaching.

Apart from the great effect on moral construction, what's more important, the life encouragement mechanism may inspirit all people to make contributions to the great undertaking of mankind's becoming immortal, and maximize all people's initiative and creativity, which is an important guarantee for realization of the great ideal. For the pursuit of eternal life, followers of a religion can abide by all moral rules strictly, keep a philanthropic mind, be a vegetarian, even never marry, never drink wine, sparing no price at all for kindness. Once we begin to practice the merits and virtues system, we shall surely motivate people's worship of life, cause them to work whole-heartedly, and compete with each other for contribution to the immortal-becoming enterprise. Meanwhile, when the merits and virtues system becomes a basic social system, and eternal life becomes everyone's common goal, all members of the society will be in one mind and make concerted efforts, so the society will have a greater centripetal force and cohesive force. When everything is judged chiefly by material gains and is driven chiefly by money, sharp contradictions and conflicts will surely result between persons, between enterprises, and between nations. Once the merits and virtues system is established, every social member will become a noble-minded person, will make the greatest contribution, which will certainly promote social development greatly. In the meantime, the whole world will regard life as the most valuable, so, mankind will certainly keep away from war, wipe out war and realize an enduring peace. Thus, it can be seen that the merits and virtues system will offer an optimal solution to the two major issues of the world, namely "peace and development".

All in all, the life encouragement mechanism centering around the "Bank of Merits and Virtues" as the core technology is a fair, just and optimal competition encouragement mechanism. Which is able to urge everyone to make greater contributions to the society, especially to the undertaking of mankind's becoming immortal. It is also capable of making everyone become a person of high credit standing and high morality. With such an optimal competitive mechanism, we shall certainly be able to boost mankind's progress greatly, so as to help mankind stride at an accelerated speed from the present mankind to the new mankind, the high mankind and until the super mankind (immortals).

PART IV - MEGATRENDS OF DEVELOPMENT OF FUTURE ECONOMY

夢星

INTRODUCTION

The first place in the front page of *Wall Street Journal* of U.S.A. on March 12, 2022:

A Nobel laureate Has Risen From the Dead: an Unprecedented Story and a Start of a New Era.

China Eternal Life Group Company has surpassed the early starters and its share price is soaring up, setting a new world record.

Great good news from China Eternal Life Group Company: The said group has successfully revived a human body from refrigeration, and the Nobel laureate, Professor WWW has become the first person revived after death since mankind came into being.

As early as 2009, China Eternal Life Group Company began to search for and offer a fat salary to hire top experts in body preservation all over the world, and started the Eternal Life Revival Project. The gross project investment was 888.8 billion U.S. dollars. After unremitting efforts of 13 years, they have succeeded at last, and people all over the world are overjoyed.

China Eternal Life Group Company's share price at NASDAQ, Wall Street has risen by 10000 times! Beginning from last night, the whole city of New York is turning out to vie with each other for buying the shares of "China Eternal Life". The total number of people subscribing for the shares has reached 20 million in the city. In a place for handing out the application form, people are queuing up for 10 km, and worldwide, there are 2.5 billion people waiting for subscription on the Internet. This morning, in less than 10 minutes, the stock price of the said group company rose by 300 times, and it had to stop the operation.

Ever since China Eternal Life Group Company got listed, it has been very popular among the stock investors, especially after the release of this news on human revival, its stock index has been rising again and again, jumping to the most eye-catching position in NASDAQ market. According to some person concerned, China Eternal Life Group Company will buy out Microsoft, Intel, GE...

Another headline in the first place of the front page of the *New York Times* of the same day:

"Faced with the Great Success of 'China Eternal Life', What Reflective Thinking Should We Have?"

The title of the report by CNN, USA is:

"Who Will Award China? Who Will Award China Eternal Life Group Company?"

This day, almost all mass media of the world have reported this great achievements of China Eternal Life Group Company.

Americans are elated by the news but are also doing self-criticism with their tail between the legs at the same time. As early as the 1960's, the U.S. took the lead in setting up human body freezing companies. Such corporations used the freezing method to preserve some patients that could not be cured at that time, in anticipation of effective cures to be worked out by future science so as to raise back those people from the dead. Later, many scientists and computer workers in the Silicon Valley of USA began to take a greater and greater interest in human body freezing, and while still alive, they started to raise money to make the rank of people join for body refrigeration. Along this line of development, those who made complete breakthroughs in human body freezing and reviving for eternal life should undoubtedly be Americans!—Just like the information industry, America is far ahead in the world. However, since 2002 when a Chinese girl wrote a book, *The Second Declaration*, the Chinese government, Chinese entrepreneurs, Chinese scientists and almost all the Chinese people have been awakened. They tried their best to catch up, spending a huge amount of manpower, financial resources and material resources in the research on human body freezing technology. No wonder Napoleon said: *"China is a sleeping lion. Don't awaken it!"* Once the sleeping lion is awakened, it will bring about earthshaking achievements for the world!

Many US scholars begin to study Chinese culture in order to find out why China has come ahead this time. According to investigation, ancient poets of China were very keen on feeling sorry of the short human life from the eternity of the universe. Su Shi's *"Mourning for my transient life, and admiring the ever-running Yangtze River"*, Qu Yuan's *"Only the universe is boundless, but our life is too short"*, and Chen Zi'ang's *"Facing the eternal universe, I cannot help shedding tears"*... All of them expressed the wish for living forever. Chinese ancients always thought: The bright moon is shining forever, the red sun is sweltering forever, and the Yangtze River and Yellow River are running day and night without stop. If humans can be so, how wonderful it will be! Now, China Eternal Life Group Company has finally got the power to realize their forefathers' wish, enabling mankind to live forever just like the bright moon, the red sun or the Yangtze River and the Yellow River!

Originally, right after the establishment of China Eternal Life Group Company, it started an Internet-based discussion on the topic "What Reward Should We Give to Nobel laureates?", with the participation by all people in the world. The discussion centered on "Should we award them with money, or life?" Thereafter, Eternal Life Group filed an application to the Nobel Prize appraisal organization. They were ready to offer a prize more valuable than the Nobel Prize money to the Nobel laureates, that is "Life Award" namely providing the full range service of body freezing and future riving free of charge. It aroused a great reverberation then, most people praised China Eternal Life Group Company for its great undertaking with boundless benefit to mankind, but there were also some people doubting whether the body could be revived after freezing-preservation.

Now, the great success in the revival of the Nobel laureate has removed all doubts about it. The "Revival Academy" set up by China Eternal Life Group Company has a research scale even several fold greater than that of the US Space Headquarters, and 500 top scientists of the world have been working here in cooperation for 13 year. Finally, they have worked out a "Super-intelligent Nanometer Reviving Robot", which is as small as one billionth of a needlepoint, can go in and out of all the capillary vessels and all the brain cells of the frozen body! The Nobel laureate got his new life in this way: When his corpse was in the frozen state, the "Nanometer Reviving Robot" began to do all kinds of repairing work for the cells in his body; after the repairing work was finished, the corpse was unfrozen, and fresh blood and a newly invented therapeutic agent were injected into him. Thus, the revival became a success!

China Eternal Life Group Company's brilliant achievement has become the hottest topic in every nook and corner of the world. You see, Alon is chatting again with his colleagues.

"Alon, you are so great! Congratulations! In buying Eternal Life's stock, you have profited by ten thousand times. You should stand treat now."

"No problem! I have already ordered an undersea life villa from China Eternal Life Group Company. Next month I will move in, then, I will hold a party in the villa, and invite Ajian, Acheng and all my good friends, of course, including you."

Alon opened his microcomputer on his left wrist, murmuring triumphantly to himself: "Really surprising: I rushed to buy some shares of China Eternal Life Group Company last month, the price should have jumped by ten thousand times by this month. I have become a billionaire, too! Tomorrow, I must go there as early as possible to queue. God bless me, oh, I should say the CEO of Eternal Life Group bless me that I can buy some more shares of theirs."

Just on the day when Eternal Life Group realized human body revival, there are thousands upon thousands stories similar to Alon's.

XXVI. HIGH TECHNOLOGY AND HIGH HUMANITY

In the future economic times, only enterprises of both high technology and high humanity can become winners.

The old economy, namely the traditional industrial economy, has already passed its peak years and is going down slope, whereas the sun of the new economy has already risen, and is covering the world with its thriving power.

Since the 1990's, the US economy has shown a supernormal prosperity for as long as 110 months, breaking the historic records in supernormal growth. In the meantime, all its indicators that should grow have grown substantially, such as the gross national product (GNP), the capital investment, the revenue level, the stock market, the consumers' and producers' confidence, the employment rate, the export volume and so on. And all the indicators that should drop have dropped, such as unemployment rate, inflation rate, interest rate, fiscal deficit and so on. Such an unprecedented miracle has stunned traditional economists, making them speechless. Especially in a time when European economy is teetering, Asian economy is locked into financial crisis and Japanese economy is flagging, American economy is showing its unique charm so striking! This is just the "New Economy" that has aroused a great attention from the whole world and triggered a great unrest in the globe.

High technology is the internal agent and major support of the New Economy, being the fundamental impetus to the sustained growth of the New Economy. America's New Economy chiefly resulted from the unmatched science and technology input. As early as 1997, the US' expenditure in R & D already amounted to 196 billion US dollars, exceeding the total amount of Japan, Germany, France and UK. By relying on its unparalleled information equipment investment the growth rate of the US' information equipment investment is 8% higher than the average growth rate of all the equipment investment. By relying on its unique environment of scientific-technical progress and technological innovation... The essence can be seen through the appearance: the inner core of the New Economy lies in high technology. Without high technology, there would be no New Economy.

After the New Economy, what economy will appear? It is a question very attractive to people.

Few people have ever given a thought to it, as they say, the new economy has just come, how can we find time to think about future economy?

Any economy has a life cycle from incubation and growth to maturation and decline. The new economy also has a cycle, and it will also decline someday. After the new economy, what should come next? We are eager to know that, and we should also know that Why? Along with the social progress at an accelerated speed, the transition from one sort of economic form to another is so fast and so surprising. Maybe few decades later or even a dozen or more years later, we shall witness the birth of a brand-new economic form with our own eyes.

Just as Toffler predicts: *The world is undergoing a fundamental revolution, and the consequences of this revolution have not been seen clearly by anyone so far; The present new economy is nothing but a prelude of this revolution.*

High technology has given birth to the new economy and the new economy is essentially just a high technology economy. Along with the flying development of the time, people are increasingly aware of the great boosting effect of technology on economical development, but they have not realized that humanity is also increasingly becoming the fundamental driving force of economical development.

What is humanity?

Humanity refers to the characteristics of the humans, being the common and most universal properties of all people. Gorky said: *"Everybody has a heart, and in each heart we can find humanity."*

The magical power of science and technology has made modern people prostrate themselves in worship and bow their heads submissively. But the huge energy of humanity has not got the equal treatment. In such a state of affairs, we need to listen respectfully to Einstein's words:

> "Remember your humanity and forget all the rest. If you can do so, what is displayed before you will be a road to the new paradise. If you fail to do so, what is before you will be the danger of universal death."

In those days, there were few people who could really understand Einstein's theory of relativity. Likewise, there cannot be many people today who can really realize Einstein's profound insight about "humanity". Many people are still far away from Einstein's level of cognition, they have not "remembered humanity and forgotten all the rest", but quite the contrary, have remembered everything but forgotten humanity and they do not know anything about humanity, do not understand the significance and value of humanity, needlessly to say the close relationship between humanity and economic development.

The essence of the new economy is high technology, whereas the future economy will take high humanity as its base, being a perfect unity and harmony between high technology and high humanity. The more we unite and combine high technology with high humanity, the more we shall be able to take the lead in future economy.

In the new economy era, if an enterprise parades itself as a high-tech enterprise, it may get an overwhelming advantage and be favored by the public. In the future economic era, only enterprises of both high technology and high humanity can become winners.

Demand is the fundamental impetus to economic development, and demand originates from humanity, from the most fundamental humanity. Without demand there would be no economy. In his monumental work *Economics*, the great master in global economics, Paul Samuelson says: **"You can turn a parrot into an economist, but the precondition must be: Let it understand 'supply' and 'demand'."** We often mention the market. The market is composed of the buyer and the seller. Trading cannot happen only with the buyer or seller. The buyer mentioned here is the demanding party, and the seller is the supplying party. Only by meeting the human demand can we win the market, and to meet the human demand, we must get clear about humanity and grasp humanity. An economy that has forgotten humanity or lost humanity will surely come to the end of the road.

According to "the cask principle": what is the most lacked, is the most important. High humanity has been neglected in both old economy and new economy, but it is just high humanity that will play the most important role in the future economy.

To grasp humanity, we must understand it first. The most fundamental humanity includes five major aspects, namely: liking life, liking pleasure, liking wisdom, liking kindness and liking beauty. Liking life and liking pleasure are natural properties of a human being (common properties of both humans and animals) Liking wisdom (truth), liking kindness and liking beauty are essential properties of a human (human characteristics distinguished from animals).

High humanity means a high-leveled human demand. A high-leveled demand for life is not a mere pursuit of a few decades' survival, but the pursuit of longevity of thousand years or ten thousand years or eternal life. A high-leveled demand for pleasure is not a mere pursuit of joy for a moment, but a pursuit of daylong ecstasy, the permanent state of happiness at every hour. A high-leveled demand for wisdom is not merely trying to become a little more clever, raising the intelligence quotient by a few or scores of percentage points, but seeking top wisdom, omniscience and omnipotence. A high-leveled demand for kindness and a high-leveled demand for beauty are not merely seeking a little better mood and a bit prettier looks, but seeking the supreme kindness and the supreme beauty.

Seeking and protecting life is a human instinct, as human's primary need is survival, or life. Marx said: *"In order to live, first of all, we need clothing, food, shelters and other necessary things. Hence, the first historic activity is production so as to meet such needs"*. Engels' expression is even clearer: *"History has been placed on its true foundation for the first time. A truth that is very evident but has been neglected completely is, first of all, people have to eat, drink, dwell, and dress..."* For a long time, the foundation of economic development has always been centering around human's survival demand and life demand.

With the present level of economic development, the low-leveled demands of life such as clothing, food, dwelling and traveling have already been met. If today's economy still spins around the basic necessities of life, however hard we try to stimulate the domestic demands, we cannot extricate ourselves from a bearish market. In the future economy, we must try our best to meet the high-leveled demand for life. The high-leveled demand for life is a demand for health and for longevity.

Liking life and hating death is the most fundamental human nature. Human's greatest agony is nothing more than the impending death. Before the threshold of death, no one hopes to open that door. The cry for a thousand-year life or ten-thousand-year life is the common wish of mankind. In the ancient times, people tried every means to seek longevity, even by abandoning all fame, positions or wealth. Mankind's most essential demand was shown to the best degree here.

The ratio of America's health industry in the GDP has risen from 12% to 18%, which equals the fact that all the GDP of China is used as health expenses in the USA... Americans are so, but do ordinary Chinese too have such a demand? So long as a person has no worry about the basic needs, anyone will be ready to spend money on health and life. Suppose a powerful high-tech enterprise can provide a product that enable people to live to the age of 200, or 500 or even 1000, what will it be like? It will surely open up a great market with a greatest business opportunity, the strongest conquering power, the largest consumer group and the vastest prospect.

Meeting the basic needs within the natural life limit is only a low-leveled satisfaction, whereas meeting the needs for a super natural life limit is a high-leveled satisfaction. Today, the continuous change and improvement of high technology has made it possible for us to break through the natural life limit. The life economy in the future economy, especially 'The 1000-year Longevity Industry' is to meet human life demand at a high level and in a high prospect. Meeting the low-leveled life demand, that is, eating one's fill, dressing to keep warm and living for a few decades, can be achieved without using the power of high technology. However, to meet the high-leveled life demand, namely living to the age of a thousand years or living forever, we must rely on the magical power of high technology.

Tending towards benefit and evading harm and tending towards pleasure and evading sufferings, are bases of ethnics. Man has an innate tendency to seek pleasure and evade suffering, which is a self-evident truth. Joy is man's ultimate pursuit. Unlike other things, which are all means, it is man's purpose, man's final aim in itself. Buddhism specializes in sufferings, but what Buddhism discusses about sufferings is for extrication, for the final aim of Elysium. Bearing sufferings, working hard or struggling hard, is just for joy essentially, namely for tomorrow's joy, and a long-term joy.

Pleasure economy is increasingly becoming a bright spot in economy, and represents a trend of future economy more and more obviously, just because seeking joy and evading suffering is a fundamental demand in human nature. Jeremy Rivkin, Chairman of Washington Economical Development Foundation, said: *"The major means to earn money now is to transform cultural resources into personal experience and amusement that requires payment of money."* the US journal *Times* predicts: *"Around 2015, developed countries will enter a leisure age, and leisure consumption will account for a half of the gross national product (GNP) of the USA."* As a matter of fact, leisure industry has already become the primary economic activity of America. According to the statistic of the department concerned in the USA. Americans have $1/3^{rd}$ of time for leisure, use $2/3^{rd}$ of their income in leisure and use $1/3^{rd}$ of their land for leisure activities. According to experts' forecast, in the next few years, in America's economic structure, the employees in the leisure industry will account for 80% - 85% of the total work force of the society. The leisure industry or entertainment industry that can meet human demand for pleasure will become an industry with the greatest potentialities, and will sweep the whole globe.

In his book *Pleasure Economy*, Wolf points out, in whatever the consumer buys, he is seeking the pleasure element from it. Driven by such a trend of "fun-oriented consumption", products and services in the market will also offer an amusement function or be combined with amusement activities, forming a pleasure economy. And he believes, the key to success of most future industries will lie in whether they can combine their products or services with amusement or not.

The show world full of flickering stars, the game industry as glorious as gold ore and the Hollywood like a dream factory can achieve great successes and earn big money. Why? Just because they have met people's demand for pleasure. But all those have only met the low-leveled demand for pleasure. The future economy is a high humanity economy, and the demands it meets are high-leveled demands for pleasure, and what it seeks is a transnormal pleasure and enjoyment. The flying progress of the virtual reality technology enables us to have all kinds of wonderful experience vividly true to life. Nowadays, some US corporations have already worked out a technique called "digital odors": You may smell an attractive aroma when seeing a film or watching TV. Other aspects, such as sense of taste, sense of touch, imagination, intuition, and information of numerals, lettering, sound and image can also find a boundless prospect through the virtual technology. So, in the new era, we may feast our eyes, ears, nose, or mouth to a miraculous effect! On the wall of Bill Gates' private room, there is no artwork at all, only an ultrathin screen connecting to the art museum, but you can enjoy the thousands upon thousands of art treasures in the art museum at your will. In future, everyone may move a world-class art museum, cinema, or amusement park into his own house, so the enjoyment level will be far and way above the previous level. What is more important; by dint of high technology, we may keep a person happy for all the 24 hours of a day, especially at work and study (after all, work and study occupy most parts of our life). It is one of the most challenging topics of high technology research, and is also where the very promising market of future economy with great potentials lies.

Respecting wisdom and despising folly is also fundamental humanity. The Greek-rooted word "philosophy" just means "loving wisdom". Mankind loves wisdom, sometimes their impulse for seeking wisdom is even stronger than the impulse for seeking joy, and they have an insuppressible respect, admiration and worship for wise people. Our love for the mastermind Zhuge Liang and the wise guy Afanti in literature is actually our admiration for their wisdom. But at the bottom of our hearts, we naturally look down upon those fools, numskulls or morons. Plato said very clearly: *"If you are wise, all people will be your friends and relatives; for you are useful and helpful to them. However, if you have no wisdom, nobody will be your friend, either your parents or your relatives, or any other people."* Why? The reason is so simple: Only by wisdom can we understand, grasp and use objective laws, the greatest energy in the world, and we realize our aim or help others to realize their aim and win a success.

Wisdom makes people respectable and honorable, and everyone wishes to be a clever or wise person. An industry that can meet the human demand for wisdom will surely have a great run. For example, all such products have become popular, as "Impossible to Forget", "Brain Relaxing Agent" and so on. The effect of these edible products is not necessarily obvious or true, but they have won a great market just because they have met the human demand. Suppose, by dint of high technology, we can meet people's high-leveled demand for wisdom by producing a product that can really help people themselves or their off-spring become extremely clever or even omniscient and omnipotent, it will certainly have an extremely vast market and an extremely bright future. It goes without saying "omniscience and omnipotence", just a pill that can improve our attention or an intracerebral chip that can improve our memory, or a gene food that can enhance our creativity... will attract people greatly.

Loving Kindness and Hating Evil

Everyone loves kind and righteous people and hates villain-ous and evil people from the bottom of their hearts. Judging from people's feelings at the time of reading a novel, seeing a movie or watching a TV play, we can see, they hate maleficences extremely and love kindness extremely too. The actor playing the part of Huang Shiren (the despotic landlord) in the Chinese drama "White-haired Girl" was nearly shot down by an angry soldier in the audience. Such kind of tragedy also happened abroad. When an American was watching the Shakespearian tragedy "Othello", he was so angered by the vil-lainy of the dramatis personae that he raised his pistol and killed the actor. In the 1990's, the kind-hearted figure Liu Huifang in TV brought tears to uncountable people all over China. In the past, in the storytelling of "The Three

Kingdoms", when we heard how Liu Bei carried the day, everyone would feel cheerful, even small kids danced for joy. When hearing how Cao Cao played the bully, everyone would look angry and indignant. Although those cannot prove human natural propensity is kind, it may show everyone likes kindness and wishes others to be kind from their inherent nature. Out of the self-benefiting nature of mankind, people also hope others are kind, which may make them feel safer, hence, people's natural propensity is for loving kindness and hating evil.

In ancient China, people's requirement to their children was described as "not afraid they will not be wise enough, but only afraid they will not cultivate their virtue". In the present period of social type transformation, many people have lost kindness and good faith. According to the scarcity principle, the more scarcer something is, the more precious it will be. Kindness and good faith have already become the most scarce resources in our market economy, so, they seem to be specially valuable. Whether to an individual or an enterprise, credit or good faith is of extremely great value. The more we long for success and great achievements, the more we should pay attention to credit and morality. In the near future, people will fully realize the power of kindness and the power of credit. Then, a high-leveled demand for kindness will appear certainly: longing to become very kind persons and holy persons, and longing for businesses to provide holy products and holy services. Holy services have the greatest power to win people's hearts. So, the future economy will not only seek customer satisfaction, but also seek moving the customers. To move the customers from the heart is the goal of a high humanity enterprise, and is the development trend of future economy.

Everyone loves beauty. Everyone is overwhelmed by the power of beauty, including all heroes. As the famous Chinese saying goes, *"I love the throne, but I love the beauty even more"*. According to a research, everyone tends to judge people by their looks, so, it is easier for beautiful people to get others' favor and help, it is easier for them to meet with success than ordinary people. Even if they have committed a crime, the penalty to a beauty is often several times lighter than ordinary people. That is the result of a survey by the American court. It is the case for common people, but even big men cannot free

Loving Beauty and Disliking Ugliness

themselves from such a human nature. Both Sun Quan (king of State Wu) and Liu Bei (king of State Shu) closed the door on Pang Tong because of his ugly appearance although he was a great wise man. In the presidential election race in the West, they will offer a fat salary to invite image consultants whose duty is design the finest image for the candidate. A public opinion test organization of America carried out a survey of 500 young females. The topic of the survey is: "Rich women, talented women and beautiful women: If your fate is up to yourself, which kind of women do you wish to be?" Very few chose rich women, and those choosing talented women were not many either, but 90% of girls said without hesitation that they wished to be beautiful women.

Loving beauty is a basic human demand, and is also an eternal demand. A great demand will certainly bring about a great market. Today, people's investment in beauty treatment and health beauty can never be underrated. According to statistics, American females spend about 30 billion US$ in buying cosmetics each year. In Japan, on a average, of the basic cosmetics used by a female in her life, beauty water is 980 cubic liters, all kinds of cream is 150 kilograms, lotion is 125 cubic liters and lipstick is 400 grams. A tiny bottle of cosmetics may be as expensive as hundreds of dollars or even a thousand dollars. Even the charge for beautifying the nail for one time is as high as five or six hundred dollars. But consumers are scrambling for it. Botulinus toxin, being called the most toxic substance in the world, flowed into the faces of 1,600,000 American females last year, just because it could remove wrinkles. Its one injection costs 500 to 1000 US$, but the wrinkle removing effect can only last four months, so, you have to take an injection anew every four months, otherwise the wrinkles will creep on your face again. Even so, American females still run after it like ducks to water. In the business of "wrinkle removing by toxin" alone, the annual total sales in the USA amount to 0.31 billion US$. There are hardly any Korean star singers or film stars who have not done face-lifting operation, thus, nowadays we have many "artificial belles" and "artificial handsome men". According to a forecast, **in the future top 500 enterprises of the world, one third will do business related to beauty treatment research, production and sales.**

The present beauty treatment industry has not met the high-leveled demand for beauty. The future economy is a high humanity economy, seeking to meet a high-leveled demand for beauty, and the beautifying effect will be far incomparable to the present cosmetics or nail beautifying products. So long as we utilize future top means of high technology, for example, means of molecular biology, bionomy and cytology, we may help people meet their high-leveled demand for beauty, helping them reach "the supreme beauty": having godly looks much more beautiful and more enchanting than today's stars. The market prospect in this aspect is also very attractive.

All such attributes as liking life and hating death, tending towards pleasure and evading suffering, respecting wisdom and despising folly, loving kindness and hating evil, loving beauty and detesting ugliness, are most basic humanities, and can transcend national boundaries and transcend ages. They belong to both Chinese and Americans, both Orientals and Occidentals, both the past and the present, and of course, belong to the future as well.

The future economy is high humanity economy, which can help people to meet their high-leveled demand for life, high-leveled demand for pleasure, high-leveled demand for wisdom, high-leveled demand for kindness, and high-leveled demand for beauty. Hence, **the future economy can realize "Three Major Leaps", namely, causing the meeting of all the three demands of mankind (demand for survival, demand for enjoyment and demand for development) to rise from the low level to the high level.**

The low-leveled demand for life such as clothing, food, shelter and traveling has already come to the end, which means that the traditional material economy has come to the end. The new way out for economy just lies in meeting people's high-leveled demand for life, pleasure, wisdom, kindness and beauty. High humanity has opened up an unprecedented, endless, perpetual and boundless market.

The greatest secret of economic development lies in seeing through humanity, respecting humanity, utilizing humanity and taking humanity as the foundation stone and starting point. Judging from human demands, humanity is characterized by "loving of life, loving of pleasure, loving of wisdom, loving of kindness and loving of beauty". Judging from human relationship, the fundamental humanity is shown in self-benefiting and morality.

Self-benefiting is human's natural property. Morality is human's essential nature, being the property distinguishing human from animals.

First, we must face up to the self-benefiting nature of humans. Self-benefiting is man's first inherent nature. Self-benefiting is different from selfishness. If the self-benefiting property expands incontinently, it will result in selfishness. Once a person is born, he (she) knows self-benefiting without teaching or learning. Of course, man also has morality. There is a root of kindness in everyone, and everyone likes kindness and hates evil naturally. At the bottom, liking kindness and hating evil is also rooted in the self-benefiting human nature. Therefore, both good people and bad people like persons with a high moral naturally. If anyone wants to succeed, he must keep moral to a certain degree. Even if you want to be a successful robber, you should also "be moral" in a certain scope, for example; within your fellow gangsters, the so-called "Even robbers have their moral". Otherwise, no one can achieve any big thing.

Human morality originates both from human self-benefiting nature and from human rationality. His rationality makes him realize that only by being moral can all people's interest be ensured, including his own interest. Without the constraint from morality, when everyone's self-benefiting nature expands extremely, it will certainly infringe upon others' interest, and finally no one's "self-benefiting" can be guaranteed. Being moral and cooperative is better than being immoral and uncooperative. Once such a conclusion has come out, both the society and the individual will be willing to be moral. Hence, humans can be moralized through education, so they will keep both the self-benefiting nature and the moral nature.

As soon as a child is born, he (she) wants things to eat, and this is an example of human' self-benefiting nature. Later, he learns to share food with his little pals, or let his pals eat first. Such a morality results from the postnatal education. So, we see, in humanity, what comes first is self-benefiting, and morality comes next. Only by such a profound understanding of humanity can we utilize humanity, follow humanity and improve humanity, so as not to defy humanity, go against humanity and distort humanity, leading to blunders.

The human nature of self-benefiting is the starting point of all economic behaviors. In understanding humanity, we should neither idolize it nor debase it, but should be practical and realistic. China made a great mistake in history, in economic construction, which just resulted from going against humanity. As the slogan in the Cultural Revolution says, "Fight ruthless against any selfish whim", that is, refusing to admit the self-benefiting human nature. Thus, we deprived people even of their right to cook privately, confiscated all private properties, even including their pots and cooking stoves, for the sake of collectivism. On the surface, the united force was great, but actually it led to a nationwide great famine, with people starving to death all over the country. Later, our economic reform led to many miracles, and fundamentally speaking, it was just because we recognized humanity, respected humanity and admitted the human nature of self-benefiting. Very soon, much more grain was yielded in China, and our economy jumped from an abyss to a great boom. Going against humans' self-benefiting nature or respecting it has resulted in such a vast difference.

Morality is also closely related to economic development. In the eyes of many people, economy and morality are incompatible just like fire and water. Such a notion is in fact entirely wrong. Morality is the soul of economical development. Upholding good faith and raising credit standing are both elementary moral rules and essential conditions of market economy. Adam Smith's saying *"The Invincible Hand"* in his *The Wealth of the Nation* has become a classical saying, being widely spread. However, people have neglected another book written by Smith, *Theory of Moral Sentiments*, which stresses repeatedly the *"The Person in the Heart"* corresponding to the *"The Invisible Hand"*. *"The Person in the Heart"* just refers to such virtues as "conscience", "sympathy", "sense of justice" and so on, and it is also essential to economic development. Smith said: *"When we take our personal interest as more important than the public's interest, the person in the heart will remind us at once. If we over-value ourselves and belittle others, we shall turn ourselves into targets of defiance and indignation of our fellowmen."* Max Weber, great thinker in modern times, gave a complete comment on Franklin's concept on "time" and "credit", thinking it was the "complete capitalism spirit". Time is money, but credit is also money. Without the basic honesty and credit, without the basic market morality, the transaction cost will certainly rise greatly. Mr. Wu Jinglian said: *A young boss said to me: "Nowadays, everyone cheats everyone. By a little inattention, you may be cheated. So, now I have money and projects, but I dare not invest... Even living on my past gains without any profit is better than being cheated of all my money."* Once deception becomes a common practice, and the whole society's credit system collapses completely, economic development will surely be threatened seriously. A moral crisis will ultimately trigger an economic crisis for certain.

The future economy will be a humanity-respecting economy, specially respecting the humans' self-benefiting nature and morality, combining both commendably. The future economy will not only value humanization in products and services, but also stresses self-benefiting and morality in the internal relationship between economic entities, thus, it lays great emphasis on the establishment of "free complexes" and "interest commonwealths". Only by following self-benefiting and morality can we guarantee the acquisition of the best interest of each person, mobilize people's initiative and creativity to the best effect, build up the most ideal inter-personal relationship and the optimal work team, and achieve the best result.

In general, **the future economy will rely on the power of high technology, respect the self-benefiting human nature and morality highly, and meet the high-leveled human demands completely, so as to bring a great progress from the new economy covering information economy, Internet economy and biologic economy, to the future economy characterized by life economy, eternal life economy, ecstasy economy, human-centered economy and universal economy.**

Only by relying on high technology can we realize high humanity. If high humanity does not depend on high technology, the principal high humanity demands, namely eternal life, daylong ecstasy, omniscience and omnipotence, supreme kindness and supreme beauty, will become absurd illusions, without possibility to come true.

High technology needs high humanity, and high humanity needs high technology. The two must be combined with each other perfectly. High technology is the tool whereas high humanity is the purpose. High technology is the bridge whereas high humanity is the beautiful and mysterious other shore. High technology and high humanity promote each other, and neither can exist without the other. Hence, the future economy should be a perfect combination of high technology economy and high humanity economy. The new economy only stresses high technology, whereas the future economy has transcended the new economy in that it has integrated the two greatest powers, high technology and high humanity. The development of high technology has provided a great possibility for meeting the high-leveled human demands, and meanwhile, only by using high technology to realize high humanity can we maximize the value of high technology and find the largest application scope for high technology. If there are only high human demands without high technology, there can be no future economy. If there is only high technology without high humanity demands, it will be impossible for the future economy to come into being, neither.

A megatrend of the future economy will be the perfect combination between the continuously growing and limitlessly powerful high technology and human natures such as "liking life and hating death", "tending towards pleasure and evading suffering", "respecting wisdom and despising folly", "loving kindness and hating evil", and "loving beauty and disliking ugliness", which will bring the greatest benefit to mankind, too.

XXVII. THE 1000-YEAR LONGEVITY INDUSTRY AND LIFE ECONOMY

Life economy is the representative of future economy, and is the leading power of future economy. Only by developing life economy can we win the future.

Life economy is the economy that prolongs people's life, and raises the quality of people's life.

In life economy, the 1000-year longevity industry is the leading industry. The 1000-year longevity industry is a great industry, covering the health care industry, medicine industry, hygiene industry, health preserving industry and so on.

If we had proposed the 1000-year longevity industry before the announcement of the completion of mankind's genome map in June

2000, it would have been regarded as an absurd dream, or would have been declared a fantasy that can never by realized. But today, the 1000-year longevity industry enabling everyone to live up to the age of 1000 has got solid scientific foundation.

Scientists asserts: Owing to the great leap brought about by breakthroughs in mankinds' gene research, mankind's future lifespan will be several times longer, and it will not be a hard thing for everyone to live to the age of 1200.

We shall usher in a biotechnology revolution, whose importance and influence power will be more than that of the past industrial revolution.

Scientific Foundationstone

This new industrial revolution will come soon. The British genetics professor Peter James has given a very specific prophesy on the breakthrough in the 1000-year longevity technology and the realization of the 1000-year lifespan dream. *"Judging from the progress of gene deciphering, we may expect, before the year 2030, mankind will be able to master the age controlling gene so as to prolong man's life and realize the wish to live to the age of a thousand years or even an eternal life"*.

Doctor Grebo, geneticist at Cambridge University, said: "I **can bet; among those who are living now, there will certainly be many people whose lifespan can reach 1000 years or so.**"

The American inventor Ray Kurzweil, reputed as "Edison of the Present Age", predicts in his new book, *A Wonderful Trip: Living to Eternity*. By the end of the 2020's, mankind will be able to prolong their life unlimitedly.

That is, realization of the wish to live to the age of 1000 will not take 10000 years or 1000 years, nor will it take 100 years or 50 years, but will be the near future from us: within 30 years.

In "Subversion of Tradition and Fetal Movement of Future", I have expounded the great possibility and inevitability of eternal life from many aspects such as life sciences, high and new medicine, dialectical philosophy and so on. The research on human gene is progressing with each passing day. The research institutions and gene research corporations all over the world are racing against the time, with their computers and robots for gene research running around the clock... All sorts of scientific discoveries are fascinating and exciting to us. Scientific breakthroughs have not only aroused people's limitless aspiration for the boundless life prospect again, but also laid the foundation for the development of life economy and the 1000-year longevity industry. Many high and new technologies have already been put into operation, and many new technical breakthroughs have already shown great possibilities. Everything has displayed an attractive prospect before us, and has offered us unprecedented great chances.

Market Analysis

Modern marketing science defines the market this way: "The market refers to customers' demand and the degree of satisfaction of it." Without demand, there would be no market. All tenable commodities have a general character: to be able to meet human demand.

Demands fall into general ones and individual ones, and general demands are more important than individual demands. The reason is so simple: the products or projects worked out for meeting general demands have a large covering area and a long duration, and can never grow out of date. Commodities that can meet mankind's general demands can lead the tide of the market and sweep the whole world easily.

The demand for life can transcend space and time, transcend national boundaries, and transcend classes, being the strongest and most extensive demand of mankind, that is, it is the greatest common demand of all demands. All people, whether wise or foolish, whether rich or poor, whether beautiful or ugly, have to become ash. Even kings, princes and all kinds of VIP who can shake heaven and earth cannot avoid such a fate. From ancient monarchs and ministers to literary giants and common people, all of them were infatuated with living forever and becoming immortal. So, we can see people's general strong demand for life.

The demand is the logical starting point of the analysis of general economics, and is also where the dynamic force of mankind's economy lies. Demands do not only fall into general and individual ones, but also fall into absolute and relative ones. The demand for life is not only mankind's general demand, but also mankind's absolute demand. Life is characterized by absolute fundamentality, absolute one-time nature and absolute preciousness (See "Absoluteness of Life"); hence, the demand for life is not relative, but absolute, and overwhelming.

If we judge from the three major elements of the act of purchase and sale in the market, namely the consumer, the purchasing desire and the purchasing power, we may see the 1000-year longevity industry and life economy are characterized by the most consumers, the strongest purchasing desire and the greatest purchasing power.

First, the consumers of this market are numerous. Craving for life is a necessity from humanity. The ordinary longevity can never meet human need completely. Even if we can live to be 100 years old, our life will be only a short century (in this short period of one century, if we cross off the naive childhood, the senescence phase that has lost vitality, the time of disease or injury, and the feelingless sleeping time equivalent to 1/3 of the life, there will be little left). In comparison with the universe counted in astronomical figures and the human history counted in terms of millions of years, such a short period will be nothing! After one hundred years, what awaits people will still be death, and ever-lasting sorrow. The 1000-year longevity industry will break through the general limit of human lifespan, guarantee everyone may live healthily to the age of 200, or 500, or 800, or 1000, which is the greatest and invincible temptation for any one. So, the consumption group of life economy and the 1000-year longevity industry will be the largest.

Secondly, the purchasing desire for life is the strongest of all desires. When the shadow of Azrael is approaching, anybody in a sudden retrospect will find, that what he (she) has been striving for in all his life should be worthless, and only the life that can never be recovered is so worth treasuring.

I have told this true story in many places: A man plunged himself into business to build his "economic base" and achieved what he had wished, wallowing in money, but now he has been knocked up, suffering from an incurable disease, and is on the brink of death. He begged one of his bosom friends, who worked in the medical circle, saying: "I can give you one half of my wealth; can you prolong my life for one year?" His friend shook his head. He begged again: "I'll give all my wealth to you, please give me a half year's life at any cost!" His friend shook his head again. For his disease has attacked the vitals beyond cure.

Only now can people realize: In this world, only the life of a person is absolute. Someone is ready to use all the big amount of wealth gained through his struggle in all his life in exchange for even half a year of life or even half a month! So, we can see how strong the desire for "buying" life is!

Thirdly, the market purchasing power of life economy and the 1000-year longevity industry is the greatest. The purchasing power is of crucial importance. Only those who have the purchasing desire subjectively and have the purchasing power objectively can take some purchasing action.

The universally applicable "80%:20% rule" is displayed most obviously in people's possession of wealth. The richest 200 persons in the world have a total income as much as the 41% of the total income of world population! Li Jiacheng's assets alone are more than the wealth of several hundred million people of west China. Let's enter the thinking of these "moneybags as rich as a country". They have struggled for all their life, and possessed inexhaustible wealth. They can satisfy any desire in the world and get anything that can be got. How wonderful it will be if they can keep all these forever! However, all such fine things gained through hard struggle will vanish instantly along with a mere death. Can they be ready to accept such a fate? "No, I must live forever! So long as I can live on, I will not spare any cost! "So you see, the American zillionaire, CEO of Oracle Company, Ellison, is trying his best to find an eternal life fountain (It is said this fountain is in West Indies, and by drinking the water from it, you can cure your disease if you have; if you have no disease, it may help you rejuvenate). In order to find the secret recipe for eternal life, the 61-year-old Ellison has secretly put a huge sum of money in this research. The Russian pharmacy magnate, Wladimir Bryantsalov has planned to invest 2 million US$ to establish a private "Lab for Research on Longevity Drugs". The "Longevity Science Foundation" set up by Russian Rich Men's Club has also begun to use a huge sum of money to do research on "longevity drugs" and Russian scientists have started their invention race in "longevity drugs". Up to now, the foundation has received more than 300 suggestion letters and prospectuses.

At all times and in all countries, the greater the purchasing power people have, the greater is the desire for life they will have, that is, the intensity of the demand for life is almost proportional to the magnitude of the purchasing power. A person's fear of loss goes up along with the rise of his position, his fame and his wealth. Fear of death is human instinct. Those who enjoy high position and great wealth in the world dread death especially, and are obsessed with life. So long as you can guarantee that they can live to the age of 1000, they could use their lifetime's wealth of billions of dollars in exchange for it.

Thus it can be seen, whether we analyze from economic demand or judge according to the three major elements of market purchase, the market of the 1000-year longevity industry and life economy is matchlessly great and matchlessly superior.

Nowadays, biologic economy has aroused a great attention, and it seems to be overwhelming the information economy and Internet economy. The key point of biologic economy is life economy and only the life economy is taking human life as the starting point and ultimate objective is the basic feature of the future economy.

The future economy will sublimate from biologic economy to life economy, applying the high and new technology achievements of bioscience to the solution of human life problems in all aspects: lifes' duration, lifes' quality and so on. Life is priceless. Life is absolutely precious. So, life economy will have a matchlessly fine future for certain.

Shapiro, head of Monsanto Biotechnology Company (MMB), has a typical saying: *"Nowadays we often discuss the three major industries in the world, agriculture, food industry and health product industry. They are still at the stage of separate operation now. But a series of revolutions in biotechnology will soon result in the mergence of these three into the life science industry."* This means, **life economy may integrate the traditional industries and promote the present economy. The tentacles of life economy will infiltrate extensively into many industries such as pharmacy, medical treatment, health care, agriculture, grocery trade and even real estate industry, causing a fundamental revolution in traditional economy covering clothing, housing and traveling and so on.**

For example: clothing industry. The macrobiotic intelligent clothing can monitor the physiologic condition of the human body at any moment and place, record and transmit scores of major physiological indexes, such as heart rate, electrocardiogram, breath and blood pressure. It can read each of your heartthrobs and the condition of your excitement, and record them at any time. It is one of the high-tech products in "The First Medicine", i.e. preventive medicine, and is also a transformation and upgrading of the traditional "clothing" industry.

Another example: food industry. As the Chinese saying goes, *"Food is God to the people."* Life economy will cause a profound revolution in agriculture and livestock husbandry, so the "God" will also change accordingly. Then, not only the rich varieties and cheep prices of food will stun you, what is more important, the future food will have all sorts of incredible supernatural efficacy. For example, the rice can be eaten as a "drug", the turnip can be sold as "ginseng", and there will be all kinds of wonderful products, such as memory improvement food, hypnotic food, hypertension curing food, cold protection food, sunstroke prevention food, wound healing accelerating food, beauty treatment food and so on. Scientists have begun to research on a memory noodle, which can improve your memory greatly after it enters your body. Some biologists are setting about changing the gene structure of banana so that banana may have a hepatic effect... We often say, *"Medicine and food come from the same source"* and often say *"Genius prodigies result from eating"*, "Good health results from eating" and "Beauty results from eating". Life economy will turn all those into realities.

Another aspect, housing industry. The present concept of "intelligent house" is very popular, but the future "life house" will be much greater than the intelligent house. The life house will help you realize your dream to live to the age of 1000. All the conditions in the house such as the air, sunlight, water, temperature, humidity and so on, will be adjusted to the state most suitable for human life. The floor will have a bactericidal function and can do self-cleaning. The colored health preserving wallpaper can be changed all the time at your will. The mural in the room will be vivid like the Mother Nature, emitting a spring-like flavor all the time. Everything will be fresh and agreeable, better than the level of the five major longevity villages of the world... We can imagine: such a life house will be very popular for certain.

Today, while E-commerce is becoming hotter and hotter and the new economy is in its peak years, life economy is approaching us. Although life economy will still take some time to enter our life, its steps are getting quicker and quicker. For the economy most representative of the future, we must be well prepared in thought, emotion and strategy, so as to seize the best chance first. **Life economy is the representative of the future economy, and is the leading power of the future economy. Only by developing life economy can we win the future.**

The 1000-year longevity industry's value lies in its ability to let everyone enjoy a long life of 1000 years. It is a great and complicated system engineering, hence, it may generate an enormous industry group covering many trades and with a huge profit prospect.

Judging from life science, we have gene technology, stem cell technology, "life clock adjustment" technology (cell telomere technology), "death hormone" removing technology and so on. Development of any of them will greatly help us to make breakthroughs in longevity, and may result in a great industry covering many trades.

Just take gene technology for example. One gene alone may generate an industry (trade). So long as it has ten thousand uses, it may generate ten thousand trades. So long as it has a hundred thousand uses, it may generate a hundred thousand trades. For a gene with an important function, the price of patent transfer alone is often as high as millions or tens of millions of dollars; some may be as high as a hundred million dollars. For example, the transfer fee of the "obesity gene" is as high as 0.14 billion US$. The transgenosis hogskin developed by our country can be used for curing fire burns, so, the price of a single piece of hogskin is as high as 20,000 RMB. The industrial development and product development of gene technology are not only very profitable, but are also hot spots in stock market investment. Although some gene technology corporations have not profited and even have no product, their share price can grow by 10 times or 40 times. For example, No sooner did West Quark Therapy Company declare they had cloned the gene for asthma than their stock price went up by many times. It is only an asthma gene in the gene technology. If it is a longevity gene, it will certainly bring about an unprecedentedly huge profit. Moreover, we have the stem cell technology, telomere technology and so on, and each great technology may generate a lot of new industries.

Life science may generate a lot of industries, and so may high and new medicine. By conquest of incurable diseases, organ transplantation, gene therapy, development of the first medicine and so on, we can produce a series of industries and bring about illimitable treasures.

Overcoming incurable diseases alone may bring about stunning great wealth. Recently, a noted US consulting corporation, Land Company, found out the following fact in its investigation on the impact of medicine and biology on the society and the economy: **If heart diseases can be cured, we shall have an increase in wealth by 48 trillion US$. If cancer can be cured, we shall have an increase of 47 trillion US$. Even if the cure rate of cancer is only 20%, we may have an increase of 10 trillion US$, which is even higher than the GDP of the USA in one year.** Many incurable diseases can be conquered, not only heart diseases and cancer, but AIDS, hypertension, apoplexy, diabetes, parkinsonism, scarlatina and so on as well. The conquest of each "incurable diseases" may bring about a huge wealth worth scores of trillion US$. What else can generate more wealth than this? Now, some British and American pharmaceutical companies and biotechnology magnates, such as Pfizer, MRK, SmithKline and so on, have aimed at this huge market, and are pouring huge money to develop or purchase technology, and are vying with each other to register their patents.

Organ transplantation can also become an industry, and bring about a huge wealth. Some biotechnology companies such as Next Ran, Alex Ion and so on are trying to move human gene into the bloodline of animals' embryos so that animals' organs may match mankind's genome better so as to reduce xenogenic repulsion. The commercial value of each hog organ after gene transformation is as high as 18,000 US$. In the USA alone, there are more than 100,000 patients dying because they cannot get human organs timely for transplantation, so, this commercial market is very great. Many companies have already begun to design and produce human organs and tissues; hence, mankind has transcended organ transplantation and entered the organ producing age. They have begun to research on the making of heart valves, breasts, ears, cartilages, noses and other organs.

Biochips can also become an industry, and bring about a surprisingly great wealth. By implanting the biochip into the human body so as to learn about a person's health condition via the chip, we may cause a great leap in the first medicine, preventive medicine. So, we can find and cure diseases timely, which may greatly prolong human life and raise the quality of life. On June 29, 1998, the United States announced its start-up of the biochip program, and many countries in the world have also begun to increase their investment in this area. Industries related to the biochip are growing up all over the world. So far, the stock of 8 biochip companies in the USA has been listed, with an average rise of 75% each year in stock price. According to experts' statistics, the production value of the biochip industry of the whole world is 1 billion US$ or so and it is generally predicted that biochips will be industrialized around the year 2010. Gene chips and related product industries will take the place of the micro-electronic chip industry, becoming one of the greatest industries in the 21st century.

Nowadays, there are no longer only "365 industries" as the traditional saying goes, but there are 3650 industries, 36500 industries, 365000 or even more. The disparity among industries is so great as to make you speechless. In some industry, for example, the stardom industry, Jin Xishan (Korean film star) will earn thousands of dollars easily just by one smile. In some other industries, for example, a sanitation worker cleaning street sweepings at midnight while braving the rain can only earn a few coins, which makes people sigh: *"Making money needs no hard sledding and hard sledding does not make money"*. In terms of industry categories, there are vastly different ones, such as "declining industries" and "sunrise industries", "industries with meager profit" and "industries with huge profit".

Lv Buwei can be said to be a mastermind rarely seen in history. As early as two thousand years ago, he realized the great differences among trades. After he saw Yiren (the later Emperor of State Qin and First Emperor of Qin Dynasty), his dialogue with his father was very meaningful.

He asked his father: "How many times of profit can we get from tilling the land?"

The answer: "10 times."

He asked again: "How many times of profit can we get from doing business, for example, selling jewelry?"

The answer: "100 times."

He asked again: "If we help a man become a king, how many times of profit can we get?"

His father was surprised: "This! this... This profit can not be counted, a thousand times, ten thousand times, or even more! I cannot say."

This story can be said to be an excellent description of "industry variance". It was just because Lv Buwei chose the industry with a huge profit resolutely that he worked a wonder later that shocked people of all ages. He became a prime minister because of helping a child become the king.

Generally speaking, an industry in a position to become a highly profitable one often has the following characteristics.

Its Market Accession Is Hard?

Once the market accession becomes less hard, those entering it first will have very great chances to get windfall profit. Before great progress is made in the genome project and in science and technology, the 1000-year longevity industry is always an area where no market competition has occurred, and it is very hard to enter this market. The footstone of this industry is the breakthrough in top life science and technology, so, the expenses for its first-stage research are extremely great, and the threshold of this industry is very high. However, once a breakthrough is made, the production cost will be very low but the selling price will be very high, so, the profit will be very fat. Anyone who can enter this area first will get windfall profit.

Its Long-term Potential Demand Has Been Neglected

Once a product corresponding to a potential demand has appeared, the supplier will have a high chance to make big profit. For a long time, before the genome project was completed, all people thought living to the age of 100 was already extremely hard, and they never dared to expect to live up to the age of 1000. Their strong demand for life has never been met from ancient times till now. Now, it is already possible for this potential bottom most demand to be aroused, triggered and met. Once this demand is met, colossal profits will come naturally.

Its Technological Progress Is Fast

Once a product with advanced technology appears, the chance for the leading developer to make colossal profits has come, too. Although our progress in biotechnology and gene technology has not reached the level to be able to uncover all mysteries of life yet, biotechnology and gene technology are growing at an increasing speed, especially the successful completion of the map of human gene has provided a great impetus to the development of related technologies. In an area with fast technological development, those who take the first chance will surely get huge profits.

Its Market Supply Is Short

Entire present economy is surplus economy; characterized by the buyer's market, and supply shortage has become very rare. The 1000-year longevity industry is the scarcest industry that has never appeared throughout history. Once a technical breakthrough is made and it is industrialized, a huge seller's market will appear for certain, characterized by demand exceeding supply, and a sudden rise of share prices.

It Contains a High Intelligence

The higher the intelligence content of the industry is, the more possible it is for the industry to make big profits with a small capital. The 1000-year longevity industry must depend on high technology, and its demand for high-tech manpower is much higher than ordinary industries, as it is a typical high intelligence industry. A high intelligence will surely have a high return.

It is an Industry with Monopolized Products

A monopolistic industry is essentially different from monopoly. Monopoly refers to relying on power and wealth to control the production or circulation channel in a certain scope to snatch a high monopoly profit. A monopolistic industry means: by virtue of wisdom, developing unique products that are hard for rivals to catch up, so as to have a unique advantage in market competition. The 1000-year longevity industry takes the development and production of monopolized products as its lifeline. At first, you need to bear some risk, but once you have succeeded in the R & D, you will be able to occupy a large market, forming a sweeping power, and getting a high return.

So long as you meet any one or two of the aforesaid conditions, you may make colossal profits. The 1000-year longevity industry can meet almost all the above conditions, so, it can become the most eye-catching profitable industry of the 21st century naturally.

The 1000-year longevity industry chiefly depends on the market, and meanwhile we should also see many factors that are closely related to the super long life, such as pestilence, natural calamities, war, air quality, water quality, environmental quality, life style and so on. Mankind's average lifetime is the highest standard in judging the level of civilization, and it is closely related to all aspects of life as well. Hence, the 1000-year longevity industry is not only a huge economic system project and technological system project, but is also a great cultural system project, political system project and social system project.

The 1000-year longevity industry and life economy need a group of ambitious and resourceful entrepreneurs taking the lead and uniting scientists and statesman together to devote themselves to this undertaking, so as to develop it into a great industry full of boundless hope.

XXVIII. GOD-SLEEPING INDUSTRY AND ETERNAL LIFE ECONOMY

Information from all aspects such as humanity demand, technical conditions, market forecast and so on has shown that the god-sleeping industry has a wonderful prospect, ,and will certainly grow at a tremendous speed.

What is the God-sleeping Industry?

In both the mythology of the orient and that of the occident, the time in the god's world is always slower than in the human world, such as the saying goes, *"Just three days have passed in the godly cave, but it is already one thousand years in the human world"*. In *Pilgrimage to the West*, there are many stories like this: a god just took a nap, but his horse or slave-girl took this chance to steal the god's treasure and went down to the human world secretly to make a great trouble for five hundred years or a thousand years. Judging from the time concept of the human world, the god's sleeping time is very long, at least several hundred years, or even thousands of years.

God-sleeping; just as its name implies, is sleeping like a god. The god-sleeping industry means the industry in which we use a high-tech method to make someone to enter a sleeping state, like a god so that his body can be preserved perfectly, waiting for someday to "awake" and get a new life. In other words, the god-sleeping industry means industrialization of the undertaking of becoming immortal in a tortuous way, namely the industrialization of the body preserving technology for becoming immortal in a tortuous way.

The god-sleeping industry is no longer a mere assumption or forecast, or something existent only in the laboratory, but has become a business practice that has been put into commercial operation already. In the USA, there are four corporations specializing in human body freezing-preservation, they have a certain customer group, and can earn quite a considerable service charge. People ready to accept this service not only include Americans, but also include Australians, Taiwanese and people from all over the world.

The target of this service may be someone with an incurable disease, or someone who is weighed down with age, or someone who is on the verge of death, or someone who has just died... It is hard for anyone to escape death, and the human body freezing-protection technology has offered people the great hope and possibility for curing their incurable diseases, conquering ageing and getting a new life. Hence, anybody can be a potential customer of body freezing protection.

Although the public is not so familiar with the body protecting technology yet, information from all aspects such as humanity demand, technical conditions, market forecast and so on has shown, the god-sleeping industry has a wonderful prospect, and will certainly grow at a tremendous speed.

Pascall once described mankind's fate of survival like this: *"Let's imagine, some people are executed under the very nose of others. Those who have survived see their own plight from their fellows' fate, so they are full of grief but just look at each other in speechless despair, waiting for their own death.*

That is just a miniature of mankind's condition."

Refusing Death

Faced with death, there is no escape for anyone. The fear of death is threatening us beyond description. The mummy is a symbol of the ancient Egyptians' "refusing death and seeking eternal life", and other nations of the world also have similar pursuits. Just as many customers of the body freezing protection say: *"I like the feeling of being alive, so, I do not want to die for better or for worse."*

Whenever we think of the picture that a person will never regain life after death, just as if falling into an abyss beyond redemption, we tremble with fear. God-sleeping is the most powerful weapon against incurable diseases and ageing. It hopes to save mankind from the edge line of the abyss of death, and enable people to say goodbye to death. At this point, just like the 1000-year longevity industry, the god-sleeping industry can meet people's absolute and common demand. Hence, we can predict, there will doubtlessly be more and more people ready to receive this service, so, the god-sleeping industry has a tremendous market potential.

Eternal life economy is the economy that enables people to win life time and life quality forever. The god-sleeping industry (industrialization of the body preserving technology) is the core industry of the eternal life economy. Moreover, there are some related industries helping people to reach eternal life, such as the mind-saving industry (industrialization of the mind-preserving technology), the revival industry and so on.

Great Technical Breakthroughs

The history of American human body freezing companies is not short, but its development is quite limited, chiefly because the technical conditions were not so mature. Now, the technical conditions of the god-sleeping industry are getting mature, much better than the past.

In the past, people's greatest worry was whether the body could revive, and even if they believed in the "revival", what they expected about the new life was no more than a dozen or more years or scores of years.

But now everything has changed!

Quantum jumps in science have convinced people that future

technology may develop to the level to "raise people from the dead", and the hope for revival or rebirth has increased by many times. At the same time, people are also convinced that, after revival, they can live for not only scores of years more, but for a thousand years more, hence, the value of revival or rebirth has increased by many times as well. Therefore, the magnetism of the products and service of the god-sleeping industry has increased by tens or hundreds of times, especially for those oldsters on the verge of death, the god-sleeping industry will be their greatest blessing and only savior. Once we realize scientists' prophesy, *"Before 2030, mankind may break through the limit of one thousand years"*, for those who are already too old to live to the year 2030, their death will mean a narrow miss from the unprecedented great happiness. How can they not feel a great pity! Yet, the god-sleeping industry and eternal life economy have offered them an ensuring buffer, helping them to be able to get the "rebirth" in 2030 or even later, and after the rebirth, their new life will not only be scores of years, but a thousand years or even an eternal life.

The god-sleeping industry will have continual great leaps in the body-preserving technology and revival technology. As far as the present is concerned, the god-sleeping industry chiefly depends on the human body low-temperature storage technology. But the current technological breakthroughs have made people discover there is not only one way for keeping the human body, but there are also some other possible ways, such as human body vitrification, gene hibernation technology and so on. The largest human body freezing institution of America at present, Alcor, has suggested a new preserving method, that is, human body vitrification, instead of refrigeration, which can prevent human cells from being frozen, so as to ensure the low-temperature freezing particles may not damage the cell membrane and keep the activity of the cell better.

What is more important, the discovery of the hibernation gene has brought about a new twilight for the god-sleeping industry. Mankind has studied animals' hibernation for more than 100 years, and has found that coldness and diseases have no effect on hibernating animals. When an animal is in hibernation, owing to the drop of temperature, its metabolic activity can decrease by 98%, the whole physiological system is in a "sleeping" state, and the life process is prolonged accordingly. Can we use animals' hibernation mechanism to put human beings into an "artificial hibernation"? In recent years, science has finally found out: Just as animals can hibernate, mankind can hibernate, too. Recently, US' and German's scientists found two hibernation-inducing genes, PL and PDK-4, which seem to be able to induce mankind into a hibernating state. Hellmayer, professor of zoophysiology of Marburg University of Germany and chairman of International Hibernation Society said: *"In the past years, we just studied aimlessly, trying to find out the door handle or button for starting up this mode of metabolism. Now, we have finally found the related genes."* The American military source offering financial aid to the study is very interested in it, hoping to apply the hibernation technology to wounded soldiers afield, for the wounded soldier could be stopped in a "dormant state", so it would be easy to carry him safely to the hospital for treatment. The US Outer Space Headquarters also hopes the study on hibernation may contribute to mankind's long-range space travel. If the god-sleeping industry uses the artificial hibernation technology, we may control the clock spring of life, putting people into a "hibernation" state, so as to preserve the body for a long time in order to "revive" at one stroke some day.

The success in eternal life economy and god-sleeping industry will also depend on the breakthrough in the revival technology. The present startling progress of mankind in nanometer technology, stem cell technology and other high and new technologies is also convincing more and more people that those receiving the body preserving service will get their "new life" some day in future. Now, American human body freezing corporations have begun to feel the number of people signing up for freezing-preservation after death is "unprecedentedly great". One of the responsible officials said: *"Our business has got recognition and acceptance from more and more people, and it is an indisputable fact."* All such results have greatly benefited from mankind's accomplishments in the domain of science in recent years.

The combination of biotechnology, nanometer technology and even new energy technology will not only greatly reduce the cost and raise the quality of eternal life economy and god-sleeping industry, but also make the realization of rebirth much earlier. The present charge for human body freezing storage is something like this: In Alcor, the charge for freezing storage of the whole human body is 120,000 US dollars, and that for partial storage (only the head) is 50,000 US dollars. In the

Human Body Freezing Research Institute of Clinton Town of Michigan State, the overall freezing charge is only 28,000 US dollars. Along with the deep going of the research, the cost of human body preservation will continue to drop, so the charge will get lower and lower.

One important phenomenon is noteworthy: Those who are fond of body refrigeration are just the elite of the times: high-tech personnel or brainpower.

Why are the consumers science and technology workers represented by the people of Silicon Valley instead of people not knowing their ABC? The reason lies in the fact that "Silicon Valley's people are greatly loyal to advances in science and technology. They believe, so long as you can use plentiful money and technology, you can solve any problem. So, why can't we live forever?"

Believing in human body freezing storage, believing that future science will revive people, and believing in the feasibility of the god-sleeping industry, are just the same as believing in the possibility of accelerated development of high technology and believing in the magical power of high technology. The technology elite standing on the top of the tide can feel the invincible might of high technology more than any other group of people, and they take a more optimistic attitude to the future and are full of confidence in the future. Thus, most of the people who sign the human body freezing storage agreement are computer scientists, software developers and other professional technologists.

So it is evident that, judging from the consuming group, human body preservation is not only a universal human demand, but is also characterized by progressiveness and is avant-garde. In judging the prospect of an industry or project, we need a looking-ahead way of thinking. The key point of the looking-ahead way of thinking is: in decision-making, we should not only consider the "money", but also look at the "prospect". Insightful entrepreneurs are very good at grasping the future trend, especially grasping the trend of change in future. The god-sleeping industry is just an avant-garde industry representing the future.

When the famous Japanese entrepreneur, SONY's founder, Akio Morita, launched his tape recorder into the market at first, he did not get a smooth going in the market. One day, he came to an antique store and found a customer buying an old flower vase with a high price. It startled him: Why was the customer willing to buy an old flower vase with a price even higher than a tape recorder? For the buyer knew the value of the flower vase. Thinking of that, he became enlightened at once: *"Yes, to sell a commodity first of all, you should let the buyer realize the value of it!"* Whenever we develop a new product, we should educate the consumer first.

Before selling your own new product to your consumers, first of all, you should sell your new idea, turning the consumers' eyes to the demand that has not been understood and met. Did Chinese drink coca-cola in the past? Did they drink the colorless and tasteless pure water? Even water can be sold for big money, really unimaginable! Thus, we can see that the market can be cultivated and can be taught!

Two salesmen from a shoe factory went to an island in the Pacific Ocean to build up a market. After arriving at the island, each of them sent a telegram back to the factory. One was: *"The people on this island do not wear shoes, so, I will return tomorrow by the first plane."* The other was: *"It is so wonderful! The people on this island have not begun to wear shoes yet. The market potential is so great. I plan to settle down here to do business."*

The market of eternal life economy and god-sleeping industry is far from being exploited yet, and just like the people on the island who have not put on shoes yet, has an extremely great potential. Moreover, the god-sleeping industry has a greater need to educate the consumers than any ordinary industry and product. The god-sleeping industry needs to guide and kindle the great potential human demand (the greatest fear of death and the greatest thirst for life), turning their eyes and attention to this demand that has not been fully understood and satisfied before. Once we have educated the consumers successfully, people's potential demand will be triggered for certain, and then, no force can stop it.

Culture decides ideology, ideology decides demand, demand decides consumption, and consumption decides production, that is, there is a chain like this: "culture—ideology—demand—consumption—production". So, it can be seen that the most fundamental is culture. To transform the production mode, we must renew human demand and revolutionize human ideology, and the key to the revolution of human ideology just lies in culture. Hence, culture may determine the economic form. "Economization of culture", "culturalization of economy" and "integration of culture and economy" are the megatrends of economic development. Culture and economy infiltrate into each other, blend with each other, promote each other, and decide each other. This will be the future trend.

The hamburger with two pieces of bread holding a piece of meat, does not taste very good, frankly speaking. It is even not as delicious as the flapjack. But why did it become popular all over the world? At first, it was labeled as the food for the middle class in the affluent society, so, eating hamburger became a vogue of Middle America. Later, it marched into the international market, and was packed as a representative of American culture, making people eating it believe firmly that eating hamburger can prove they are modernists. Thus, it can be seen that the worldwide spread of the hamburger depended on culture, the culture that marks American vogue and fashionable.

To develop eternal life economy and god-sleeping industry at a fast rate, we must rely on the wide spreading of the life culture. Human life problem is mankind's most important problem. Once the society has accepted the culture that values life most, the life culture has become the leitmotiv of the society and every citizen has built up the idea that **"One life is more important than a solar system"**, then, the products and services of eternal life economy and god-sleeping industry will be accepted by everyone and everyone will be vying for them.

To sum up, **to open up a broad market for eternal life economy and god-sleeping industry, we need to shake off the bondage of traditional culture, including ideas, concepts, morality, customs and force of habit, and need to let such a culture go deep into the hearts of the people. That is, "esteem life, love life, and life interest is above all things".**

In nuclear physics, there is a concept called "threshold quantity", referring to the minimum quantity of radiating matter that can lead to a nuclear reaction. Once we reach that threshold quantity, the whole changing process will be self-sufficient. We may also introduce this "threshold quantity" concept into social science areas: We need a certain number of people who take the lead in adopting a new product, new method or new concept, and only after a series of chain reactions can it become the vogue to be followed by everyone. That "certain number" is just the threshold quantity.

American scholar Rogers has proved in his study: If 13% of the people have accepted a new idea, sooner or later, at least 84% of the people will accept it. He also says: *"Once a new invention has reached the threshold quantity, it will be adopted naturally."* In the social field, the threshold quantity is about 5%-20%.

"All things may change from small to big, so long as you can trigger the vogue." Threshold quantity is just the shooting point. Mastering the secret of the threshold quantity equals mastering the way to trigger the vogue, thus, you will be able to lead the vogue and sweep the world.

As an enterprise, it may consider the threshold quantity for eternal life economy and god-sleeping industry. What is the threshold quantity for accepting this new trend of thought of life culture and god-sleeping? By trying to reach the threshold quantity and reach the shooting point, you may trigger the vogue and make the fashion. The key is: First of all, we must make some people to use or accept it, so as to lead the public in their imitation and propagation, letting the public follow the public.

At first, it is necessary to take some intentional propaganda strategy and marketing strategy so as to concentrate the fire to reach the threshold quantity, then, change will become self-sufficient, turning automatically from a small trend to a megatrend. Once the threshold quantity of the god-sleeping industry is reached, receiving the god-sleeping service will become a vogue for certain, growing into a new vogue in the 21ˢᵗ century.

As a very promising eternal life economy industry, the god-sleeping industry has a broad market space and huge profit return. Meanwhile, the god-sleeping industry is never an isolated economic system, as it is closely related to many areas such as philosophy, ethic, moral, law and so on.

The god-sleeping industry should boost relevant philosophical innovation, ethical progress, moral improvement and legal establishment. For instance: how should we ensure the rights and interests of the people accepting the god-sleeping service (patients of incurable diseases or old people on the verge of death who have the economic condition)? How should our society treat them? How should their descendants treat them?

Do the service providing private companies' credit and commitment have a proper legal assurance? ...

Like many large industrial projects, the god-sleeping industry is in dire need for cooperation between the enterprise and the government, and in dire need for the dual force of the market and the government. The government can offer a awards to scientists, entrepreneurs, educationists and so on who have made great contributions to the whole society and mankind. The award will be life that is more valuable than money, that is, the government will provide the high-grade god-sleeping service to them free of charge...

As the world's top prize, the Nobel Prize takes money as the reward to those excellent personages who have made great contributions to mankind. If we offer god-sleeping service to Nobel laureates at the same time, such a reward will be more important and attractive than money for themselves, and moreover, the result of such a reward, namely their living forever, will contribute more to the society, and meanwhile will boost the development of eternal life economy and god-sleeping industry to a better effect.

XXIX. ECSTASY INDUSTRY AND PLEASURE ECONOMY

The rise of pleasure economy has just begun: In our age, an unprecedented revolution is happening, In the past, only products in kind could increase wealth, but now "joy" can also boost economic development.

Embracing Pleasure Economy

According to the definition in *Contemporary Chinese Dictionary*, "pleasure" means "making people happy" or "joyous activity". Pleasure economy is meant to make pleasure, being an enjoyable economy. Seeking pleasure is a human nature, and people are ready to give alms liberally in order to get pleasure. America's Hollywood, Japan's Nintendo and Hong Kong's entertainment circle have been operating pleasure as an industry for a long time, and their rich profit is amazing! In Japan's economy, 20% of the earning comes from the electronic game trade in the entertainment industry. America's audio-video product is the second greatest export commodity only secondary to plane export. A drama is a market and an industry that may bring profit for several years, one decade or several decades. The global income from one musical opera, "Miss Saigon" has exceeded 1.3 billion US dollars. "Lion King" earned 0.74 billion US dollars in the world. "Space Fighter 7" game was sold in 6 million copies all over the world, bringing a sales amount of 0.3 billion US dollars! On the day when "Space Fighter 8" was put onto the market, as many as 2.5 million copies were sold, with a sales amount exceeding 0.1 billion US dollars. In Britain, the economic income from arts has exceeded that from the automotive industry. Korea's game industry has built up a production value approximately 20 billion US dollars, with a growth rate as high as 40%, and will become the most profitable industry in the country. SONY, Intel, Microsoft and some other famous companies are marching into the field of digital entertainment with one accord coincidentally, just as the saying goes, "Great minds think alike". Bill Gates said, seeing that the game industry is flourishing increasingly, he would list the game industry development as the central task of Microsoft's future development. Microsoft's Asian Academy set "digital entertainment" as their major direction in research long ago, which has made Microsoft transform from a software empire into an entertainment consumption empire.

Pleasure economy will play a very eye-catching role in future. Any enterprise with an acute sense to the market will be aware of a new flying chance is coming, and pleasure economy is increasingly becoming a focus point attracting entrepreneurs, economists and futurists. The rise of pleasure economy has just shown: In our age, an unprecedented revolution is happening: **In the past, only products in kind could add to wealth, but now "joy" can also boost economic development.**

Among the majority of Chinese people, few have a good impression of pleasure, entertainment, or games. People tend to think games are only things for children, and the entertainment industry is usually not regarded as a regular occupation. But the fact is quite different. Today, 3/4 of computer entertainment software users are adults above 18, and 82% of them are in the age range of 25-55. Moreover, the players' educational status is generally quite high.

We really need an ideological revolution, and thinking liberty!

Seeking pleasure and avoiding suffering is fundamental humanity. At the bottom, all human activities are for pleasure. With regard to our understanding of playing, let's lend an ear to what the famous economist Yu Guangyuan said: *"Playing is one of the basic needs of human life, and is a human instinct"*, *"We should have a playing culture, should study the science of playing, master the techniques of playing and develop the art of playing"*. The German thinker Shiller had a famous viewpoint in his *Letters on Esthetic Education* as early as 200 years ago: *"Only when a man is a man in the full sense can he play, and only when a man is playing can he be regarded as a full man."* Along with the continual development of the society, our material life is getting extremely rich, so our human nature of pleasure seeking will be shown more and more clearly. When the basic human needs of survival have been met, people will certainly seek pleasure more eagerly and increase their expenditure in amusement.

> Only when a man is a man in the full sense can he play, and only when a man is playing can he be regarded as a full man. —Shiller

We often hear Internet economy is an "attention economy", and an "eyeball economy", but in fact, essentially, Internet economy should be called a "pleasure economy". Whether to steal the spotlight or keep the attention, we need to depend on amusement. Ding Lei, Chairman of the Board of Directors of Netease, one of China's three major web sites, made a prompt decision: To develop on-line gaming and take on-line gaming as Netease' key point in forthcoming development. Ding Lei said: *"Last week, I went to some Internet bars inland for investigation. I took a look at 45 Internet bars. It is hard to imagine: 99% of the people were playing OICQ, they took a great delight from the chatting, and many people can chat whole night. Chatting could give people illusions.* One man could chat with two girls at the same time, and also fool another boy, which could meet his desire for fancying to a great extent. Some others were playing games, especially online games..." According to a survey; most of the netizens access the Internet just for chatting and playing games. Of course, the pleasure from the Internet is never only from chatting and gaming, but we can see that people's greatest interest lies in pleasure, and their greatest wish is that the net will bring pleasure to them, the pleasure that is hard to come in reality! So, it is evident that the key to whether a net company can profit or not just lies in whether it can "bring pleasure to people".

The effect of pleasure economy on future economy will be all-round, and the key to the success of most future industries will just lie in whether they can bring pleasure to people successfully. Nowadays, from performance to marketplace shopping, from finance companies to stock market information, from traditional dining and drinking to new-styled mobile phones, the amusement element is infiltrating into each sector of the whole economy. **The greatest function of amusement is to make people ready to accept. If a product has an amusing effect, everyone will like to use it. If a brand has an amusing effect, everyone will be ready to a take to it. All kinds of industries and trades must try their best to offer pleasure functions, pleasure services and pleasure activities, so as to meet customers' pleasure demand to the greatest extent.** Mcdonald's president said: *"Remember: We do not belong to the dining industry, but belong to the amusement industry."* A great many children cry for eating Mcdonald's, just because they want to collect the set of novelty toys offered by Mcdonald's. In the economic activity from now on, amusement will never be a specified or limitary industry, as any industry or trade must be transformed by the amusement element. All traditional basic necessities of life should have an amusing function if you want them to have a greater competitive power. The future clothes will be high-tech clothes that can bring a happy mood to you when you put them on, the future food will be amusing food; the future house will be a "pleasure house": a joyful paradise in every aspect, such as the modeling, design, color, temperature, humidity, smell, sounding and so on...

Ecstasy" here refers to utmost pleasure, or greatest joy. The ecstasy industry is an industry that offers people the utmost joy, making both "quality" (from the low-leveled joy to the high-leveled joy) and "quantity" of the joy (from the low-degreed joy to the high-degreed joy; from temporary pleasure to daylong pleasure) reach an unprecedented height.

> Remember: We do not belong to the dining industry, but belong to the amusement industry.
>
> —Mcdonald's President

The ecstasy industry is a great industry with limitless charm, and is the mega trend of pleasure economy development. The ecstasy industry is more advanced, more forward-looking and more humane than the ordinary amusement industry. In order to seek the utmost pleasure, the ecstasy industry depends on the magical power of high technology more than the ordinary amusement. The more we let high technology participate, the more possible it is for us to help people get the utmost pleasure. Such an amusement has an unprecedented and enigmatical charm. So to speak, whoever has invested in the ecstasy industry and mastered the high-tech amusement will hold the key to the door of wealth.

"High technology will play a shocking role in changing our pleasure experience."

Combination between biotechnology and information technology may bring us more vivid virtual experience.

In toda's amusement, 4 kinds of information take the predominance; digit, word, sound and image. Information may also come from many other forms, such as olfaction, gustation, touch, imagination, intuition and so on. In future, high and new technology development of all the other information forms such as olfaction, gustation may be put into commercial use, too. So far some corporations have already worked out digital odors, for instance; the basic factors of an odor may be expressed in digits, then loaded onto a moderately priced chip. Just imagine: You may get an electronic greeting card emitting a fragrant smell; You are watching a digital film with a unique smell effect; Or even you can touch the petticoat of the heroine in the teleplay and have the real tactile sensation, and you may dance together with her... Everything will show a wonderful virtual world that is "more realistic than the reality".

Virtual reality technology can simulate the real world vividly, transcending space and time, and can put people in it completely.

The famous director Spielberg named his filmmaking company "Visional Factory", which can offer people a surreal visional world, enabling them to get a lot of special experience and pleasure. However, all such traditional amusements as seeing a movie, watching TV or read-

ing a novel have great shortcoming. You cannot participate in it or interact with it, but would be just a passive onlooker and recipient. By dint
of high technology, we shall not be passive, shall not just receive, but can
participate and experience directly. For example; nanometer technology
may result in a very tiny machine or chip occupying a place in the sen-
sory area of your brain, which may produce a virtual reality effect so that
it is hard for you to distinguish between the real and the simulated. The
best food you have eaten may be enjoyed repeatedly in the company of
different people. Going to Fuji Mountain or visiting Louvre will become
meaningless. The intracerebral emulational means can do or sense anything people can do or sense: You can become a knight of the Dark Ages
for a while, or change into a chivalrous swordsman of ancient China in
Jin Yong's novels for another while, or turn into a princess in the Castle
of Beauty for some time, or have a taste of the feeling of being a world
plutocrat... Such pleasures are unimaginable even to the kings of ancient
times.

More and more people will use the broadband to "order" amusement.

According to a survey, in families with broadband Internet, their
average consumption in amusement is far higher than those without
broadband. One of the five major public accounting firms of the world,
Ernest & Young did a global survey covering 128 CEOs of large multinational companies, such as Nokia, Timewarner, SONY, American
AT&T and so on. The result shows that most of the CEOs, namely three
fifths of them, believe "The broadband will change people's amusement
experience". More and more people will use the broadband technology
to "order" amusement. Nowadays, by dial-up access, you can download
only 1/10 of a song in one minute. In future, in one minute you can
download 300 songs. In the near future, one hard disk can store all the
songs in the 150,000 CDs released in the times. In 2000, the global capital investment in broadband amounted to 100 billion US dollars. All
the movies, teleplays, novels and albums of paintings stored in store-
houses in the past may be digitized and put onto the broadband network, for you to choose any time, which will certainly be a huge profitable business opportunity.

Using high technology to work out amusing devices and amusing agents.

Do you still remember that experiment with a mouse? After the scientists found the pleasure centrum in the mouse, they designed a lever: Each time the mouse jammed on the lever, it would stimulate the mouse' pleasure centrum, so the mouse felt extremely happy, so, the little animal was very fond of jamming on it and never gave up even until tired to death. The mouse has a pleasure centrum, so does a man. Through in-depth research on human pleasure centrum, it will not be hard to work out a device or drug that can make people feel extremely happy. Of course, we should ensure no side effect will come out when bringing pleasure to people. Once such a product is worked out, its market profit will be too great to estimate.

In the 20th century, invention of the contraceptive and condom smashed the shackles on females, freeing mankind from endless worry and anguish. If there had been contraceptives in Marx' times, he would not have said, **"For a person with a great aspiration, there is nothing more foolish than marriage."** Marx had six children and three of them died because of abject poverty. The death of each child was a heartbreaking pain and a heavy blow. Had there been contraceptives, Marx' would have been much happier, and so would have his wife Jenny. The pleasure such a tiny pill has brought to mankind, especially females, is beyond estimation, and of course, it has also brought about huge economic interests to its inventor and producers. Once a device or pill able to bring extreme pleasure to people is worked out, its impact on mankind will be even greater, and at the same time it will bring economic benefit beyond comparison.

In conclusion, the overall and in-depth participation of high technology is the great difference of the ecstasy industry from the traditional amusement industry.

The ecstasy industry has another great feature: It not only stresses the "quantity" of joy, but also emphasizes the "quality" of it. The ecstasy industry is to help people get pleasure anytime, anywhere and in any case, not only the momentary pleasure from playing electronic games or watching digital films. The ecstasy industry will also help people to get a more profound and higher pleasure of the soul, the pleasure of self-realization, instead of a superficial, low-leveled sensual pleasure. The ecstasy industry lays special stress on the time effect of pleasure, not only considering the present, but also caring about the medium-term and long-term effect, that is, considering the negative effect of the present consumption on the future, and considering the general effect of the whole process.

We should not vulgarize pleasure, thinking only watching TV, seeing a movie, playing games, visiting friends or relatives, eating and drinking, or window-shopping is pleasure, but should also see that we must seek the pleasure in learning, the pleasure in creation, the pleasure in success, the pleasure in self-realization and so on. Maslow's theory of need hierarchy has also shown that the highest pleasure of mankind is the pleasure in self-realization, and self-realization tends towards the "peak experience" most easily. Man's demand for seeking spiritual satisfaction and self-realization is limitless, which has also opened up boundless possibilities for economic development.

In our life, study and work take up the primary position. Now most people's study and work cannot be said to be joyful, even most of them are crying to heavens. **"If work is a pleasure, life will be a paradise! If work is an obligation, life will be a hell!"(Goethe) When we judge according to this saying, I'm afraid that many people are still living in an abyss of suffering.**

The ecstasy industry is just to help us turn both work and study into pleasures, making everything in life full of pleasure. Work will becomes man's primary necessity, and he will not just work for making a living, but work for his own pleasure, for his self-realization. Through the use of many kinds of high and new technology, our future work will become full of fun and fascination. It needs one to go all out, just as a child plays games. When he spares no efforts, he will get the greatest fun. Competition, risking, imitation and all kinds of interesting jobs are just like gaming, and it will be just the essence of future work.

The ecstasy industry should also make our study full of fun. On one hand, reform human gene; and on the other hand, use high technology to reform education. A brand-new amusement-styled study will replace the present infusion-typed education, and children will learn many fresh things from their environment in a very free state. The essence of amusement is infatuation. When someone is madly infatuated with something, this thing must be very attractive. As this is the case, why not use high and new technology to turn learning into a way of amusement so that children can become infatuated in it? Moreover, a joyful learning does not tend towards fatigue and repulsion easily, and one can enhance his thinking and raise his quality in spite of himself at ease...

Build Up the Elysium

Elysium is mankind's greatest ideal. The high development of the ecstasy industry and pleasure economy may contribute most directly to the build-up of Elysium. But the build-up of Elysium is complicated system engineering. To let everyone get the utmost pleasure, we must consider scores of social factors such as educational level, health index, environmental protection, traffic, equality, democracy, political stability and so on. Only when our society has become an ecstasy society can we get the real utmost pleasure. Just watching movies, playing games, and letting everyone have plentiful money, cannot be called "utmost pleasure" yet. If there is no progress in our environmental protection, we cannot breathe fresh air and cannot see blue sky, we will not feel happy. If our society is full of evil, people have no sense of security, and malfeasants often set up our bristles, we will not feel pleasure, either. Some wrong concepts, ethics or morals may also impede our pleasure. For instance, Chinese ancients thought *"One woman should never make love with two men"*, *"Even if you starve to death, never lose your chastity"* are moral virtues, and they absolutized such a moral, which caused a surprising loss to Chinese females' pleasure for quite a long time.

The high progress of the whole society and the fast development of the productive power are the basic keystones for building up the Elysium and letting everyone live in a paradise.

In the future society, we shall have a full automation, such as production automation, office automation and home automation, leaving us more leisure time at our disposal, so that we may do self-satisfaction and self-realization, we even can *"Go fishing for two days and dry the nets for three"* (Chinese proverb, meaning working in a leisurely manner). After realization of production automation, we can work at home and everyone may become a SOHO, without the need to read the boss' faces or crowd ourselves in the bus. We shall have more free time to stay with our family. Housework automation can liberate women greatly, adding a lot of time for pleasure and enjoyment. The invention of the washing machine has made a great contribution to the emancipation of women all over the world from the onerous housework. Not long ago, mothers had to struggle around the cooking stove, splitting firewood, burning coal, with their nostrils filled with black smoke. After having the gas burner, they were greatly liberated. From now on, all boring family chores, such as cooking, clothes washing and so on will be trusted to the robot, hence, we shall be able to hold our youth and retain our pleasure. Engels had a famous saying: *"Women' emancipation is a natural yardstick of social emancipation"*. Likewise; women's pleasure is also a natural yardstick for judging the happiness of the society. High technology should be women's friend, and all weak persons' friend. High technology should bring liberation and pleasure to the disadvantageous group.

To build an Elysium, we must rely on the over-speed development of the whole world's economy and the magical revolution of productive power. The key to the advent of the Elysium as early as possible lies in boosting the mushroom growth of the pleasure economy and the ecstasy industry. We should carry out amusement education and ecstasy education, cultivate amusement personnel and ecstasy personnel (high-quality personnel who are able to use high technology to increase pleasure for people), and always turn out new concepts, new ideas, new inventions and new techniques for the products and services of pleasure economy and ecstasy industry. Meanwhile, the government and the society should offer a correct public opinion orientation, a good policy support, especially doing strategic thinking and long-term planning for pleasure economy and ecstasy industry. So long as we do so, a human paradise or Elysium will appear before long.

XXX. EUGENIC INDUSTRY AND HUMAN-CENTERED ECONOMY

The eugenic industry can not only lead to the appearance of numerous genius prodigies, but also enable them to surpass all past genius prodigies in every aspect.

What Is Eugenic Industry?

First of all, let's trace back to mankind's history of eugenics.

Sir Francis Galton, Darwin's cousin, put forward the concept of "eugenics" for the first time in history in 1833, but in fact, mankind has never been subject to the mercy of nature since the ancient times. Even in the ancient times characterized by a very low productive power, people also took all kinds of measures to ensure their descendants would be healthier and brighter.

People of the ancient Sparta bathed their new-born infants with wine. If the baby was safe and sound, it showed the baby was strong enough, and they would keep him (her) alive. Some babies could not bear such a torture, which showed they were weak, then they would put them to death and eliminate weak ones, and in this way they kept their nation strong and prosperous.

In many primitive tribes, there is an unwritten stipulation: Whenever they find a new-born is deformed or handicapped, they will put him to death without exception, and strictly prohibited intermarriage within directly-related members of a family, so as to guarantee the clan's prosperity.

We have already entered the 21st century. In such a great new century, we must go beyond the traditional ways of eugenics, and go all out to develop the eugenic industry. The eugenic industry means using high-tech eugenics technology, namely biologic technology and life technology represented by gene technology, to optimize and reform human gene directly, and applying such an eugenics technology to industrialized operation.

In not so distant future, mankind may generate perfect descendants through gene reform, which has become the consensus of many scientists. Eugenics by way of gene technology can not only avoid genetic diseases, but also improve the person's appearance, physique, intelligence quotient, conduct, character, temperament and so on. Such eugenics is unexampled to any previous eugenics of mankind

Eugenics through technology has become an inevitable trend. Here, we do not discuss the technical feasibility of eugenic industry, and just analyze it in terms of economy.

It goes without saying; a new industry's prospect and fate will depend on the market demand. People's eugenic demand is far stronger than ordinary demands. "Children First" is the paramount idea of many families, and parents place their greatest hope on their children. **According to an investigation, when people are asked what they live for, most of them answer: For children!**

For children, they can spend anything. For children, they can sacrifice anything. For children, they can bear any great suffering. For children, they can accept any hardships. Especially in the present China characterized by the single-child family, people's attention to their children is greater than any time in the past. For the child, they save a huge amount of money; for the child, they are ready to spend a lot of money; for the child, they do great investment. The whole life of Chinese parents is just for the child. Whether a male or female, for the child, they spare no cost.

A man wants to have a child. When his pals invited him to play in the bar at the weekend, he shook his hand again and again: "No. You go yourself. I've got to care about my future son." Now the chief reason for the call to let women return to the home is "for children". You see, the children's future can drag the greatest feminist back to her home. People's desire for investment in their children's education is much stronger than their desire for investment in their own education. Just take a look at the roaring business of the "noblesse" kindergartens and schools, the booming sales of home training books and home teaching reports, you may see parents' great enthusiasm in educational investment and the great popularity of the concept "for the child". Then, the eugenic industry that is closely related to the fate of the child will surely trigger an even greater demand!

Which father or mother does not hope that their own child will be the best? Which nation or country does not wish its own next generation to become the best? Do you wish your child to become a virtuous, wise, able, strong and good-looking person, or a vulgar, selfish, lazy, stupid, imbecile, weak and ugly person? The answer goes without saying. If you know there is such a method that can help your child have genes for becoming more beautiful, healthier and more intelligent, you will certainly take it. It is out of our common humanity. And it is just where the market power lies, and it determines the greatest market prospect for the eugenic industry.

The eugenic industry that conforms to the strong wish and great consumer demand of the general public has no reason to be rejected, and has no reason to be decried. It is totally different from the genocide driven by horror and animosity in the disguise of eugenics in the last century, One is a cruel fascist political massacre while the other is a fine and voluntary business behavior. One is an extremely terrible human tragedy while the other is a great matter with an ever-lasting value. The difference is as vast as between Heaven and Hell.

Education Industry vs Eugenics Industry

There are two major factors deciding a person's quality: nature and nurture. Along with the continual improvement of people's understanding of the gene, scientists are getting more and more surprised at the great role that gene plays in deciding a person's quality and fate, even to the extent that many scientists think "Inheritance prevails education". According to scientists, all the seven original sins listed by the Christianity of the Dark Ages, namely; anger, greed, theft, avarice, raping or seducing, false pride, and homicide, have much to do with genetic elements, at least 50% comes from the DNA.

Both education industry and eugenics industry concern "human production". Education aims at postnatal improvement while eugenics is meant for congenital optimization. Both determine people's quality and fate.

Nowadays, most people have a great fervor for educational consumption. According to a survey by China Zero-point Market Research Company, in urban families of China, the educational expenditure for children has become the item that grows fastest among all their items of expenditure. Families like educational investment very much, and so does the state. The government is becoming more and more aware that education is the most important, and prosperity of the state chiefly depends on education.

Judging from the rising education industry and the accelerated growth of educational investment, it is easy to imagine that the eugenics industry will be prosperous for certain, and the eugenic market will flourish for sure. In future economy, the eugenics industry will, like the education industry, unfold and establish a series of brand-new concepts of different levels, such as eugenic service, eugenic demand, eugenic choice, eugenic opportunities, eugenic rights and interests, eugenic market, eugenic business, eugenic exploitation, eugenic contribution ratio, eugenic trust, eugenic savings, eugenic stock, eugenic banks, eugenics industrial groups and so on.

Whether for the individual or the family or the state, their purpose in stressing education and investing in education is only one: for improving personal quality. Personal quality is the base of all things. The level of the quality is the primary cause of the fate of the individual, the family or even the whole nation. Personal quality not only hangs upon postnatal education, but also depends on the innate gene. **The congenital is more fundamental than the postnatal. In the past, we were always powerless before the "congenital", so, we belittled or disregarded "the congenital", focused all our attention and time onto the postnatal education. Now, we have already had the efficient path and feasible program for changing "the congenital" and optimizing "the congenital", that is, we can use gene technology and eugenic industry to enhance personal quality. Why not take action at once?**

In comparison with the education industry (postnatal project), the eugenics industry (congenital project) has its unique advantages. Education not only needs to spend a lot of monetary cost, but also needs to pay precious and long life cost or time cost. It involves both personal cost and family cost, especially the life cost of the person himself for more than a dozen years or even scores of years, the parents' great deal of time, energy and life cost, and the risk cost (For example, if you send a small child abroad for study, you will have to bear the risk of "losing your child"). Moreover, it also includes social cost (the input by the state and the society for every citizen's education). But the greatest advantage of eugenics just lies in its saving of time cost, you can get it done once and forever, without any time cost or life cost of the person himself and his parents, then you can benefit from it greatly all your life. Meanwhile, eugenics may also greatly shorten the education term so as to enhance the educational efficiency greatly. Now, one person needs to spend a dozen or more years or even longer time in education before he is able to serve the society. In the future, as human natural predisposition has been improved through eugenics, the term of years for education may be greatly shortened, so, all the present "wonder child classes" will become "ordinary classes". In the ancient times, Gan Luo became a prime minister at the age of 12. In future, with eugenics, everyone may become Gan Luo after an education of three or five years.

Therefore, we should not only invest greatly in education but also invest greatly in eugenics. The investment in eugenics can by no means be replaced by investment in education. Whoever invests in it will benefit. If the family invests in eugenics, the family will get benefit. If the state invests in eugenics, the state will benefit. The state will be the greatest investor and beneficiary of the eugenics industry. Just like education industry, eugenics industry is a great social undertaking, and can also bring a huge economic benefit, being a special industry able to contribute greatly to economy. It is a basic industry indeed, full of strategic significance.

Production of Humans and Human-centered Economy

Economic activities are chiefly production activities. Production activities are the foundation for the existence and development of mankind and human society.

Production falls into two categories: one is the production of materials, and the other is the production of humans. Production of humans is more important than production of materials. Engels points out in his *Origin of the Family, Private Ownership and Nation*: "*Production can be classified into two kinds. On one hand, it is the production of means of subsistence, namely; food, clothing, houses and the necessary tools for themand on the other hand, it is the production of mankind itself, namely breeding of the race.*"

In comparison between these two kinds of production, it is easy to

see; Production of mankind itself is more fundamental. On one hand, the ultimate objective of production of material objects is for people, for people's happiness and for people's development. On the other hand, human is the most active and most decisive factor in production, or we may say, "The most essential condition for a production to go on is the workman", and the workman is just the result of production of mankind itself.

The concept "production of mankind itself" is never static and unchangeable one, but should develop in step with the times. If we just rest on Engels' understanding and regard the production of mankind itself only as "multiplication of the human race", we shall only come to such a conclusion: what we need is just to control and adjust the population according to the need and possibility of production of material objects. Of course, such a conclusion is far from enough for us. Nowadays, there are already some men of insight pointing out that education is "human reproduction", only by education can we make a human become a human in the real sense, and education is just "shaping humans" and "producing humans". **In my opinion, production of mankind itself never only refers to multiplication of the race, nor should only indicate education development. In the present time, its full connotation should chiefly refer to overall optimization of mankind itself by the aid of high technology so as to make people develop most perfectly, becoming the real, complete, ideal and perfect people.** The aforesaid "birth optimization through technology", "body optimization through technology" and "education optimization through technology" have already covered all aspects of human production through high technology.

Human-centered economy is more important than material economy. As there are two kinds of production, there are also two kinds of economy accordingly. What corresponds to the production of materials is the material economy, whereas what corresponds to the production of humans is the human-centered economy. Human-centered consumption is more important than material consumption. For example, between buying a car and buying a product that can enable people to become younger for ten years, and at the same price, which will you choose? I think most people will choose the latter. We can say definitely, the benefit and pleasure brought about by buying a car cannot be compared with the benefit and pleasure brought about by making people younger for ten years. The human-centered consumption that makes people young forever, omniscient and extremely beautiful and so on is far better than material consumption such as automobiles, not only in benefit and happiness, but also in resource consumption and environmental protection, as it costs the least resources and is most environment-friendly. In future economy, we shall shift gradually from material consumption to human-centered consumption, then enterprises investing greatly in human production will become mainstream enterprises, and human production will also become the main content of economy. The human-centered economy taking eugenics industry and education industry as the main will take the leading position in future economy.

Geniuses are torches and deities of mankind. Someone has done a counting and found. In mankind's civilization history of five thousand years, those who really deserved the calling of "genius" were no more than 400. Geniuses are so rare!

Geniuses are boosting social development and human progress. By developing the eugenics industry, we can produce geniuses in large quantities, and our age will also become an age full of geniuses hence. Then, thousands upon thousands of people will become Aristotle, Shakespeare, Newton, Darwin, Einstein, Curie, Malthus, Keynes, Marx, Hegel, Floyd...

The eugenics industry will not only be able to turn out a lot of geniuses, making the quantity shoot up, but also make the geniuses much greater than before in every respect. Most of the geniuses of the past lived a very miserable life: either died in their peak years, or could not be accepted by the society. Pascal said: *"From the age of 18, I did not have a single day without anguish."* When Pascal was 23, he became paralyzed because of apoplexy. He often fainted down, suffered from many chronic diseases, and often had a nervous breakdown. Mozart was a great musical genius, but he was never happy. In his latter half of life, he lived like a beggar, and died when he was only 35. Von Gogh was extraordinarily gifted in painting, but he suffered from delirium, cut off his own ear himself, and ended his life in the peak years in a sad report of the pistol. They were geniuses, but at the same time they were also pitiable men. By optimizing the gene through technology, we may turn pitiful geniuses into happy geniuses and long-lived geniuses, so they can make greater contributions to the society.

Through the eugenics industry, we shall turn out super geniuses continually, not only surpassing the past geniuses far and away in quantity, but also surpassing them far and away in level and quality. At the same time, we may also avoid all sorts of sufferings and worries of the past geniuses. Thus, we shall have an unprecedented great breakthrough both in "quantity" and in "quality".

Wonderful Prospect of Overall Development

The overall development theory is the pith of Marxism, which is also a great contribution to mankind's education. Overall development, that is, raising human quality in all aspects, is mankind's fine wish, but just relying on education is far from enough, and we must rely on eugenics as well.

In human overall development, there are at least three aspects, namely; moral, intelligence and body. Moral is the first. Gene improvement can play a great role just in this aspect. Some opponents of eugenics chiefly worry that if we produce people of super intelligence and let them rule the world, it will bring a great threat to the world. But the eugenics through technology I am talking about is not only to optimize the intelligence, but also to optimize the character, especially the moral character. Such eugenics will not bring any threat to the world, but will help the social progress. Scientists have found, all kinds of antisocial behavior such as shyness, misanthropy and crimes are related to the gene. Some research center has estimated the affecting degree of inheritance on the character as follows: anxiety 55%, creativity 55%, compliance 60%, aggressiveness 48% and extraversion 61%. According to the report from the scientists of London Children Health Institute, girls who have inherited the father's X chromosome are more expressive, easier to make friends, more harmonious with their families and teachers, and easier to teach than those who have inherited the mother's X chromosome... the DNA decides every aspect of a person, even including smoking, risk taking and so on. Eugenics will be greatly helpful to optimization of all human aspects.

The eugenics industry never means optimization of only intelligence, but an all-round optimization of moral, intelligence and body, being the full optimization in all the six major aspects, namely moral, intelligence, ability, body, beauty and pleasure.

An often used example by some opponents of eugenics is the renowned scientist Hawking: You see, Hawking is a handicapped person but can be a scientist. Do we need to improve the gene? But have they ever asked Hawking: "Do you like to be a handicapped man?" If gene optimization can offer him a pair of healthy legs that can walk like flying, will he refuse it? Han Feizi was a genius and the book he wrote made First Emperor of Qin Dynasty strike the table and shout bravo. But he had the trouble of stammer. Did he wish to be a stammering man? If his speaking ability had been as great as his writing ability, maybe he would not have lost his life. People "produced" by the eugenics industry will be not only extremely intelligent, but also optimistic and outgoing, healthy and long-lived, kind and noble. Such people will have all kinds of noble virtues and peerless attributes, such as a kind character, a wise mind, a robust body, a tenacious willpower, a modest temperament, an optimistic mood and so on. To sum up, only by the eugenics industry can we provide a solid congenital foundation for overall development, so as to realize Marx' ideal and enable everyone to have a high quality in all aspects.

Eugenics Industry and the Individual.

A person's quality and fate depend on the integrated result of nature and nurture. The proportions of the two factors are not unalterable, but changeable in fact. Now, the proportion of postnatal education is greater, but along with the gradual application of gene technology for improvement of the congenital gene, the congenital factor's effect on a person's fate will become greater and greater, or even take the leading position. If you are born with a heart disease, or born with a deformity, or born with weak intelligence, what a life will you be faced with? On the contrary, after a eugenic operation and you have got a good gene so that you are much cleverer, much stronger, much prettier and much more optimistic, then, what a life will you spend? Needless to say, the eugenics industry is closely related to the individual's interest and fate.

Eugenics Industry and the Family.

The family is a cell of the society. All parents in the world hope their children will be excellent, and it is their most lasting and strongest wish. As far as the family is concerned, the hurt brought about by the child born with some genetic disease is indescribably cruel. Zhou Guoping's (Chinese writer) child Niuniu was diagnosed to have an incurable disease soon after her birth, and from then on, sadness and pain would intertwine his family forever. On the contrary, bringing a most excellent child through the eugenics industry will be the family's greatest interest and greatest blessing.

Eugenics Industry and the Nation.

A nation's fate consists in its children, and children are future citizens. We often hear such sayings "Football training should start from childhood", "Computer training should start from childhood", and so on. Every undertaking should start from childhood. Children's quality concerns the nation's prospect and the nation's future. Only when personal quality has changed fundamentally can the face of the country change by root. If we do not lift the citizens' quality, China can only continue to be the largest supplier of cheap labour in the world, and our dream to become a great power can never come true.

The eugenics industry will be the greatest foundation engineering for turning out more than one billion "great Chinese".

Eugenics Industry and Mankind.

The eugenics industry is mankind's challenge to God, being the greatest revolutionary breakthrough in human history of development. In all the past ages, mankind's progress was only shown in transformation of the outer world, but what is more fundamental and most important than transformation of the outer world is transformation of the internal world. The eugenics industry is a great undertaking of mankind's transformation of its own internal world. **Through the eugenics industry, mankind can dominate its own evolution, which will help mankind enter an epoch-marking stage of quantum leaps.**

Development of the human society is increasingly dependent on the driving force of high technology. High technology originates from high-quality people. The eugenics industry can turn out millions upon millions of super-quality people, thereby speeding up the development of the human society by 100 times, 1000 times or more, and most profoundly deciding all mankind' overall interest and overall fate.

All in all, the eugenics industry is greatly significant to the individual, the family, the nation and all mankind. It is great system engineering, being the key project of all key projects, and the foundation project of all the foundation projects. "The eugenics industry" can also be called "the eugenics engineering". In talking of the greatest engineering in human history, people often think of the "Manhattan Engineering" and the "Apollo Engineering". But the "eugenics engineering" is far more meaningful than the "Manhattan Engineering" and the "Apollo Engineering", as it concerns mankind's future all the more. It is not only decisive to the person's or the family's happiness, but also determinative to the development of the nation and mankind. It is most beneficial to the individual, the family and the country, so it will surely win the largest market.

The eugenics industry, as the core of the human-centered economy, needs the great market power for business operation, and needs generous investment from far-sighted entrepreneurs. Meanwhile, the whole society should build a supporting system for the eugenics industry, the government should also draw up corresponding development policies and measures, and organize all forces to speed up the research, so as to develop the eugenics industry and human-centered economy as fast as possible.

XXXI. OUTER SPACE INDUSTRY AND UNIVERSE ECONOMY

I think, mankind may face the danger of extinction before the end of this century unless we develop our living space in outer space.

—Hawking

The outer space industry is one of the important parts of the future economy. The eminent futurist Arvin Toffler became world famous because of his prediction of information economy and Internet economy as the "Third Wave". Now he regards outer space industry and outer space economy as the "Fourth Wave".

The outer space industry is a great and attractive industry, with huge economic interests and strategic interests. The tellurian overlord America plans to invest hundreds of billion US dollars in its ambitious outer space industry in the coming decade. Recently, America's NASA (National Aeronautics and Space Administration) drew up a draft entitled *How to Strengthen the Outer Space Industry Strategy*, preparing to develop the outer space industry on a large scale.

Outer Space Tourism

Since May 2001 when the Russian "Alliance-Y" rocket carrying the 60-odd-year American moneybag Dennis Tito flew to the outer space, who became the first space tourist in mankind's history, a space tourism wave has been started all over the world. People touring the outer space will increase very fast. It is a great progress in civilization. Just as the historian from NASA, Aleksej Rollander, said: *"Our space flight cause is just to serve all earthlings. If everyone can queue up to take a trip to the outer space some day, it will be too wonderful."*

At present, many corporations in the world have hit at the outer space tourist market, and are going all out to develop reusable rockets for that. Some companies have even accepted reservations in advance. Many people are fervent about outer space sightseeing. According to a survey of Englishmen, Japanese, Americans and Germans, most of them have shown a great interest in an outer space tour. Since the American billionaire Tito's successful outer space travel by the Russian spacecraft, the young moneybag from South Africa, Mark Shutler has also taken a space trip smoothly. Many big stars and zillionaires are longing for a space trip. Among them, there are James Cameron, director of "Terminator" and "Titanic"; Lance Bass, coryphaeus of the band "Super Boys"; the great beauty, Cindy Crawford, and so on. Mankind is just ushering in a brand-new era of outer space traveling.

Outer space tourism is so unconventional, full of attraction. By outer space tourism, you can taste the unique fun of living in the outer space totally different from the earth, can do space sightseeing in the "space hotel" (space station), can land on another star for exploration, can spend a holiday on the moon or Mars or farther stars, can do space walk while enjoying the sunrise and sunset in the outer space, can play special games there, such as space baseball, space golf, space bungee jump, and so on...

Outer space tourism can not only give people an unprecedented special experience, but also revolutionize their ideology and educate mankind.

From more than 400 km away, the three astronauts living in the international space station have seen what we cannot see on the earth and felt what we cannot feel on the earth, too. What the American astronaut Frank Culberts, the commander in the international space station, has seen about New York caused him to have an unprecedented love for the earth. Culbertson said: *"The space travel has opened my eyes greatly, so I no longer consider myself always right." "We should take a look at our earth from the universe. It is such a tiny and fragile world, just like a pebble. If we cannot learn to live in amity, what faces the world will only be a disaster."* From the vast outer space, have a look at the blue star breeding us mankind, we shall have a strong feeling of "Since we are in the same boat, we must help each other, advance and retreat together". If all people on the earth can take a trip to the outer space and have a good look from there, it will be a great education to mankind.

The expenses of outer space tourism will drop gradually. The present price is 20 million US dollars. We can forecast, one decade later, it may drop to 10,000 - 20,000 dollars, and finally it may drop to the level that every commoner can afford. Now in America there are 20-odd companies working hard to develop rocket aircraft or airships for outer space tour. By such a new airship, you have no need to go through any training, just as convenient as taking a plane. And, the price is much lower than what Tito paid, being about 20,000 US dollars. The tourists will have 3 hours for roaming in the earth orbit, and by the huge rotating cabin, they can enjoy the beautiful scenes of the earth and outer space. According to estimation, not long from now, those who wish to take a walk in the outer space may line up for that. According to a research firm's survey, 20% Americans like to spend their income of four years to take a 6-day space travel to the moon. Some experts estimate, even if it is 50,000 US dollars per capita, the outer space tourist market will increase by 60 billion US dollars each year, and moreover, "commoner-oriented outer space tourism" will become the major trend, so, the final picture will be like this, just as Isozaki Mitsuteru, chairman of Japan Rocket Association, predicts: *"In the 21ˢᵗ century, outer space tourism will become a pastime every ordinary person can enjoy."*

Outer space tourism will become a newly arising industry in the new century, with a huge profit potential. America, Russia, Japan and many other countries have thought high of the outer space tourist industry, and a space commercial war is on the verge of breaking out.

Space Agriculture and Space Industry

By utilizing the peculiar environment of outer space, we can develop new-typed agriculture, new-typed industry; by use of the special conditions of outer space, we can turn out all kinds of special foodstuff and special products, and industrialize and commercialize them.

According to scientists, the micro-gravitation, high-vacuum and intense-radiation environment of the outer space can cause many helpful variations of biologic cells and chromosomes. Hence, starting from the 1970's, mankind has been sending all kinds of biologic seed into the outer space via spacecraft, causing them to have beneficial property changes, thus, we have bred many new fine species with composite properties. So far, mankind has already done outer space breeding for more than 400 kinds of crop, such as rice, wheat, maize, sorghum, cotton, watermelon and so on, and the results are amazingly good. Under such conditions, not only the seed of space rice and wheat have a shortened growth period, but also their average yield per acre is 20% higher than the seed on the earth, and moreover, their protein content has increased by 8% to 20%. Have you ever seen a 2-meter-long turnip, or a one-kilogram tomato? Why are these plants so big? It turns out that they have come from the outer space. In comparison with common vegetables, "space vegetables" have a high nutrient content, are easier to grow, have an obvious stronger capacity against cold, insects, virus, drought, water-logging, and diseases, and have a greater yield per acre. For example, "No. 1 space tomato" has a yield per acre as high as 10,000 - 20,000 kilograms. So far, space breeding has already brought about 4 billion-ton space grain for mankind. So, it can be seen that, by utilizing the specific environment of outer space, we can revolutionize our agriculture, work out breed varieties that are better in quality and more high yielding. So, the grain problem puzzling mankind can be solved this way.

Space agriculture contains many great business opportunities, and is increasingly growing into industrialization like the space industry. Mankind has already moved some momentous industrial projects from the earth onto some star in space. The outer space environment is very special, characterized by micro-gravitation, super cleanness, high vacuum, deep cooling, strong radioactivity and so on. It is hard for any natural or man-made environment on the earth to have such characteristics. Hence, by development of space industry, we could refine or manufacture industrial products with very high quality in purity, shape, strength and life, which is hardly possible for the traditional industries and manufacturing industry on the earth. **So, a great many scientists have regarded outer space as an ideal production base for getting new industrial materials and new bio products.** "The rarer a thing is, the more it is worth" and "The finer a thing is, the more it is worth" are general principles in the economic field. At first, space products "are valued because of their rarity", and along with the gradual drop of the cost, they will be "valued because of their fineness". In the whole process from the incubation period of the industry to the development stage and the mature period, they will result in generous profits. So far, Russian astronauts have finished 14500 experiments on manufacturing of industrial products in their space stations. America has listed scores of products that can be produced in the outer space. According to statistics, industries related to the outer space grow at the speed of 20% each year, and the profit of international space industries has exceeded 80 billion US dollars.

Thus, it can be seen that both space agriculture and space industry will be great industries with huge commercial value and economic potentials.

Energy crisis and environmental pollution are great problems confronted by mankind. Space industry can show its great advantages in this respect.

For a long time, people have been dreaming to go the outer space to collect solar energy, transmit it to the earth, turning it into electric power, so as to solve the energy crisis of mankind.

Asking Energy from Outer Space

In the near future, the first solar power plant constructed jointly by US National Aeronautics and Space Administration and Ministry of Energy will be assembled in space and begin to supply power to the earth soon.

Solar electric power generation has an unparalleled superiority. The solar energy received by the earth only accounts for 1/2,000,000,000 of all the energy emitted by the surface of the sun, and this energy is equivalent to 30,000 - 40,000 times of the total energy needed by the globe. So, it can be said an inexhaustible source. Moreover, utilizing solar energy will neither result in the "greenhouse effect" nor changes the global climate, or will cause environmental pollution.

The chief way to get energy from the outer space is like this: Use the photoelectricity board, maser, receiver, solar satellite and its transmitting apparatus, etc. to collect solar energy and convert it into electric energy, then convert it into microwave so as to transmit it onto the earth. The ground receiving station uses a purpose-made metal sheet to convert the microwave into electric current, then uses a rectifier to change the alternating current into direct current. According to measurement, a standard receiving station's electricity generating power may reach 5 billion KW, equivalent to the generating capacity of 5 large nuclear power plants, so the efficiency is quite high. In future, we can also do mining on the moon to get the silicon and aluminum used for making the photoelectricity board and the shelf, which will greatly reduce the weight of goods to be sent into the orbit from the earth, so, the cost of solar electric power generation will be further reduced. Thus, it will be easier to popularize and industrialize it.

Asking Homesteads from Outer Space

There is tremendous space outside the earth.

In the immensurable cosmic space, the earth is only one out of the 8 planets of the solar system. The sun is only an average star among the 100 billion stars or so in the Milky Way galaxy. The Milky Way galaxy is only one of the 100 billion galaxies or so in the universe. In the vast universe, the earth is as tiny as a mote indeed. If the earth is no longer suitable for mankind's living some day, will mankind only await its doom?

As the ancient Chinese saying goes, *"A wily rabbit has three burrows"*. Today, mankind has only one homestead, the earth. If the earth goes wrong some day, everything will perish! Besides, the earth is never a bed of roses, and we face all sorts of known hazards and all sorts of unknown hazards here.

From now on, we must take precautions, to find some more "caves" for mankind. Now, American Space Administration is quickening its steps of exploration, so as to enable mankind to land on Mars in 2015. Recently, at the seminar held in California on the physical and biologic characteristics suitable for human living on Mars, some scientists said, if the atmosphere of Mars is full of ultra greenhouse gases, it will become a star suitable for human living. As the story goes, we have already found some stars enwrapped with air outside the solar system, where life may exist, or even there may be an environment suitable for human living. By the present technology, the spaceship can help us realize our dream to live in the outer space. In addition, we may also use the genetic engineering to "improve" mankind's gene so that humans can live on other stars. Maybe many people just regard settling in another star as a science fiction, but just think, many stories in science fiction have already been turned into reality. Tomorrow, mankind should be able to live in the outer space.

Famous scientist Hawking also goes along with the idea of emigration to the outer space: *"I think, mankind may face the danger of extinction unless we develop our living space in outer space. When we only rely on one star (earth), there may be too many tragedies that cannot be foreseen indeed. Of course, we don't have to be pessimistic; after all we can go to other stars for survival."* By marching into the outer space, mankind can open up more living space. What we shall possess will not be just one "earth", but also the second earth, the third earth, and the Nth earth.

Marching Towards Universe Economy

More than 20 years ago, Wernher von Braun, one of the originators of the modern rocket, predicted: *"The 21st century will be a century for scientific activities and commercial activities in the outer space, and a century for interstellar navigation and for beginning to establish permanent footholds for mankind outside the mother star earth."* That is to say, the 21st century is a century for the outer space industry. And, in the 21st century, we shall also usher in an era of universe economy.

The universe is a limitlessly vast world. The universe

economy takes the space industry as the core, covering all kinds of economical projects carried out by utilizing extraterrestrial resources. In the vast universe with limitless resources, there are boundless potentials and limitless business opportunities. It will not only become an ideal super tourist base and super production base, but also become a super research base, super military base... The universe economy will free us from spatial limit and spatial bondage in the real sense. We shall be able to open up new military domains and new economic domains in numerous new homesteads.

The exploitation of the cosmos technology will arouse the attention of more and more countries. For example, the competition between America and Russia in aerospace technology is still very keen. The outer space or the universe can not only be commercially utilized, but is also closely related to national security. If a country wants to get the commanding point for future development, it must go all out to develop the space industry and universe economy that are closely related to national security and international competitive power.

Establishing Contact with High-intelligence Lives

Space industry and universe economy also include establishing contact with high-intelligence lives. **In the vast universe, we cannot rule out the possibility of existence of some other high-intelligence lives. Maybe they have already conquered many hard problems puzzling mankind, including living forever and becoming immortal.** If mankind can get their help, and get the secret recipe and technology for living forever and becoming immortals, we shall certainly be able to speed up mankind's steps of evolution into immortals. Hence, establishing contact with high-intelligence lives is extremely valuable to mankind.

Scientists of all countries are very zealous in their endeavor to search after high-intelligence lives in the outer space. American scientists have started up a new program to look for extraterrestrial intelligence, and got great financial aid from the two millionaires of Microsoft. One is Paul Allen, one of the originators of Microsoft; the other is Myhrvold, former Chief Officer of Technology of Microsoft. The former contributed 11.5 million US dollars, and the latter donated 1 million US dollars, to American Silicon Valley's "Institute on Searching after Extraterrestrial Intelligence". American institute for the "Program of

Searching for Extraterrestrial Civilization" has begun to search for laser signals sent out by extraterrestrial intelligence, and have already checked about 300 stars so far. The United Nations is also planning to spend 1 billion US dollars to search after extraterrestrial intelligence, that is, to work out the largest radiotelescope in the human world, which will cost 1 billion dollars at least. China is also preparing to construct the largest radio antenna in the world in Guizhou, to monitor the cosmic radio waves from the outer space. It can also send electric wave signals, so, our "extraterrestrial friends" tens of thousands of light years away may receive greetings from China. This "FAST" is Chinese scientists' first formal searching after extraterrestrial intelligence.

The reason for spending so much money in searching after extraterrestrial intelligence is just that it has a great value for our mankind's future.

Solving Problems That Cannot Be Solved on the Earth

By utilizing the universe or the outer space, we may do a lot of things that cannot be done on the earth and solve numerous hard problems that cannot be solved on the earth. The noted futurist Toffler said: *"The outer space may help us solve a great many problems impossible to be solved on the earth or even it may change our understanding of things by the root."* Toffler's company is just working together with NASA, chiefly doing research on bioastronautics. Through the cosmic station, we may learn about some biologic changes. In the outer space, owing to the very small gravitation, the pressure on the cell is also small, very favorable for cell experiment, so, the results of medical science produced this way will be quite different from the experimental results on the earth. That is to say, in the sky, we can do what we cannot do on the earth. In the universe, we can do all kinds of physical tests and biologic tests, develop space materials, space medicines, space electronic technology, space biotechnology... After the space vegetables and space melons and fruits come to the table, space drugs will also step into our drugstores. At present, there are already many pharmaceutical enterprises hoping to send their biological samples into the outer space for a tour. Any medicine developed by using the outer space conditions will have features incomparable to anything on the earth, so, we can make "wonder drugs" and "godly drugs".

The law changes with the change in the conditions. Cosmic conditions are completely different from telluric conditions; hence, in the god-given peculiar conditions, we might solve many hard problems. We shall borrow the heavenly power to work wonders. Using the space environment, we can do better study on life and longevity. For instance, some scientists have begun their bold assumption about childbearing in the outer space. They are studying on impregnation and giving birth to children in space, so as to get "space humans" with super intelligence and super physical strength...

By using the macrocosm, we can work out all sorts of wonderful high and new technologies. The macrocosm has also left a huge space for scientific workers to give play to their imaginative power and creativity, giving us a boundless commercial space and uncountable business opportunities...

XXXII. DEVELOPMENT MODE OF FUTURE ECONOMY

From global "heart pooling" to global "intelligence pooling" and global fund pooling, we may win over all the people in the world, all the elite of the world and all the capital of the world.

Financing: Staking for the Future

As the ancient saying goes, *"Food and fodder should go before troops and horses"*. In the past, the war depended on food and fodder, but the war in the present marketplace is a war for capital. The capital is the lifeline of economic activities. Many men of insight agree to such an idea: We will be in a future framed by the three major economic forces, namely; multinational capital, national capital and private capital.

Financing is of vital importance to future economy. All such industries as the 1000-year longevity industry, the god-sleeping industry, the ecstasy industry, the eugenics industry and so on, are large projects with high risks and high return, so, they need large-scaled operation and ample investment. The thinking mode for operation of these industries is totally different from the old way of thinking characterized by primitive accumulation and keeping expenditures within the limits of income, and the management philosophy of placing hope on self accumulation for winning the world is destined to fail in the future economy. Only by dint of the support of rich capital can we get brilliant achievements in life economy, eternal life economy, pleasure economy, human-centered economy and universe economy.

Usually, the sources of fund fall into the following channels: state allocation, bank loan, listed financing, and venture investment. Venture investment can be called the "engine" for development of high-grade technology industries. Traditional investment, such as banking fund, bonds and so on, usually prefers the rich to the poor, hence, projects that need fund most and have the highest rate of return on investment usually cannot get the loan because of its high risk. In contrast, venture investment only thinks high of newly arising industries characterized by a great growing potential, and, in order to get the huge return in future, it is ready to take a great risk.

In the highly fluctuating capital market, the 1000-year longevity industry, the god-sleeping industry, the ecstasy industry, the eugenics industry and so on are just suitable for venture investment, so they are most promising to get venture investment.

Venture investment just prefers investing money in hi-tech industries. Mr. Cheng Siwei, Chairman of the Central Committee of China Democratic National Construction Association, and the earliest advocate of venture investment, said: *"Venture investment means such a trade investment behavior: putting funds in the area of high and new technology development with a great risk of failure, in the hope of getting a high capital return after the project succeeds."* No industry or project of the future economy represented by life economy can get away from high technology, any of them is a top new science and technology industry and hence, they are most suitable for venture investment.

Venture investors just prefer investing money in the future. They are good at bottom pouring for the long term, so, people call venture investors "people going ahead of the times". Life economy, eternal life economy and so on are economies facing the future, representing the future development megatrends, which just suits venture investment.

Venture investors are most willing to put funds into green field projects with great potential profit. Once the project prospers, the venture investment may get scores of times or even thousands of times of return.

Such areas as life economy and human-centered economy are not only high-tech economy, but also high-humanity economy, and once any project hits a success, it will trigger a huge market demand, and its profit return will be hundreds or thousands of times.

Such future economies as life economy and human-centered economy give mankind a wonderful future to use high technology to realize high humanity, which is enough to impel, attract or encourage more and more investors and financial institutions, so, the financing will go smoothly, getting the sufficient funds needed by the 1000-year longevity industry, the god-sleeping industry, the ecstasy industry and so on, ensuring their over-speed development.

Intelligence Pooling is More Important than Fund Pooling

People are usually very thirsty for capital, as if craving for the manna, so, it is not hard to understand the great strategic position of fund pooling. However, perhaps not many people are not really aware, Intelligence pooling is more crucial than fund pooling.

American economist, Schurz, has found: **"What constitutes the wealth of the present countries with high incomes? Chiefly human ability."** Thus, he has put forward the concept "Human Capital", and won the Nobel Prize in economics hence.

Capital falls into two kinds: one is physical capital, and the other is human capital. In comparison with physical capital, human capital is more important. Human capital may also be further subdivided into physical strength capital and intelligence capital. Intelligence capital is the most central, most essential and most valuable part of human capital. So, it is not hard for us to reach the conclusion that in comparison with fund pooling, intelligence pooling is more important.

From ancient time till now, any great accomplishment ha depended on intelligence pooling. In the period of Three Kingdoms (220-280 AD), all the three kings, Cao Cao, Liu Bei and Sun Quan, knew this truth. Cao Cao's maxim is *"I use the wisdom of the world, guided by Dao (the Way or Law of the universe), so I am bound to triumph everywhere"*. Liu Bei's maxim is *"Anyone who wishes to achieve something great must take human as the core"*. Taking human beings as the core just means taking wisdom as the core. Liu Bei visited the cottage thrice in succession to ask Zhuge Liang to help him conquer the world, and when he got him at last, he felt "like a fish that has got water". Sun Quan's maxim is "*If you can use the forces of the people, you will be invincible under the sun. If you can use the wisdom of the people, you will have no need to fear saints*".

The times have changed, and the battleground has changed, too, but what remains unchanged forever is the contending for brainpower. The three major multinational magnates have already come to our home to contend for wisdom. Microsoft has spread its net to Zhongguancun of Beijing, setting up its China-based Academy and planning to invest 80 million US dollars in the academy within six years. Intel has regarded China as its own "new brain", establishing the China-based Intel Laboratory to exploit the global market in China. Bell Laboratory's CEO said, "*We have chosen Beijing as our head office besides America*", borrowing China' brain directly.

In the future, high technology will become the most crucial production factor in economic development. Life economy is a high-tech economy, and is also a high-intelligence economy. Hence, the position of intelligence pooling is higher than any economy, and intelligence pooling is playing a greater decisive role in project breakthrough and enterprise development.

Since intelligence pooling has braught a lot of "most clever brains" together, what can not be accomplished? Moreover, in fund pooling, we must also do intelligence pooling first. **Only with highly intelligent personnel can we expand our financing channels better, expand the total amount of fund pooling and raise the efficiency of fund pooling further.** While making a decision many venture investors specially check whether you have a high-intelligence team or not.

"*So long as I have done fund pooling, I can get intelligence: with money, what can I not buy?*" There are many people holding such an idea. As the saying goes, "*He who has wealth speaks louder than others*". Of course, fund holding may contribute a great deal to intelligence pooling. Without money, it will be hard to hold any wise persons. However, in comparison with "fund", "intelligence" has its own characteristics, so, **"intelligence buying" does not equal to "intelligence pooling".**

The great difference between "intelligence" and "fund" is like this:

Monetary capital can transfer from your hand to another man's hand, or flow from your purse into another man's purse. But intelligence can only remain in the individual's brain, being completely possessed by the person himself. No matter what means others may use, they cannot take away his intelligence, and his intelligence can only belong to him, until the end of his life. So, it is not hard to see: In financing, so long as you have moved the fund into your project or enterprise through a transaction, you have succeeded completely. But intelligence pooling is not so simple. Attracting talented people to join your enterprise (so that you may use the intelligence of others legally within a certain time) is only the first step in intelligence pooling.

In intelligence pooling, of course we hope to get high-intelligence people who are both "diligent" and "loyal". However, a wise person can be a "workaholic", doing creative work madly, but may also "live in the Cao camp but with his heart in the Han camp", and not contribute any intelligence. The reason is so simple: only the holder of the intelligence can devaluate his own intelligence or even shut it off, making it into nothing. If a person is forced to labor, he will repulse from the heart, so, he will pretend ignorance, work to rule, or even watch his\her time to do destruction. To let people be both "diligent" and "loyal", only using money is not enough. What means should we use? Offer him shares, which is equivalent to giving him hope, and paying him a long-term rich profit. Offer him power, so that he can put his intelligence to good use to meet his demand for self-realization. Moreover, we have to meet his other psychological needs. For instance, when an enterprise invited an able person, in order to retain him, the enterprise used his name to name the technology building when they built it, which moved him so much that he said he would devote his wisdom to this enterprise all his life. In addition, whether the employee is both diligent and loyal, has much to do with the personal charisma of the leader of the enterprise. Whether the high-intelligence people are willing to accept the leadership, and what kind of people's leadership they will accept, concerns the result of the intelligence pooling directly.

In intelligence pooling, the emphasis lies in blending human intelligence into the enterprise or the project. Human intelligence is different from physical strength as physical strength can be controlled just by transaction or enslaving whereas intelligence can be called into play only by encouraging and exciting. Sometimes, even if you think hard all day long, it is still hard for inspiration to come out if you do not have a good mood or state of mind. If you just buy the wise people with a fat salary, but you lack an agreeable free working platform and lack a good enterprise system and corporate culture to stimulate or guide them, their wisdom may not be of use. Why is "even the air of Silicon Valley full of originality"? Silicon Valley allows failure more than any other place, and it protects failure, so that people's innovative impetus can be protected and their intelligence can be kindled to the best effect. On the contrary, a corporate culture that disallows failure and despises failure may only make the most intelligent personnel keep on the straight and narrow, and their talent can only be submerged in a mediocre environment...

We must also realize that intelligence pooling is not a temporary act, but a long-term strategy, that is, we must keep a lasting magnetism to high-intelligence people. Intelligence is a private property and may run away itself. In intelligence pooling, we must learn to "treasure this running capital". According to an investigation, of the people with a yearly salary of 100,000 Yuan in Shenzhen, about 30% wish for job hopping. Even 100,000 Yuan cannot hold them. Are they for money? No. The chief cause is only wants for change of an undertaking platform that may be more favorable for their self-realization and display of their abilities. So, it can be seen that we cannot only depend on money in intelligence pooling.

Intelligence pooling is an alliance and mergence of intelligence, that is, we cannot only rely on some single man's wisdom to win the world. As far as the enterprise is concerned, the work team's intelligence quotient is more important than the individuals' intelligence quotient. If every one of the able persons employed has a high intelligence, but their high intelligence is used for internal friction, that is, for intriguing with each other inside the enterprise, it will be just against the purpose of intelligence pooling. Intelligence pooling is to blend all people's intelligence together, blend it together with the enterprise, letting it belong to the enterprise or the collective, so as to build up a wise work group.

In intelligence pooling, we should also make good use of the magical power of high technology, let human intelligence combine with the intelligent network and artificial intelligence, realize a large-scaled man-machine integration and man-machine interaction and realize intelligence explosion and intelligence revolution. This is the great feature of the future economy, and also the intelligence pooling prospect the future economy has been pursuing by all means.

Heart Pooling Is the Base

Fund pooling is important, but intelligence pooling is even more important. However, whether fund pooling or intelligence pooling, we must take "heart pooling" as the base.

What is heart pooling?

Heart pooling means getting all people's mind, will and wish together. That is to say, to form a common ideal, aspiration and goal, have a common faith, belief, philosophy, concept of value and have a common thought, wish and aim. In other words, heart pooling means building a shared vision, making everyone in the enterprise think alike, and melt many people's "individual ideal" into the "common ideal".

There is nothing more powerful than the will of the people. The key to success in everything lies in striving after the will of the people, seizing the will of the people, changing the will of the people and winning the will of the people.

Only by heart pooling can we succeed in fund pooling, getting ample money. Originator of the web site "Alibaba", Ma Yun, did not earn a cent at first, but why did many international zaibatsus and venture investors vie with each other to throw money to him? For his agitative speech attracted the investors, so they were willing to untie their purse strings for him. Not only winning capital depends on heart pooling, but attracting able people also depends on heart pooling. After listening to his speech, a Harvard's MBA expressed his wish to join in at once. Although the reward he got was not even enough for making a phone call to his girl friend abroad, he was most willing to do so. Ma Yun said: *"Alibaba does not depend on a fat salary to attract able people, but depends on the enterprise' "nine major concepts of value", and on the "thought unification". Isn't "thought unification" just the same as heart pooling? If you have conveyed the confidence in the future undertaking to the investors and able people, and conquered their mind, you will be able to conquer everything ahead."*

The ups and downs of the capital market are actually decided chiefly by the investment confidence. Junior Superman Li Zekai could earn 30 billion Yuan a day, which was even more than the income of his father Major Superman in all his lifetime. It was just because his "Digital Harbor" won the will of the people, all the investors felt confident in Junior Superman, and in the future of his enterprise.

Heart pooling is not only very important in attracting capital, but also more important in intelligence pooling, that is, alluring able people and their mind. Liu Bei only emerged from obscurity as a man selling straw sandals and weaving straw mats, so he had no way to use high posts and fat salary to tempt people, like Cao Cao, and what he could rely on was just heart pooling. Through the "sworn brothers" relationship, he won the hearts and ability of Guan Yu and Zhang Fei who were famous for their great skills in martial art. What he won was people's hearts, but Cao Cao failed in winning over Guan Yu despite his all kinds of efforts, such as "one small feast every three days and one big feast every five days", treasured horse and beautiful girls, high posts and fat salary. Xu Shu, as famous as Zhuge Liang in wisdom, was determined not to offer any advice all his life although he was deceived into serving Cao Cao. Liu Bei's greatest intelligence pooling act in his life was his winning Zhuge Liang as his military counselor, after "visiting Zhuge Liang's cottage thrice in succession" despite all troubles. From heart pooling to intelligence pooling and from intelligence pooling to conquest of the world: That is just the root cause of Liu Bei's success in winning one third of China from a humble origin.

A large joint-venture enterprise in Shanghai with ten thousand employees invited an American managerial expert for its training work. Upon mounting onto the platform, this expert drew four curving red lines on the screen of the projector. When people were wondering what these red lines meant, the expert said: *"These four curving lines are the routes of the Long March of the Red Army, which was a miracle in the world! ... At that time, the Red Army's life was indescribably hard, but no one demanded an allowance, and they defeated the powerful enemy at last. What did they rely on? They just relied on a belief and a spirit!"* Chinese Communist Party's wonder of using the weak to defeat the strong and using the small to conquer the big, resulted chiefly from the heart pooling chiefly. Even in the hardest time with the least prospect of victory, in the poorest and hardest conditions, they could make all the people in the Party and the Army work with one heart and one mind, making all the soldiers believe in their future ideal, so as to struggle bravely for their beautiful common ideal. If any enterprise can reach such a state, what ideal cannot be realized?

As the ancient saying goes: Those who have won public support will win the world! Today, this "world" is the market, the vast market. In heart pooling, we should not only try to win the hearts of the employees, the middle and high-leveled leaders, the shareholders, the investors and the cooperators, making all of them work with one heart and one mind, but also strive to win the hearts of the customers and the public, so that the general public, the customers may have a good opinion on the enterprise' projects, products and services, ready to accept them.

Life economy and human-centered economy have natural advantages in heart pooling. The future economy values culture more than ordinary economic forms. Life culture and life economy interact with each other. Life culture will win the will of the people, give people new concepts and new beliefs and help them build up new ideals, so that life economy may do heart pooling more effectively. Especially, the human-centered culture can help the heart pooling of human-centered economy greatly.

In the future, heart pooling will be carried out via advanced means and tools, namely by making full use of the Internet, we can do heart pooling throughout the world. By dint of the magical power of the Internet, we may break through restrictions in geographical location, national boundary or spatial-temporal limit to exert a cultural influence, break the government or news media's monopoly in information, so as to make information release, information transmission and information acquisition very easy. An ordinary middle school pupil can make his own web site so that readers from another side of the earth may read his articles just by a click of the mouse, and be influenced by his ideas this way.

Future economic forms such as life economy, human-centered economy and so on, can take the greatest advantage of the Internet in heart pooling, as they have the global mental foundation for heart pooling. Life economy is rooted in mankind's common needs and common humanities, as the thirst for life has transcended the restrictions in race, skin color, nationality, national boundary, age, sex, and occupation. With this widest humanity foundation and mentality foundation, the

Internet's heart pooling effect can be called into full play. Through this powerful and free information release platform, we can spread the earth-shaking and amazing life culture to people all over the world in the shortest time and at the highest efficiency, so that we may find out people with the common thought, common aspiration and common goal as many as possible, get supporters and appreciators from every corner of the earth, win the hearts of all people in the world, and build up the vastest market.

Famous Chinese merchant Hu Xueyan had a well-known saying: How great one man's business can become, chiefly depends on how great his vision is. If his vision is in Hangzhou, he can only do business within Hangzhou. If his vision is in China, he can do business all over China. If his vision is in the whole world, he can do business all over the world. Hu Xueyan had such a brilliant viewpoint even in those days of old China, and it is a pity that he did not live in the Internet age. The Internet has offered us chances and conditions open to the whole world. The future economy—(life economy, eternal life economy, pleasure economy, human-centered economy and universe economy) has both the ideal tool of Internet for its global heart pooling, and the humanity and mentality foundation for global heart pooling.

From global heart pooling to global intelligence pooling and global fund pooling, we may win over all the people in the world, all the elite of the world and all the capital of the world. Once we have achieved the great mergence of global hearts, global intelligence and global capital, the fine future of life economy and human-centered economy will come soon.

Future economy will take the "three-in-one mode" of heart pooling, intelligence pooling and fund pooling, and all the three will grow together and promote each other. Heart pooling is the switch: First of all, we must revolutionize the ideology and win the will of the people. Once we have won the will of the people, we may go smoothly for intelligence pooling and fund pooling. Intelligence pooling is the core. The future economy is a high-technology and high-intelligence economy, chiefly relying on intelligence for breakthrough. Once we have succeeded in intelligence pooling, we may do well in heart pooling and fund pooling. Fund pooling is the guarantee: Only a high investment can lead to great development of high technology. With ample funds, we may also do better in heart pooling and intelligence pooling.

In using the "three-in-one mode of mutual pooling", we should lay special stress on the interest drive.

The reason why our world is both full of dangers and full of splendors is just it is a world governed by lots and lots of interest. Seeking maximization of interest is a perpetual motive force for enterprise development.

As the ancient saying goes, *"Everyone is busy going and coming for interest"*. We have a new saying now: *"There is no eternal friend, but only eternal interest"*. An enterprise is an organization striving for interest, and is a typical interest group. Russell said, barely talking about interest has already approached the truth. Everyone tends toward benefit and avoids harm, so, the interest lever is the universal lever driving people forward.

> The future economy will be driven chiefly by three kinds of interest: The first is material interest, the second is spiritual interest and the third is life interest.

Material interest chiefly refers to remuneration, such as salary, welfare, allowance, bonus, share option and so on.

Spiritual interest refers to the sense of competence in work, sense of accomplishment, sense of duty, sense of mission, sense of sublimity, sense of challenge and sense of innovation, including the degree of being valued within the enterprise, the influence power in the work group, the condition for personal growth and self-realization, and so on. Spiritual interest sometimes has a greater driving effect even than material interest. Money only has an alluring power, but a sacred undertaking has a cohesive power. Even if you have got a reward of 1 million dollars, if you have no stage for displaying your talent in this corporation and you are just an ornament like a flower vase, you may also flounce out.

What is more important than material interest and spiritual interest is that life interest that is characteristic of the life economy. Economics is in fact a science on minimization of life cost and maximization of pleasure and happiness. Mankind's all economic activities concern pay-out of human life cost, and all economic activities are for saving life cost. Just as Marx said: *"All saving, in the last analysis, is saving of time"*. And *"time is just life"*. The traditional economy is meant to make the best use of the limited life cost whereas life economy is chiefly not to save life cost passively, but to prolong life actively, that is, it can give the greatest interest, life interest, which cannot be obtained from any other economy.

With the triple driving by material interest, spirit interest and life interest, we may achieve a better effect in heart pooling, intelligence pooling and fund pooling.

The "three-in-one mode" of heart pooling, intelligence pooling and fund pooling can not get away from the best organization form of future economy.

Future economy's business organizations will be both free complexes and interest commonwealths.

The "free complex" is Marx and Engels' fine conception in their design of the communist society. The best future enterprise form should be the ideal prospect of communism, the free complex, which is unexpected to many people.

In human nature, there is a strong thirst for freedom, and only freedom can kindle intelligence. For a high-intelligence person, "Give me freedom, or death!" Freedom means "the possibility for everyone to do anything by the most suitable means according to his own judgment and rationality", then, a free complex just means using one's own judgment and reason to make the decision to unite together.

Only a high freedom can kindle people's inspiration, trigger people's creativity and make people's devotion and fervor surpass what is expected in the strict supervision of traditional enterprises. In a traditional enterprise, the following phenomena would be hard for people to tolerate. Technical personnel are not restricted by the business hours: when they like to come, they come; when they like to go, they go. Some people do not keep their clothes tidy, some put their legs on the table... But all these will be encouraged, the working mode will get freer and freer, and every worker will be a free individual. Their joining is based on the common idea, common ideal and common pursuance; they come together because of their common interest, aspiration and fervor in life economy and human-centered economy. All of them are willing to devote their own time, youth, talent and wisdom to this great undertaking as they think it is worthwhile.

Meanwhile, future enterprises will also be interest commonwealths.

Practice has proved that the more an enterprise becomes an interest commonwealth, the more possible for it to grow fast. More than 90% of listed companies of America have practiced the employee stock ownership plan. In comparison with the time before practicing the employee stock ownership plan, the labor productivity has risen by 1/3, the average rate of profit has risen by 50%, and the employees' income has become higher by 25% to 60%. Bill Gates' Microsoft has more than 27,000 workers, and 80% of them have stock option. Bill Gates' idea of employee shareholding is not only a welfare measure, but is also a competitive one, as the obtaining of the stock option is based on the employee's contribution to the Company. It is just for this reason that working in Microsoft has become more challenging and more attractive. "You have never seen such vigor, ability and enthusiasm from people elsewhere." Hence, Microsoft has grown so fast, and when Microsoft made a great success and Gates became the richest man in the world, his Company had also turned out many millionaires.

A scholar has described employee shareholding as "gold in four aspects". Gold handcuffs, golden dream, golden handshake and gold bowl. Gold handcuffs mean that the rich income brought about by enterprise development "handcuffs" (locks) the employees so that they can be willing to contribute to the development of the enterprise. Golden dream describes the stock option system as a long-term reward. Golden handshake refers to the fact that all the managerial personnel and common workers fight in the same "trench". Gold bowl refers to the fact that the workers treasure their own work much more. The gold handcuffs, golden dream, golden handshake and gold bowl melt the interest of everyone in the enterprise together, let them stand together regardless of any situation, thus, the enterprise may develop at a super speed, and the employees themselves may also get huge interest.

The appearance of the interest commonwealth has challenged the traditional economics. Judging from the traditional economics, profit is the surplus of the salary, and the salary and the profit are in contradiction to each other. Such a contradiction has not only resulted in the tense labor relation, but also led to class conflict and hostility in the whole society. Yet, the appearance of employee shareholding has transcended the interest relationship in the traditional economics, making interest contradiction into interest harmony.

The future enterprise will be both an interest commonwealth and an undertaking commonwealth, or to be more exact, it will be a superposition of undertaking commonwealth and interest commonwealth. The common undertaking represents the common interest that integrates material interest and spiritual interest together, integrates immediate interest and long-term interest together, combines the individual's interest and the collective's interest into one, and combines the enterprise' interest and the society's interest into one. When such a common interest contained in the undertaking of the enterprise is accepted by every member, this enterprise will have a very stable base, so it can build up a "Spartak square matrix" with a great cohesive strength.

We regard the "free complex" as the software system for the construction of the future enterprise' work team, and consider the "interest commonwealth" as the hardware system. Only when they combine closely with each other can we build up the most perfect and most harmonious work team. **This organizational structure of a free complex and an interest commonwealth will give the enterprise the two most needed forces: "vital force" and "composite force".** The free complex and the interest commonwealth will be the sources of both vital force and composite force of the enterprise.

The future high-technology and high-humanity enterprise will not only build up a "free complex" and "interest commonwealth" within itself, but also seek "free joining" and "common benefit" with dealers, customers, the public and the society, striving for mutual credit, mutual benefit, mutual participation and joint development, and trying to get the most extensive cooperation and harmony, and establish an extremely wide and stable free complex and interest commonwealth. Only on the basis of heart pooling, intelligence pooling and fund pooling can we establish a good free complex and interest commonwealth. Meanwhile, so long as we have set up a wide and solid free complex and interest commonwealth, we will be able to do the best heart pooling, intelligence pooling and fund pooling. Once all people in the world have joined into a solid free complex and interest commonwealth, we shall be able to do the best heart pooling, intelligence pooling and fund pooling globally.

Export, investment and consumption are usually called "the three major impetuses of economic growth", or "the troika". Heart pooling, intelligence pooling and fund pooling will become the "three trains" of the future economy, as they will work together, becoming the inexhaustible impetus for the flying development of future economy.

All in all, using the "three-in-one mode of heart pooling, intelligence pooling and fund pooling" and taking the road of combination of free complex and interest commonwealth, will be the inevitable choice and power source of future enterprises driven by the billow of high technology and high humanity.

XXXIII. TRANSCENDING THE TRADITIONAL INDUSTRY THEORY

In the 21ˢᵗ century, every country, especially our China that has never taken the leading position after the agricultural economy, is faced with a more profound industrial revolution, and a new great industrial structure adjustment is awaiting us, which is full of opportunities and challenges.

On September 26, 1792 in the name of congratulating on Emperor Qianlong's birthday, Great Britain sent a great diplomatic corps headed by Lord Macartney to "The Central Country", in five ships. They sailed for 10 months and reached Dagu Port of Tianjin at the end of July 1793. On September 14, they presented themselves to Emperor Qianlong, and submitted the letters of credence and all kinds of industrial products to the emperor. However, China posed itself as a great central state then, so it did not take this unusual visit seriously, paying little attention to the industrial products from Britain, as it never realized the leading role of industry in the future economy.

Chinese economy used to rank No. 1 in the world, with GNP accounting for one third of the world, but then it slid gradually to the verge of backwardness. Of course, there were many causes, but among others, a very important one was: from that moment in the Qianlong Period, China missed the opportunity to catch up with the industrial revolution and industrial structure transformation that opened up the new era of mankind. For so many centuries, China had been far ahead of the West, but it was cast behind out and away by the world just because it failed to transform from agriculture to industrial revolution. This great strategic misplay of that time has put China into disgrace and disaster for more than one century.

A backward industrial structure may lead a very advanced economy towards decline, but an advanced industrial structure may lead a backward economy to prosperity.

Let's turn our eyes to the 1980's. Japanese economy shocked America

From the Heyday of Qianlong to Silicon Valley Wonder

violently, plunging America into a great panic, and every American was worrying about "Japan's possible purchase of America". For a time, economists all over the world were lost in the study on the Japanese mode, crying up Japan's wonder. However, in a short period of one decade, everything changed. Japan's economy got into a period of wandering and sagging, whereas America worked a "new economy" wonder lasting as long as 110 months, which not only released the terrible alarm of that year, but also consolidated its throne as the overlord of global economy. What helped US economy turn defeat into victory was just their industrial structure transformation. US economy underwent the greatest industrial transformation in the century. Manufacturing industry has shifted to hi-tech industry, the former manufacturing industry has almost degenerated as an appendage of economy, whereas the high-grade technology industries represented by network technology and information technology have become the greatest powers boosting the sustained growth of American new economy. "The Silicon Valley Wonder" has replaced "The Japanese Wonder". Yet, Silicon Valley is merely a valley 48 km long and 16 km wide in America, even smaller than a county, but its annual production value is equivalent to 1/4 of China's GDP! The wealth in astronomical figures generated from it has shocked the whole world, thus, Silicon Valley has become a target of worship and imitation, hence, Singapore, Japan, Taiwan, Hong Kong, Korea, China and so on are vying with each other to build up their own "Silicon Valley".

History has proved repeatedly: Transformation of the industrial structure and change of the leading industry play a decisive role in economical development.

From Trio Industry to Quintet Industry

At present, the generally accepted theory on industrial structure is the "trio industry" theory, namely; the first industry (agriculture), the secondary industry (industry) and the third/tertiary industry (service industry). In 1940, British economist C. Clark began to use the concept of "tertiary industry" widely. From then on this theory was put into extensive use in the world, and many countries have classified their sectors of national economy in terms of this theory.

Today, we can see: In the developed countries, the structure of the three major industries has undergone some significant change. Generally, the work force in the first industry (agriculture) has dropped, only accounting for 5 - 6% and in the USA it only accounts for less than 2%. The proportion of workers in the productive industry has also greatly dropped: In most developed countries, it has dropped below 30%, and in America, the proportion of productive workers has dropped from 33% to 17% in the recent 30 years. If we are still tied down by the concept of worker-peasant alliance, America's base of worker-peasant alliance is only less than 20%. More and more work forces are turning to the flourishing tertiary industry that is on the upgrade all the time. Meanwhile, in the proportion of the GDP, that of agriculture and industry is getting lower and lower, and the service industry has become the major factor for the growth of GDP in many countries.

The traditional economy theory classified all that does not belong to the first industry and secondary industry into the tertiary industry. It is just like taking the tertiary industry as a big bag, letting everything be put into it. Apart from agriculture and industry, all the other sectors are put into this big bag, so it has become a hodgepodge.

Such a classification was proper in the time when agriculture and industry were in the dominant position of economy and the tertiary industry was still at the embryonic stage. However, in the present time when agriculture and industry are increasingly shrinking and dropping to very unimportant positions and the tertiary industry is becoming more and more diversified in economic activities and its share in the total output value is tending to be dominant, such a classification seems to be very improper.

Development of the times is forcing a change in the industrial structure. Technology industry and human-centered industry are just showing their great special power in the tertiary industry, becoming more and more important in position. These two major industries should be separated from the tertiary industry, and get their due status.

Even those who have never read Analects of Confucius are familiar with the saying *"Right names lead to right things"*, so, "rectification of the name" is very important. The name of "tertiary industry" may give people an impression that it is in a secondary position in economy, ranking as "the third". But as a matter of fact, this "third" only resulted from the historic order of industries (the order of appearance in history). If we order according to the future sequence of industries (order of importance in future economy), we should just revert the order. Meanwhile, "the quintet industry" theory should replace the "trio industry" theory, becoming the compass for future economy development.

In the historic order of industry development.

First industry—Agriculture

Second industry—Industry (in the narrow sense)

Tertiary industry—Service industry

Fourth industry—Science and technology industry

Fifth industry—Human-centered industry

In the future order of industry development.

First industry—Human-centered industry

Second industry—Science and technology industry

Tertiary industry—Service industry

Fourth industry—Industry (in the narrow sense)

Fifth industry—Agriculture

In comparison with the old theory of "trio industry", the new theory of "quintet industry" has made the following contribution: It has separated the human-centered industry and the science and technology industry from the original "tertiary industry", and takes the human-centered industry as the first industry of the economy of the 21st century and regards the science and technology industry as the second industry.

Our economic theory should keep abreast with the times, and the "trio industry" theory has lagged behind the times, unsuitable to the economical development of the 21st century. So long as we stand ahead of the time and take a great foresight of the future, we will certainly find the "quintet industry" pattern will be the megatrend of future economy development.

The human-centered industry includes culture industry, education industry, eugenics industry, life industry and so on.

What does the human-centered industry produce?

It produces humans and consummate human beings!

Human production is the most fundamental production of all productions, and the human-centered industry is the most fundamental industry of all industries. The human-centered industry is the foreground in future economy. Hence, the human-centered industry is the first industry.

Let's look at the culture industry first. **The future economy will be ruled by the culture industry. In many developed countries and regions the culture industry is a rising new force constituting a very important part in economic growth, and will become a major growth point and mainstay industry in their national economy.** Just as Daniel Bell says in his *Advent of the Postindustrial Society*, *"Economy has become the production of the life style displayed by the culture"*. Hollywood seems to be just as important as Silicon Valley. Without understanding the charm of the culture industry, we shall be unable to understand why Jordan's economic contribution should be 20 billion US dollars, why books, perfumes, underwear, movies and Nike sneakers about Jordan should be sold like hot cakes, why NBA tournament should get such a great audience, why so many countries should be vying to get the building right to Disney parks...

In his inaugural address, Korean president Kim Dae-jung declared: *"Build Our Country through Culture"*, stressing that they should develop the culture industry into the backbone industry in the 21st century. The power of "Korea-philes" is getting greater and greater in China, and Korean Teleplays have a far-reaching impact on China.

American writer J. K.Rowling used to be in abject poverty and lived on the government's relief before she published her best seller *Harry Potter* that has sold 270 million copies in the world! According to a British news medium's report, British parents have spent 40 million pounds buying products related to *Harry Potter*. That cafe frequented by Rowling has also become world famous now. Coca-Cola Company spent 1 billion US dollars to buy the right to print Harry Potter's smile on the coke can. The film "Harry Potter" has also become a hit, challenging

"Titanic", bringing about a huge profit... The report released by New England Committee highlights the promoting role played by culture in bringing about job opportunities. In the said region, there are 245,000 people engaged in this industry, outnumbering the people in the software industry that used to be considered as the leading industry in that region. So, culture economy has already reached the scale comparable to Silicon Valley's electronics and communication sector. The report stresses that culture economy has infused a vitality into the urban economy that seems to have tended towards decline, it can build up a healthier community, which is of far reaching importance to the sustainable development of England.

While using their culture to attack China and changing the Chinese people's ideology deeply or even overly (reshaping and producing people intangibly), westerners also make big money from it. Oscar, NBA, Disney, MTV, Hollywood, Madonna, Mcdonald's, TOEFL, GRE, CNN... Each of them has resulted in a huge production value and each of them has produced far-reaching impact on the world. Especially, while making big money, Americans are spreading their concept of values, which has not only conquered the populace and young students, but also overcome scholars and celebrities, even the elite on the top of the society have prostrated themselves before the American culture. Yet, on the other hand, Chinese culture's competitive power and position in the world culture market seems to be shrinking, as if there is nothing in our culture except Peking opera and the Great Wall. The Chinese have not realized that culture is also an industry, and always regard culture as a trade needing sponsorship from others, so, even the traditional Chinese story "Hua Mulan" had to be made into a film by Hollywood. The advertisement sales volume of the whole magazine market of China is only 1% to 2% of that of the USA. China is behind the West in science and technology. But should we bite the ground in culture industry too? If we do not awaken, we shall lose our future unavoidably!

The education industry is an industry that "produces" the most important production factor: highly able persons. The education industry is a basic industry characterized by all-inclusive contents and a guiding role, and the return rate of educational investment is very high as well. Education itself is a profitable industry. The education output value of the USA ranks No. 3 in all the industries of US economy, the income from education service alone is as high as 200 billion dollars each year, providing job opportunities for 2 million people. More than a half of the institutions of higher learning in the developed countries are private universities from Harvard, Yale to Waseda... Facts of economic development in the developed countries and regions have proved, "The investment with the highest return is education". Why do we call some countries "developed countries"? The top reason is their advanced education. "First-rate education turns out first-rate people, and only first-rate people can work out first-rate products". The education industry has good reason to be regarded as one of the top key industries.

The human-centered industry also includes the eugenics industry, the 1000-year longevity industry and so on. Everything should take the human beings as the base and the human beings takes life as the base. Health and life are eternal themes in mankind's pursuance, hence, life industry including the 1000-year longevity industry and so on is also a human-centered industry with top significance. I have already talked about such industries in the previous sections.

Second Industry: Science and Technology Industry

"Science" and "technology" are different concepts. The definition of "science" in *Great Chinese Dictionary* is: *"It is a knowledge system on nature, society and thinking"*. The definition of "technology" made by French scientist Diderot in his *Encyclopedia* is: *"Technology is a system of the methods, means and rules for attaining a certain goal."* The task of science is to solve the problems of "what is" something and "why it is so", whereas the task of technology is to answer questions about "what should be done" and "how to do".

The science and technology industry includes such knowledge production industries as scientific research industry, technical study industry, technological development industry, science and technology consultation industry, managerial consultation industry and so on.

Science and technology are becoming more and more important in economic development. According to statistics, in developed countries, 70-80% of their postwar economic growth resulted from advances in science and technology. The contribution ratio of science and technology knowledge to economic growth was only 5% to 20% at the beginning of the 20th century, but now in some developed countries it has already risen to 80% to 90%. In these countries, it has really become the first productive force. As the overlord of global economy, America has benefited greatly from its rich investment in science and technology. In 1994, America's investment in research and development amounted to 173.02 billion US dollars, equivalent to the total investment of Japan and Germany in this area. In 1997, it reached 205.742 billion US dollars. The massive investment in research and development has kept America's leading position in the world in almost all the basic research areas and 24 of the 27 key technology fields, especially in information technology, biological engineering, new materials and space navigation technology. In 1997, the contribution ratio of America's hi-tech industry to US economy growth already exceeded 55%. Moreover, the cycle of science and technology R & D has been greatly shortened: In 1990, the average new product development cycle in America was 35 months; In 1995, it was shortened to 24 months. It will be further shortened from now on. Thus, the speed of industrialization of high-grade technologies is rising very fast.

Scientific research achievements are the greatest assets and the most precious assets. **Bill Gates has neither factory building, nor manufacturing machines, and he became the top zillionaire of the world just by his wisdom and research achievements.** The huge energy of the science and technology industry has become evident for a long time. All such life economy and human-centered economy as the 1000-year longevity industry, the god-sleeping industry, the ecstasy industry, the eugenics industry, the space industry and so on, will depend on the power of high technology.

Science and technology will have greater and greater power in future development. All kinds of high technology; such as electronics, information technology, life science technology, new energy technology, new material technology, space technology, oceanology, environmental technology and management technology and so on, will show their magical powers in future economy. Biological cloning technology has already been able to change the traditional agriculture, traditional livestock husbandry and aquaculture on a large scale. For the human genome, a keen competition for the patent rights is going on, each patent is an incantation of Alibaba for opening a great treasure-trove. New energy sources and new materials are coming out one after another, which can almost free the economic operation from the impact of fluctuant international petroleum prices. Development of environmental technology has cast the twilight of harmonious coexistence of humanity and nature... In prospect to the new century, the science and technology industry will become the main battleground for competition among all countries. All kinds of high and new technologies will mature, climaxes of science and technology development will come one after another, causing great revolutions in the productive forces of human society.

Hence, the science and technology industry will naturally become the second industry. Whether for the government or for the enterprise they must take the human-centered industry and the science and technology industry as the priorities in drawing up their strategies for economic development. The human-centered industry and the science and technology industry are not the fifth industry or the fourth industry, but are the first industry and the second industry in status. That is the grand strategy for development in the 21st century.

Industry Reshaping and Frog-leaping Development

By great development the first industry is the human-centered industry, and the second industry is the science and technology industry. We may reshape the traditional agriculture, industry and service industry, causing fundamental changes in all traditional trades including agriculture, foodstuff industry, textile industry, automobile industry, real estate industry, manufacturing industry and so on. Even some declining industries may regain their life due to the stimulation by the human-centered industry and the science and technology industry. In other words, only by dint of the human-centered industry and the science and technology industry can the future agriculture, industry and service industry (the service industry separated from the human-centered industry and the science and technology industry) "scale a new height".

For example, by reshaping agriculture through science and technology, we may turn it into a new agriculture, which will be full of high technology and high humanity. Then, tomato, milk, meat and so on will make people healthier, more long-liveing, more beautiful and cleverer. The future agriculture will be integrated with the health industry and the medicine industry, and the link for this integration will be high technology. The future agriculture will also develop towards gene agriculture, fine agriculture and Internet-based agriculture. Then, there will be "super rice" with a per mu yield of one ton and "super pig" that can grow 1 kg a day; Agriculture will become very fine work, irrigation will become "drip irrigation", and the Internet-based agricultural database, and Internet-based technology and information service will help the farmers get any agricultural information instantly. By reforming the traditional agriculture through high technology, we shall find new methods for agriculture. And we can deduce the rest from this, such as reforming the traditional manufacturing industry through science and technology, reforming the service industry through science and technology, and so on.

To reshape traditional industries through science and technology, we must do employment structure adjustment also while doing industrial structure adjustment. China has adjusted its industrial structure into a manufacturing industry type, but its employment structure remains to be the agriculture type, thus, it is a dualistic structure now. Only changing the industrial structure but keeping the employment structure unchanged, has brought about an even greater problem: Farmers account for 75% of the whole country's population, but this "first industry" with the most labor inputs in China only accounts for 15.9% of the total amount of national economy. The "three rural problems" (agriculture, countryside and farmers) have become the "obstinate diseases" in China's modernization process. Some foreign scholars think, in comparison with Europe and America, China has lagged behind one hundred years in the key areas. Take Germany for example, the proportion of Germany's work force in industry was higher than that of the present China even as early as the 1870's; Germany's urbanization level in 1880 was much higher even than the present China. According to the statistics in 1990, 26.3% of Chinese live in towns, but the rate of urban population in Germany already reached 39% in 1877. To a great extent, China's economy is still at the stage of the fifth industry—agriculture. From agriculture to industry and to service industry, it will need a great industrial structure transformation. In such a situation, we may just use human-centered industry and the science and technology industry to boost the industrialization process, instead of waiting for the realization of the industrialization goal and then beginning to develop the human-centered industry and the science and technology industry.

Judging from the surface, today's Chinese economy is very prosperous, the manufacturing industry is progressing by leaps and bounds, the infrastructure construction is growing like a raging fire, and China has almost become a world factory. In future, China will certainly become a great power of the world.

Can we say so definitely? There is no doubt that China's economy will progress in all aspects for a long time, but if we do not pay attention to the industrial structure and neglect adjustment of industrial structure, it will also be doubtless that China will degrade into a labor country and a processing plant of the world. **If we only adhere to the "trio industry theory" do not see the predominance of the human-centered industry and the science and technology industry in the future economy, and lack a long-term general plan, a general strategy and general measures, China will get into a miserable state again.** We cannot evade the ruthless reality: Our country has not even finished the task of industrialization, and our employment structure is still at the agriculture stage largely. We need to both "make up for the old lesson" and "learn the new lesson". What is to be done? Should we only follow others?

The way of thinking decides the way out. What way of thinking should we use to face this grim reality?

The only way out is the road of frog-leaping development. If we await industrialization before starting the service industry and starting up the science and technology industry and the human-centered industry, we will lag behind in every step, and never be able to catch up or surpass others. On the contrary, if we, according to the quintet industry theory, foresee the chance in future economy timely, do strategic deployment for future economy ahead of time, go all out to develop the first industry— the human-centered industry, and the second industry—the science and technology industry, then we may use the minimum learning cost and the maximum learning efficiency to catch up with "the global co-rotation speed". In this way, we may be able to take the "later-coming advantage" and work wonders.

Perhaps someone will say that it is too unrealistic to transform our country's economic structure to the science and technology industry and the human-centered industry. But we must know, only by re-enforcing development effort in the science and technology industry and the human-centered industry can we realize a frog-leaping development. Only when the science and technology industry and the human-centered industry have grown greatly can the quality of all citizens such as workers and farmers be improved greatly, so as to realize the macro adjustment of the industrial structure. At the early stage in the history of America, agriculture was their primary industry and they had to use 95% of the manpower to support the population of the whole country. But today, America only uses less than 2% of the people to support the population of the whole country, and it has become the greatest grain exporter in the world. If we continue to bind the 900 million farmers to the land, we shall have only one way out: death. Only improving the farmers' quality as fast as possible and transferring farmers to other jobs in large numbers is the true grand strategy of national economy development of our country.

Whoever can foresee the future will win. Whoever can conform to the trend of the times and conform to the trend of industrial revolution, will become winner in the future economy, or may even leap from the present backward country to a world's advanced country. Otherwise, if we always lag behind the developed countries, we can only be losers forever.

What Confucius liked most about Yan Hui (his student) was his "not blaming others" and "not making the same mistake again". We should never attribute our own failure to others, but should always blame ourselves. **In refusing change and refusing industrial revolution, we have paid a price of one century's stigmata, and nearly two century's lagging behind. What is more important, we should learn "not to make the same mistake" again.** In the face of the new economic situation in the 21st century, we should never be satisfied with the present prosperity and peace, but be ready to bear some pains for the successful transformation of industrial structure, that is, pay the due price for the strategic victory. Otherwise, we will lose in the future, and our nation will be confronted by endless trouble for certain.

In the 21st century, every country, especially our China that has never taken the leading position after the agricultural economy, is faced with a more profound industrial revolution and a new great industrial structure adjustment is awaiting us, which is full of opportunities and challenges.

According to the quintet industry theory: laying stress on development of the science and technology industry and the human-centered industry will be a great opportunity for a great leap forward for both developed countries and developing countries.

The earlier we realize that, seize the chance, make plans and take actions, the earlier we will win.

XXXIV. TRANSCENDING THE TRADITIONAL PRODUCTIVE FORCE THEORY

Our world is transforming from "material-centered pattern" to "knowledge-centered pattern" and further to "human-centered pattern"; the major productive force is also changing from "material productive force" to "knowledge productive force" and further to "human-centered productive force".

Material Productive Force - Knowledge Productive Force - Human-centered Productive Force

Today, in a Western, developed country, one big automobile factory's annual output can almost meet the need of all countries in the world for one year. Such a phenomenon similar to surplus of productive force has already become a common phenomenon.

Is productive force really in surplus?

What is in surplus is only the material productive force. In the present times, we have to recognize the other two major productive forces: Knowledge productive force and human-centered productive force.

Since the first official use of the concept "knowledge-based economy" in the document of the international organization "Organization for Economic Cooperation and Development (OECD) in 1996, the tide of knowledge economy has swept the whole world, knowledge productive force has been taking the place of material productive force to become the main force in economic development. In the hot topic "new economy", the true connotation of the "new" is the new productive force. Both agricultural economy and industrial economy take the material productive force as the main, but the new economy takes knowledge productive force as the main, and its growth chiefly depends on the production, spreading and application of knowledge. Just as managerial master Peter Druc points out, the major function of the new economy is *"production and distribution of knowledge and information, instead of production and distribution of materials"*.

Our world is transforming from "material-centered pattern" to "knowledge-centered pattern" and further to "human-centered pattern"; the major productive force is also changing from "material productive force" to "knowledge productive force" and further to "human-centered productive force".

The victory or defeat in economic competition increasingly hangs upon the human beings, that is, hangs upon the "human capital" or "human resources". A research result by World Bank has proved that 64% of the wealth in the world is dependent on human capital. According to a recent research report from America, in America, 80% of the work posts are essentially for brainwork. If the same set of advanced equipment can have a production efficiency of 10 in New York and Tokyo, its efficiency might be even unable to reach 3 in Cairo or Rangoon. After Arabia made a big fortune because of its petroleum, such an idea occurred to some Arabian plutocrats: "Buy a modernization". As a result, advanced technical equipment has been introduced, and modern factory buildings have been set up, yet, they lack well-trained people, the work force's quality is very low, and the management level is very low, hence, their production efficiency is very low. Therefore, the program "to buy a modernization" has to be declared as a failure. On the contrary, America, as the representative of the flourishing "new economy", has greatly benefited from the human-centered productive force. Just as Chairman of the US Committee of First Lunar Landing Exploration said, *"America's greatest assets are a group of inventors and entrepreneurs, and they are the key to the wealth of the country."* America's educational investment accounts for 7% of the GDP or more, and the sophisticated education system makes it possible for the citizens to receive new education at any moment and place. Meanwhile, America has been gathering highly able people from all countries of the world. In the recent years, America's number of people winning the Nobel Prize accounted for a half of the prize winners in each year. Many American corporations carry out the employee stock ownership plan; their wage cost is the first big item in the production costs, accounting for 80% or so.

So we can see that in the current and future economy, the real impetus is the human or the human-centered productive force.

Human being is a tool-making animal. From burnishing the first piece of stone, mankind began its tool making and tool using history. The role played by tools in human progress is so great that people even use the feature or material of the tool to name the age, such as Stone Age, Bronze Age, Iron Age, Steamer Age and so on.

The great change of material productive force lies in the revolution of the tool. Making stoneware was the first great change of mankind's material productive force. The greatest change in the modern history of human material productive force should be said to be the invention of the steamer. In the 21st century, the full-automated production system represented by the highly capable robot will become the greatest revolution in mankind's tool using history.

The trend has already shown up.

Nowadays, **America's peasant population is less than 2%. The workers' proportion is only 17%, and it is predicted that by 2020, it will be less than 2%;** and by 2025, ordinary workers will lose their position completely. Along with the speedup of robot industrialization, there will be even sharper revolutions. Then farmers and workers doing traditional production of material objects will disappear from our earth. Unmanned factories and unmanned farms will become common phenomena.

In future, there will be such a picture in most factories: No one, not even a single person. No talking voice, no sneezing sound, no coughing sound, but only the sound of machines' running. These mechanical "workers" neither need rest, nor do they demand a wage, or complain about poor labour insurance conditions. They just remain loyal and devoted one hundred percent, do not know tiredness, work 24 hours every day, even do not hesitate to go through fire and water. They can replace human beings completely in doing tiring, boring and dangerous jobs that we are unwilling to do. They never complain of suffering, and keep polite and obedient forever.

Robots will not only be active on the production lines of factories, but also enter every sector of our life. Throughout the ages, what housewives have been longing for is just to shake off the onerous housework completely some day. From ancient time till present, housework has been heavy and boring, but it has to be done, even to the extent that many economists have drummed up the idea of letting women return to the home to specialize in housework. Just as an economist says, *"There is hardly anything in the world that can be more like Sisyphus' stone than housework; Clean clothes become dirty, dirty clothes are washed clean, but become dirty again, so you have to wash them again... Such things repeat again and again, day after night, without end at all. Housewives are tired in mind and exhausted in strength because of such tedious jobs, but they have just kept the existing state, and have done nothing beyond it."* In the near future, Sisyphus' stone may leave us at last. When we go to bed every night, the robots as small as matchboxes will come out one after another, working in good order, such as mopping the floor, wiping the furniture, getting clothes in order... According to the prediction by Rolfe Dittel, a German professor, 5 to 10 years later, home robot servants may be popularized to ordinary homes like the washing machine.

Factory production will be automated, and so will housework. Nowadays, housework also belongs to the service industry, so, many economists are considering that the state should pay wages to women who have returned home for housework. But all such areas of material production will be automated, without the need for manpower, and mankind will be liberated completely from this field.

Apart from automation of material production, with more robots participating, the products will become cheaper. The general development trend is: the cost of material production will drop greatly. **Once production needs no manpower, and after a great deal of "self-reproduction" of the robot, its price will become very cheap, approaching "zero price". The whole process from high price to middle price and to low price, until "zero price" finally, can be called a "zero price" trend of material production brought about by all-round automation.**

Production automation and product tending towards zero-price are great changes in material productive force. And it depends on the degree of realization of the two major indices in development of the robot: One is the working capacity of the robot, and the other is the cost of the robot.

First, let's look at the robot's working capacity. Robots may replace humans and surpass humans completely. Not only in physical strength, but also in intelligence. It will approach human intelligence gradually, and it is even possible that robots surpass humans in all aspects some day. Many scientists assert that the robot's intelligence might surpass the combined intelligence of Einstein and Hawking.

Secondly, let's look at the cost of the robot. It is the key factor deciding whether material production can be fully automated and products can tend towards zero-price. Negroponte once predicted that we would spend one dollar to buy a computer in future. In future, may all of us spend the money just enough for buying a cup of coffee, to buy a robot that is both intelligent and willing to bear the burden of hard jobs. Quite possible! According to the prediction by United Nations' Economic Commission, in the coming 5 to 10 years, robots may be as popular as today's computers or mobile phones: every family may have one or even several robots.

Production cost is closely related to the speed of product development. The computer used to be inaccessible to most people, but its developing speed is so fast that in every 18 months, its performance indices rise by one time and its price drops by one half. What about the mobile phone? In China, in 1990, there were only 20,000 mobiles in the whole country. At that time, holding a mobile was the marker of a tycoon. But nowadays, it grows at the shocking rate of 280% each year, and younglings change their mobiles as casually as changing a dress. If we develop it consciously, the developing speed of the robot will surely surpass that of the computer or the mobile far and away. What is more important, robots may do self-reproduction. Each time the robot reproduces itself, it can self-correct its shortcomings cleverly, so that the duplicates can be much better. In "producing" a person useful to the society, the parents spend several decades in bringing him up, so the time cost and social cost are very high. But the robot can do self-reproduction continually, and the cost may approach "zero" gradually.

Will mankind really retreat completely from material production?

From ancient time till now, material production has always been in the unshakable primary position. The reason is so simple: First of all, we must eat, drink, wear, and live. But does that signify material production must be primary? No! Air and sunlight are also the most necessary things for human survival, but we do not need to "produce" them ourselves. Thus, although they are absolutely necessary to every person, we never need to pay money for them. The same is true of future material production. Although such needs met by material production as food, drink, clothes and houses are basic needs for human survival, the development of robots will automate material production completely, material production will no longer need human labor, and the cost of robots will be almost as cheap as "zero", then, prices of products will also be almost as cheap as "zero", just like air and sunlight, without need to pay money, being close to free supply.

In the face of the automation trend of material production, some people have got into a panic wanting to stop it. But it is a pity that all those efforts are just like throwing straws against the wind. In time of the industrial revolution in the 18th century, a so-called "Lud Movement" broke out in Great Britain, which was a movement of artificial destruction of machines. Today, similar things have appeared: There are many people fearing robots and hating robots. Some scientists are worrying that the would-be "robot species" on the earth might threaten human life. In fact, it will not be so. As early as the time before the robot was born, American science fiction writer Asimov published a book *Robot*, which set three ground rules for the relationship between robot and humans:

> The robot may not hurt mankind.
>
> The robot may not sit watching when mankind is being hurt.
>
> The robot may not result in hurting of mankind.

In any condition or any case, the robot should never go against the three major rules. The robot is made by man, and all robots in future will abide strictly by these three major rules. Hence, the robot is bound to be absolutely obedient and loyal to mankind, and should never threaten human life.

The all-round automatic production system represented by the robot has caused an earthshaking change in mankind's material productive force. After realization of full automation and zero-price of products, man will be freed from the material production that has swallowed most of his time and energy since his birth, and he will use all his time and energy for more valuable things. So, all fine dreams that mankind dare not imagine now will come true for certain.

After the robot does everything for mankind, what should man do then? Where does human value lie?

To answer this question, please answer another question first: Initialy in America, its farming population was as high as 95%, but now it is only 2%. Where have the former farmers gone? In addition, where have the so many industrial workers gone? Most of them have turned to the service industry. From now on, even the service industry will be mostly undertaken by robots. So, more and more people are bound to flow into the knowledge production industry, namely the science and technology industry. The whole society will step into a science society and intelligence society with stress being laid on development of the knowledge productive force.

Super change in material production results from automation, whereas super change in knowledge productive force results from intelligentization. Today's knowledge production (scientific research fruits) has already piled up endlessly and become uncountable, but future knowledge production will make scientific research fruits grow in an explosive way like atomic fission.

We know, the energy from atom splitting is the most powerful and most amazing energy in the human world. There are two conditions for atomic fission to generate such a huge energy:

It is just because of these two conditions that atomic fission can produce such a surprising energy in a short moment.

Knowledge Productive Force: Intelligentization and Godly Brain's Work

> In each fission, the energy increases by one time from the previous fission.
>
> Each fission only takes one 50 trillionth of a second.

When knowledge production or scientific research grows like atomic fission, we can imagine: Suppose the achievements from the first research are 1 billion, those from the second research will be 2 billion, those from the third will be 4 billion, and those from the fifth will be 8 billion! Moreover, between each two researches, there is only "one 50 trillionth of a second"! You can imagine how fast the increase of knowledge production is. That will be the situation to be brought about by explosion of knowledge productive force.

Making explosion in knowledge productive force will depend on the increase in number of people engaged in knowledge production (once mankind is liberated from material production, all the several billion people may join knowledge production); and what is more important, that it will depend on whether mankind will adopt the godly brain system for research or not.

What is the "godly brain"?

> godly brain = super human brain + super computer + super outer brain + super huge brain

As the poem says, *"Why is the water in the channel so clear? Just because the source is running water"*. The source of scientific research achievements or knowledge production is the brain. The brain is where wisdom lives, it is the source of intelligence, it is the most mysterious and most precious part of the humans, and it is the core parent body for knowledge production.

First, we must create a **super human brain**. This depends on our knowledge of the brain. When the brain science makes great progress so as to uncover all secrets about the growth, division of work and so on about our brain, such knowledge will greatly help the research on artificial intelligence. When we use all such knowledge to develop and improve our brain, everyone may give play to all the creative potentials of the brain; What is more important is "building up the brain", that is, using gene technology, nanometer technology, computer technology and so on to rebuild the brain. The famous scientist Hawking advocates the "cyber-technology", that is, connecting the human brain to the computer directly. He says, *We must work out such a technology as soon as possible, so as to let machine intelligence make contributions to human intelligence, instead of letting it play a threatening role.* By "building the brain" through high technology, everyone of us may have a super brain, which will be much better than the brains of top scientists or people of high intelligence.

What comes next is the **super computer (electronic brain)**. The future computer will develop quickly to have super-speed, super-miniature size and super-intelligence. The future supper computer will be a quantum computer, with a speed billions of times faster than the traditional computer. It will also be a photon computer, not only with an amazingly fast calculating speed, but also with a great anti-jamming ability and a fault tolerance capacity like the human brain. It will also be a DNA computer, not only with a huge information storage capacity, but also with a self-repairing ability, and the ability to be connected to the human brain. It will also be an artificial intelligence computer, it can imitate human logical thinking, imitate human imaginal thinking, express all kinds of feelings, do creative work, and have a self "consciousness". It will also be a nanometer computer, so tiny as to be one millionth of the cell, and hardly consuming any energy source... To sum up, the present computer will grow into a super computer full of unthinkably great intelligence. Gordon Moore, Honorary Chairman of the Board of Directors of Intel Company, believes: *"Silicon intelligence will develop so highly that it is hard to distinguish the computer from the human being."* Australian scientist Garris says he is working on the first artificial brain that can imitate the human brain, and he predicts that after 2011, such an artificial brain's intelligence might surpass the human brain by 40 times, and Raymond Kurzweil, President of Kurzweil Technology Company believes.

In future, one computer will be equal to one billion human brains of the present time. Then, if we use one billion such super computers for scientific research, how can the research scale and efficiency not reach a miraculous level?

> "Around the year 2019, a 1000-dollar-worth personal computer will have the basic computing power of a human brain, including 100 billion neurons and 100 trillion nerve connectives. By 2030, a computer system valued at 1000 dollars will be equivalent to the total intelligence of 1000 persons' brains. By 2050, it will be equivalent to the power of 1 billion human brains."

The godly brain system will also need the help from the **super "outer brain"**. One of the greatest presidents in the world, President Roosevelt, was best at using the outer brain, namely the brain trust. In his brain trust, there were lawyers, economists, financiers, teachers, socialites, merchants, newspaper editors, labour leaders, state officials and so on, with various backgrounds and various thoughts. These people composed a great outer brain, the brain trust, becoming an inexhaustible source of wisdom for Roosevelt's outer brain, the practice of the "brain trust" was quite similar to the outer brain in the Spring and Autumn Annals period of China, namely the many "hangers-on" kept by an aristocrat. The future "super outer brain" will utilize the super expert advisory system and super "think bank" to make use of all kinds of outer brains. In doing scientific research, no one will be single-handed, but will make use of the super outer brain most expediently, so as to get the needed wisdom and intelligence from people of all nations, all nationalities, all trades and all ages, and get the wise support and wise help from experts from all trades at any time. It will not only reduce a lot of repeated tests and repeated studies in research, but also help everyone's intelligence and creation to reach an unprecedented height in both quality and quantity. The super outer brain will become the distributing center of human intelligence, and a solid foundation stone for human progress.

Moreover, the godly brain system cannot get away from the super huge brain. The super huge brain is an integration of many super human brains, or combination of many super computers, or summation of many super outer brains.

All things mankind has achieved till now resulted from intelligence. **When mankind has built up a godly brain system, namely an integration of uncountable super human brains, super computers, super outer brains and super huge brains, we shall be able to do godly brain research at an amazing speed.** Then, the Internet will become an intelligence network, becoming "the global nerve system"; everyone will be a point in this global information network system. On this network, what will be going on will no longer be a mere informational interchange, but will be intelligence exchange. The intelligence of the whole world and all mankind will be linked together, so the whole world will become an intelligent world, human intelligence will rise by billions or zillions of times, the speed of scientific development will be like atomic fission. Some new branches of science or new research findings may come out in every 50 trillionth of a second! Then, the progressing speed of science and technology will be billions or zillions of times of the present speed! When knowledge productive force explodes on such a scale, the future economy such as the 1000-year longevity industry, the god-sleeping industry, the ecstasy industry, the eugenics industry, the outer space industry and so on driven by high and new science and technology as the core "nuclear power", will be certainly growing like raging fires.

The product of human-centered production is human itself. All that are attached to the human body, such as human life, health, youth, knowledge, skill, intelligence, moral and so on, may bring about more special productive forces. In future economic activities, the human-centered productive force will become the most fundamental and most important productive force.

Unlike Malthus' stress on population quantity and Schurz' stress on human capital, what the future economy stresses is the human-centered production. Human-centered production is neither the control over human natural multiplication emphasized by Malthus, nor the investment in traditional human education stressed by Schurz, but means the overall optimization of the human through high technology, such as optimization of birth, optimization of body and optimization of education.

By the magical power of birth optimization, body optimization and education optimization through science and technology, we may turn out geniuses almost omniscient and omnipotent, with supreme virtues and supreme beauty. That is just the divine human-centered production of the 21ˢᵗ century.

A small change results from the way for small changes; a great change results from the way for great changes; an abrupt change results from the way for abrupt changes.

In the case of China, the disadvantage lies in human beings, but the advantage also lies in human beings. One Qian Xuesen's value used to be regarded as three or five divisions of the army by an American senior officer who was reluctant to let him return to China. But according to an American columnist's opinion, his value is far more than five divisions. He wrote such a paragraph: *"How can you compare Qian Xuesen's value in science to merely three or five divisions? The fly bomb he developed for the Communist Party of China not only balanced the gap between China and America in strategic weapons, but also added a great counterweight against the threat from USSR, and meanwhile, between the two military super powers of America and USSR, China has used the fly bomb plus its 1 billion population to create a situation of tripartite confrontation. So, we can almost say this one person equals one country in strength!"*

> Labor creates mankind—Labor is the way for the abrupt change from ape to man.
>
> Science and technology create gods—Science and technology constitute the way for the abrupt change from man to god.

"One person equals one country", that is the value of a great genius! One person may change one country's position in the pattern of the whole world. If all the 1.2 billion people of China are like Qian Xuesen, China's comprehensive national power will surely shock the whole world, and be matchless! The deified human-centered production will just mean producing omniscient geniuses with the highest virtues and supreme beauty, making all people as capable as Qian Xuesen, or, one person in future might be equal to 100 or 1000 Qian Xuesens in ability and wisdom. Imagine, what a great change or sudden change it will cause to the productive forces of the whole society.

We can use high technology in eugenics, letting ourselves play the role of God to do godly human production, that is, to improve human gene, so that people may have the basic quality of a super genius as soon as they are born. Moreover, we can do overall optimization of education using science and technology. By all sorts of high technology means, we can transcend all restrictions and build up a magical global education system with the minimum cost and the maximum efficiency. Anyone may get the optimal education at any moment and place. We may provide him the most needed knowledge, skills and wisdom, and offer him the conditions for getting such knowledge, skills and wisdom most conveniently and most efficiently. Thus, education may become highly quality-oriented and highly individualized, and can be available for all the lifetime. Meanwhile, the deified human production will also need optimization of the body through science and technology. We may use high technology to optimize all aspects of the person from inside to outside. As for optimization of birth, education and body through science and technology, there are detailed explanations in the previous sections; hence, there is no need to give any detail here.

God was really too stingy when He created humans. He gave humans some intelligence but also made them suffer from many hardships and defects, so, the created humans are far from being perfect. In the future, mankind will be able to upgrade human-centered production to a deified state. They will apply the omnipotent means of science and technology to the production of humans themselves. Through the deified human-centered production, everyone on the earth will become a genius, then, the future times will be full of geniuses. In such times and such a world, will the productive force be only ten thousand times or one hundred million times of the present one?

Whether it is a revolution in social form or in economic form, it will depend on the development of the productive force, in the last analysis. Whether the human society can march into a matchlessly bright future, whether mankind can leap into the most ideal race of gods, will, by root, depend on the great leaping changes in the productive force.

A leaping change is an unexpected sudden change happening instantly. Volcanoes and earthquakes are leaping changes. Evolution of the ape into the humans, and leap of the human beings into "the god" are leaping changes, too.

"Saltatory evolution" came into being as a brand-new science long ago, and it is called "a breakthrough in using precision tools to describe complicated phenomena in biology, social sciences and other fields", "being the greatest revolution in mathematics in the 300 years since Newton and Leibnits' invention of calculus".

China has an old saying: *Heaven does not change, and nor does the Way.* As a matter of fact, both Heaven and the Way are changing. In the new century and new epoch, every aspect of the world is faced with unprecedented changes, such as science and technology, economy, politics, thinking and so on.

We are just on the space-time point of a great change that has never occurred in history, a great abrupt change in material productive force, knowledge productive force and human-centered productive force. High automation is causing an abrupt change in material productive force, that is, production tending towards zero-cost. High intelligentization is causing an abrupt change in knowledge productive force, namely research via the godly brain. High deity-oriented evolution is causing an abrupt change in human-centered productive force, namely geniuses coming forth in great numbers. All such changes are so great and sudden that nothing in history can match them.

The magical abrupt changes in the three kinds of productive force will certainly lead to a great explosion of the entire productive force.

American futurist John Naisbitt points out: *"We are just in an unusual period in mankind's history. The two decisive factors to social reforms, namely new concept of value and new economic needs, have already appeared."*

The concept of value advocated by the human-centered culture and life culture and so on is just the new concept of value he mentions. Such economic needs as those related to the highly-humane 1000-year longevity industry, like the god-sleeping industry, the ecstasy industry, the eugenics industry, the outer space industry and so on, are just the new economic needs in his utterance. Such a new concept of value and new economic needs will also further accelerate the great explosion of the entire productive force, including material productive force, knowledge productive force and human-centered productive force.

Human development chiefly hangs on the development of productive force. Human change also chiefly depends on the change of productive force. When we have understood and mastered the principle of magical abrupt change of productive force, especially the law of abrupt change of the three kinds of productive force, the development of the productive force of the whole society will be like atomic nucleus fission. It will have a chain reaction at the fastest rate and on the largest scale, so, a great explosion of human productive force will appear like a great atomic explosion. Then, it will be certain for us to realize the most splendid ideal of mankind!

EPILOGUE

夢星

Human beings were born together, and all mankind turns out to be one enti-ty. It is just the world formed by mankind that has split mankind. What a foolish world! What a deceptive world! What a terrible world!

—Binstoc

Today, we are in an extremely precarious situation!!

After the cold war, what awaited mankind was not a peace all over the world, but instead, we have entered an even more dangerous age. While enjoying the happy moments brought about by high technology, we are also faced with numerous fatal disasters caused by mankind itself! The nuclear terror lasting for more than four decades has not disappeared, but on the contrary, the situation is getting harder and harder to control. Today, the nuclear weapons held by mankind are enough to destroy mankind a dozen of times or more. The monster made by mankind itself that is powerful enough to destroy mankind itself. It was only used once right after its birth, and then, it seemed to have been "put on the shelf" and the world has calmed down. However, in fact, such things as nuclear contest, nuclear deterrent, nuclear terror, nuclear pollution, nuclear smuggle, nuclear accidents and so on, have never stopped. On December 19, 1998 Great Britain's weekly *'Economist'* published an article, which held that the probability of a nuclear war is quite great in the coming 50 years. Along with the spreading of nuclear technology, making a nuclear bomb is no longer a very hard thing. According to a story published in America, an American college student designed a practical nuclear bomb in a homework. A 12-year-old German boy built a "minitype nuclear power plant" in the cellar of his home. Today, even poor countries such as India, Pakistan and North Korea have already made or are able to make their own nuclear weapons. Some countries seem to stay away from nuclear weapons, but have the ability to make them completely. For example, Japan can make a nuclear weapon just within a week. As we know, the more nuclear countries there are, the greater the likelihood of a nuclear war will be, the harder for us to control the whole situation. When everyone is at swords' points, even if no one provokes a nuclear war intentionally, a nuclear accident might be triggered by a wrong judgment by any one! For instance; a computer program error triggered the Early Warning System, which nearly led to pressing of the button of the nuclear-armed missile. If the button had been pressed, millions of people might have been killed and even all mankind might have been destroyed. Such an "accidental firing" is not impossible at all.

Today, terrorism has become more rampant and ruthless, as it defies any moral and any rule, never gives a thought to prevent innocent from being hurt, and is all-pervasive in this globalized age. In the past, if someone had said that the Pentagon would be bombed and the World Trade Building would be razed to the ground, it would have beeen certainly considered as nonsense. But this nonsense has become reality, turning into the earth-shaking 9\11 event. After the "9\11", people have been thinking: What else is impossible? Terrorists have already been regarded by us as evil, and once the evil kisses the fiend "nuclear", what will happen? Information has shown that terrorists are very interested in the making of the atomic bomb. Moreover, nuclear smuggling, and spreading of nuclear matter and nuclear technology, has made it much easier to make a nuclear bomb, and they might start a nuclear explosion even without nuclear weapon. For example, if you use the same way as the "9\11" to bomb the nuclear arsenal, you may trigger an unprecedented nuclear explosion, putting mankind into a fatal nuclear disaster.

Mankind cannot afford a fatal nuclear disaster, not even once. What we have mentioned is just the nuclear weapon. Along with the fast development of high technology, biological weapons, chemical weapons, laser weapons, potential energy weapons and electronic weapons... all of them are making contributions to the perdition of mankind. In making weapons, mankind is destroying itself, we are really going at a tremendous pace. Scientist Hawking is not too afraid of the "9\11", but after the panic caused by the anthrax bacterium, he thinks, this will signify a greater disaster for mankind. What is more terrible than the nuke, is the biological weapon, as it is more convenient to make, its price is much lower, it is called a "weapon of the poor", and it is more suitable for mankind's "suicide". As the story goes, setting up a biological laboratory now only takes about three or four hundred Chinese dollars (RMB), and this cost can be reduced continually. British scientists are trying to make a "super virus" using human gene mixed with a virus very similar to the common cold virus. Such a virus has an extremely great power, and is very inconspicuous. You can never know who is attacking, when the attack begins, and how powerful the attack is. Scientists say, only 20g virus might "get rid of" all the 6 billion people of the world! If Shoko Asahara had put 20g toxic heat virus in the subway of Tokyo instead of the toxic gas, mankind would have perished! ...

Needless to say, the "gift" brought about by the flying development of technology is the unprecedented rise of efficiency in mankind's self-killing. The atomic bomb has raised the efficiency of self-killing to such a degree that it can wipe out all people in an instant, and the biological weapon has further raised this power, with a great drop in cost as well. Technological theurgy may be used for destroying ourselves. Either through a nuclear war to wipe out all human beings instantly, or through unscrupulous use of atomic technology to destroy our own homestead, or through the use of a virus to poison our own heredity and cause it to degenerate...

I do not know how to describe mankind's "wisdom": While mastering all kinds of skills and methods to improve our own living conditions greatly, we have been vying with each other madly in inventing all kinds of means to destroy ourselves at one stroke!

Someone takes the earth as a ship sailing in space. All earthmen are taking this ship together. This seems to be very fine, but it is a "drunkards' ship". Many passengers on the ship are drunkards or madmen, some are shooting each other, some are digging holes on the ship, some are steering the ship madly toward a reef... On such a ship, there are all kinds of hidden crises, and disasters may occur at any moment. Isn't it the picture of the human world? There are great shadows of war, there are bloodcurdling dark clouds and there are ruinous ecological damages... Everything is so terrible, so hard to predict, and so hard to control!

The root of all such crises can be generalized into two principal aspects: misuse of high technology, and self-killing of mankind. Those are the root causes of mankind's "going down directly to Hell". All disastrous consequences can find answers from these two aspects.

To change such a terrible situation, mankind's only hope lies in making the right choice. The right choice can only be: Let high technology serve high humanity, and integrate with high humanity harmoniously. Meanwhile, turn the mutual intriguing and mutual killing of mankind into working together in sincere cooperation! All these will depend on a new culture.

Only by the new culture can we turn science and technology into the most humane science and technology, and turn inhumane and lowly humane high technology into high technology with high humanity. Becoming immortal together is the greatest pursuance with high humanity. The ideal of being immortal together contains a spirit of high humanity, sparkles with the splendor of high humanity, and gives people the greatest and most profound ultimate care. Only from the standpoint of highest respect of humanity can we have such an ideal. It means a pursuit of active evolution of mankind, that is, face the deepest tribulation and tragedy of mankind as is shown in the normal process of birth, ageing, disease and death, and apply the magical power of science and technology to the pursuit of the truly greatest happiness and interest of mankind. **In the present 21ˢᵗ century, we must turn high technology from the orientation of killing and annihilating human life to the orientation of prolonging and upgrading human life, and turn the economic center from the traditional "basic necessities of life" to more profound humanity areas such as "birth, ageing, disease and death".** The real way out for the high technology in the 21ˢᵗ century is to attain such high humanity goals as "rejuvenation", "living forever", "omniscience and omnipotence", "daylong ecstasy" and so on, namely to become "gods". It is also the real way out for mankind.

Only by the new culture can we turn mankind's situation of mutual distrust and mutual killing into a prospect of walking hand in hand and helping each other towards the same goal. Everything in the human world centers around interest. "There is no eternal friend, nor eternal enemy, but only eternal interest". That is a steadfast truth remaining unchanged through all ages. Just because of interest, we may see such things in international relations. Two parties with fierce hatred to each other suddenly shake hands and make up, whereas two parties in complete rapport turn hostile to each other in an instant. Everything can be changed, and the cause of change is just "interest". Today, human beings have not built up a common interest and a common goal, and nor have they become aware of their common interest and common goal. With the new culture, all mankind will be able to realize the common interest and common goal, that is, becoming "gods". It is also the longest (even eternal) common interest and common goal of mankind. Then, this "spaceship" of ours will have a clear course, will not be dizzy and giddy like a drunkard, but will ride the wind and cleave the waves forward towards the Paradise.

Binstoc said sadly but also hopefully: *"Human beings were born together, and all mankind turns out to be one entity. It is just the world formed by mankind that has split mankind. What a foolish world! What a deceptive world! What a terrible world!"* To change the extremely foolish, deceptive and terrible world resulting from "splitting of mankind", and to re-unite mankind as a whole, we must rely on the birth of a new culture that represents an interest higher than the national interest or the interest of an ethnic group, that is, mankind's interest. Once the god-becoming ideal representing the common interest and common ideal of all mankind becomes true, and once the god-becoming culture becomes the common culture of all human beings, then mankind will come together easily, cooperate closely with each other and live in harmony forever, and self-killing tragedies will no long occur.

Today's mankind is just at the crossroad between Paradise and Hell. On one hand, it is beset with crises, but on the other hand, it is full of hope. Once mankind changes the use of the magical power of high technology from evil purposes to good purposes, all great ideals of mankind will come true. All the goals mentioned in this book, such as optimization of birth through technology, optimization of education through technology, optimization of body through technology, longevity in a straight way or in a tortuous way, perfection of the body preserving technology and perfection of the mind preserving technology... may be attained completely.

The key to the realization of the great ideal of mankind's leap into being gods just lies in its Archimedean lever: high technology and mankind's great union. If we do not rely on high technology, all such ideals as eternal life, daylong ecstasy, omniscience and omnipotence, supreme kindness and supreme beauty, will be pure illusions, without possibility to come true at all. Becoming "god" is the attractive and mysterious "other shore" while high technology and great union of mankind are the bridges or ships for reaching the other shore stably and efficiently.

So long as we can humanize high technology and build a great union for all mankind, we human beings may "go straight up to Paradise", and become gods, so that everybody may live forever and be happy forever. Meanwhile, only by humanization of high technology and great union of all mankind can we avoid "going down straight to Hell", escaping from the perdition of the earth and mankind.

The greatest horror and greatest interest will certainly urge mankind to make the correct choice as soon as possible. Of course, it depends on mankind's understanding of such a greatest horror and interest, depending on their degree of awareness. Hence, the most important and valuable work of a truly conscientious and far-sighted thinker, statesman or entrepreneur is to help the present mankind become aware of its greatest horror and greatest interest, and find out the way to overcome the greatest horror and realize the greatest interest. *The Second Declaration* is just meant to make a contribution in this respect.

There are two groups of futurologists: the optimist and the pessimistic. These two groups of people just stand on their own side, quarrel with each other, and no one can convince the other. Wang Meng (famous Chinese writer) said, *"I am a profound pessimist"*, "The so-called profound pessimists are those who know all the weak points of mankind, also know all the heavy burdens of our nation, knowing all the problems in our culture..." I am also a profound pessimist, but at the same time, a profound optimist as well. For I not only know all the weak points of mankind and all sorts of problems in our culture, but also know all the strong points of mankind and all kinds of resources in our culture, know how to get rid of mankind's weak points, solve the problems in our culture, and how to set up a whole set of new culture that may result in the birth of a new mankind and new world. Hence, I am optimistic, profoundly optimistic. To be accurate, I am an integrator of profound pessimism and profound optimism.

If mankind keeps on walking on the old road without change, it will "go straight down to Hell". If mankind can awake to make a revolution, changing onto the new road, seeking the leap into gods, it will "go straight up to Heaven".

Now, we are faced with a great choice between mankind's total perdition and total happiness.

Mankind's ideal of being immortals may both avoid mankind's perdition and bring the highest happiness to mankind. Mankind's pursuance of being gods is based on mankind's deepest common interest, and accompanied by the loftiest functions, such as leading to great harmony and peace in the human world and controlling mankind's future.

Hence, we are pessimistic about mankind, but at the same time, we are optimistic about mankind all the more.

Mankind will gradually become aware that its greatest mistake is neglecting itself and neglecting its own life.

The 21st century will be a century for mankind to care about itself, value its own life, and strive for the greatest freedom—limitless freedom of life.

Once mankind is clearly aware that becoming "gods" is the greatest interest and highest ideal that mankind should seek indeed, it will signify an unprecedentedly great awakening and hence, it will start up the great project of re-creation of world culture, the great project of re-creation of world politics and the great project of re-creation of world economy, in short, the project of re-creation of the whole world. Only in this way can we avoid mankind's "going down straight to Hell", and ensure mankind's "going up straight to Heaven".

POSTSCRIPT

夢星

This is a book I wrote with my greatest devotion.

"Sincerely stand at China's standpoint and sincerely stand at mankind's standpoint" is my basic requirement for myself in writing this book.

I am both a citizen of China and a citizen of the world. I am both a member of the Chinese nation and a member of mankind. "Everyone's real nationality should be mankind", and of course it does not mean we should abandon our own "motherland", or abandon our own nationality. In fact, there is no contradiction between China's standpoint and mankind's standpoint. Especially in the present times of globalization, the relationship between people and between nations is getting closer and closer, and the overwhelming common fate is becoming clearer and clearer. If we just look at problems and grasp the future from our own region, it will be just like blind men trying to size up the elephant: taking a part for the whole, and grasping the small but losing the big. Only when we integrate one region with the overall situation and unite the nation with mankind, and consider everything from the standpoint of all mankind, can we find out the best way out.

Aristotle says: *"The virtue we must study is the virtue of mankind. For the benevolence we are seeking is the benevolence of mankind, and the happiness we are pursuing is the happiness of mankind."* In the study of futurology, what is most needed is the great breadth of view, great mind, great vision and great wisdom for the happiness of all mankind. *The Second Declaration* is a book written for mankind: thinking for mankind, programming for mankind, letting mankind become aware of its own merits and demerits clearly, aware of its own greatest interest and greatest happiness, and pointing out the ways and approaches leading to the greatest interest and greatest happiness.

Therefore, this book is written both for Chinese and for the whole world and all mankind.

The future has limitless possibilities, but the future can be chosen and created, too. Choosing mankind's future and creating mankind's future do not depend on any country or any group of people, but depend on all the 6 billion people on the earth, and depend on every person in the world. To the best of my belief: An idea that has not been spread cannot realize its value, and the size of scope of spreading determines directly the value of the idea. If the ideas in *The Second Declaration* are not spread, not known, not understood, not appreciated and not supported by other people, what significance and value will they have? Only when more people in the world have known, understood, appreciated and supported them can we make concerted efforts and gallop towards the most wonderful future together. When I am writing here, a story has occurred to my mind:

Long long ago, people had a great idea: Build a sky ladder through which we can climb to Heaven! All people from every quarter of the world came, gathered together for this great idea, and worked hard around the clock. When God saw this, HE could not tolerate people's efforts to ascend Heaven. From that day on, HE began to muss up human language. Originally all people used the same language, but now suddenly it turned to many different languages. People could not communicate with each other, so, naturally they could not work together, thus, their wish to build a sky ladder to ascend Heaven failed at last.

Without spreading and without communication, it will be impossible to work together with one heart and one mind. Without mankind's great union, it will be impossible to accomplish the great undertaking of mankind. Hence, I wish to have more readers of *The Second Declaration*, so as to spread these ideas to more people. I hope some people of insight will translate it into other languages so as to let more people in the world read them, know them and realize them.

Although *The Second Declaration* has been finished, judging from another sense, it is not finished yet. Everything has just begun. Mankind's new road has just started, and the spreading of mankind's new ideas has just started, too.

I like the saying by Negroponte very much: *"The best way to forecast the future is to create the future!"* In order to create the fine future for mankind "to go up straight to Heaven", we need those people who cherish a firm belief described as *"Oceans may dry up and rocks may crumble, but my heart will never change",* and need more people of high virtues, high wisdom and deep insight to join our great undertaking of creation of the future. I am longing for all kinds of opinions from my readers, whether agreeing or disagreeing, whether inquiring or oppugning, whether praising or criticizing. All opinions will be the best "gifts" to me. Besides, I hope all my friends will pay more attention, offer more support and give more help to mankind's advanced culture, to the great ideal and undertaking of mankind's becoming immortal together.

Finally, I extend my heartfelt thanks to my most beloved teacher Mr. Wang Keling, as his great wisdom has enlightened me deeply and the whole process of research and writing of this book was carried on under his guidance. I express my great thanks to my dear mother Ms Liu Xiangxiu, and my best friend Miss Wang Fei, for their help in the writing of this book. Moreover, I extend my heartfelt gratitude to all readers and friends who have shown concern for my new book.

APPENDIX

WANG XIAOPING'S ANSWERS TO THE REPORTER

夢星

1. From *Dacheng Success Theory* to *Ability Panic*, to *The Second Declaration*, from joint authorship to independent writing, from the age of 19 to 20, and to 21, from one field to another, you turn out one book each year, and each one is better than the other. The secrets of your success are very interesting to us. People wish to know how you learn or study, how you think, how you arrange your daily life. What do you like to eat? What sleeping posture do you like? At what time do you like to get up? At what time do you like to read and write? Do you have any body building method or brain training method? All these are interesting topics for people. Will you please tell us something about your everyday life? How do you spend a day usually?

Answer: My learning is chiefly research-typed. The research-typed learning has brought me up, making me part completely from the learning in traditional schools. I always thought that learning is the mother of success. If you say I have some secrets of success, the first one should be this research-typed learning. Research-typed learning has many benefits.

First, it can combine learning with thinking.

Second, it can build up one's personality and innovative quality.

Third, it can kill two birds with one stone, getting fruits both in learning and research, so the efficiency can be maximized.

As for how to think; I prefer the "Taiji three ways of thinking", that is, first, think on the positive, then, think on the opposite, and finally, do the combined thinking on the two aspects. It just coincides with Hegel's "positive subject, opposite subject and combined subject". Besides, I also like systematic thinking, often using the "system synthesis integrated method" suggested by Mr. Qian to do my thinking and studying. Many people say a great feature of my books is the strong systematic character, as the expounding is all-inclusive and systematic. I am also fond of "reading thinking", that is, taking the book as a catalyst, and thinking while reading; "writing thinking", that is, using writing to streamline my thinking, I have written a lot of thinking notes, sometimes I write more than ten thousand characters of thinking notes in one day; "oral thinking", that is, thinking through talking. Of the above "three ways of thinking", what I like best is "oral thinking", as I like thinking while chatting and communicating with others. Many of my novel and wonderful viewpoints never occurred to me before my talking, but in talking, they flew to my mind easily.

As for arrangement of daily life, I have a rough arrangement, and my activities are basically within the framework. When I want to eat, I just eat; when I want to sleep, I simply sleep; when I feel like reading, I just read; when wanting to ponder, I just go in for pondering. I do everything just as naturally as possible.

I like to eat anything delicious, especially the dishes cooked by my mother.

In Buddhism, there is a saying *"Great Ease"*, and I like it most. So, I like sleeping lightheartedly the most. Just keep as relaxed as possible, without special attention to any posture. A sound sleep may come only when I pay no attention to posture.

I like working at night, I go to bed quite late, but I do not get up too late. As for when to read and when to write, I prefer following my interest, and do not like to be bridled to any fixed pattern.

I used to do long-distance running. Now, living at the foot of Xiangshan Mountain, mountaineering has become very convenient, so I have begun to take a liking to mountaineering. I also like dancing, I am fond of a special Buddhist dance. Not long ago, I was bitten with Yoga. In Yoga, there are some movements in imitation of the fish, snake, lion, camel and so on, which have a very good effect for relaxing the body and removing tiredness. All such exercise has a good effect for keeping fit, both for the body and for the brain. Bodybuilding and brain building are closely related to each other. When you keep a fit body, your brain will naturally be very keen and flexible. Brain strengthening has much to do with food and drink. Someone said: *"Geniuses are out of eating",* and I think there is some truth in it. My mom is a doctor of traditional Chinese medicine, and also a half cuisine master. She always cooks very nutritious and delicious food for us. For instance, in the morning, I can always eat the breakfast designed specially by my mom: the three-in-one of soya-bean milk, peanut milk and cow's milk. In addition, she often cooks black rice, black fungus and black sesame seed for us. Such "white + black" "food for geniuses" is both delicious and effective for brain strengthening.

My life centers chiefly on book writing. When writing books, I spend most of my time by the computer or on the net or in piles of books.

2. The information volume of *The Second Declaration* is very great. It seems that you have read a lot of books for it, and the knowledge covers many areas, such as science and technology history of China and foreign countries, history of China and foreign countries, anthropology, life sciences, esthetics, literature, futurology, economics, political science, philosophy, modern high and new technologies and so on, both in natural science and social science. You have become quite familiar with all such branches of knowledge, as if having them at your fingertips. What methods do you use to take in such a huge amount of knowledge and information? What books have you ever read? Can you list them? When you wrote *The Second Declaration*, what books or data did you study?

Answer: When Yao Xueyin wrote *Li Zicheng* (a historic novel), he made tens of thousands of cards. I also make my cards and also have made tens of thousands of cards. But unlike Yao Xueyin's case, all my cards are "electronic cards", just as I call them. The "Electronic Card", in fact, is made via the computer database software. It is much easier to manage and use them, hence, they are much more advanced than Yao Xueyin's manually-made cards. But in making these "electronic cards", I also incurred a great time cost. Yet, it is worthwhile. It is because of them that I can master encyclopedic knowledge and use it freely at any moment.

Besides reading books and making "electronic cards", I also owe my achievements to the Internet. Without the Internet, it would be absolutely impossible for me to write *The Second Declaration* so satisfactorily. A lot of my knowledge and information come from the Internet. I remember seeing an interesting story on the Internet: "In which dynasty do you like to live most?", but I do not remember very clearly whether the final conclusion was the Tang Dynasty or the Song Dynasty. It is a typical fallback thinking of the Chinese. If they had let me choose, I would have certainly chosen living in the present times. Just the availability of the Internet will make me feel very grateful to the life in the present times.

The books I have read are too many. To list all of them, the paper in your hand is not enough at all.

When I wrote *The Second Declaration*, I studied *Great Thinking Ways of Chinese Sages, Cosmos Spirit, High Technology and High Thinking, High Technology and the Society, Century of Biotechnology, Transfer of Power, Society in Dream, Earth Times of the 21ˢᵗ Century, Systematic Perspective and Mankind's Prospect, Conflicts Among Civilizations, Death of the Chinese, Singing of the Crane and Way to Gods,* and so on. And the data I have covered in my research are even more.

3. What do you like to do usually? What books do you like to read? How do you read them? Is your reading method a systematic study or just a random study for urgent needs? Do you only read what is useful at present for yourself or place stress on build-up of your entire knowledge framework? How did you cultivate such a high learning efficiency?

Answer: I like reading the most. Especially at night, in the soft lamp light, sitting on the bed, taking up a book casually, and drinking a mouthful of fragrant tea: It is really a great enjoyment!

The books I like to read are too many. I can take a great delight from any book I happen to take up. I have many ways to read: fine reading, or wide reading, or reading through, or selected reading, or skimming through... Each has its own use. If just for contest in speed, I can read very fast. But usually I have a different state of mind in reading, and usually I take a researcher's attitude to the book, as I specially prefer this way: First of all, take a birds-eye view, to get the first impression, then digest, to assimilate its essence. Finally, criticize, that is, using a questioning attitude to criticize the book.

As for my way of reading, it is neither a pure systematic study, nor a pure pragmatic study for urgent needs, but is a new-typed style of learning characterized by combination of the two. I always give a systematic thinking to the subject I want to investigate, and I do systematic study according to the topic I have chosen. This systematic study is different from the text book systematic study in schools, as the system is constructed according to the need of the topic, and the key points are also decided according to the need of the topic.

I feel, generally speaking, first of all, we must grasp the scientific knowledge system of modern times, and have general knowledge background and knowledge eyeshot. Then, according to the need in the research, do in-depth study. I think, if you try to build up the overall knowledge framework, you must pay special attention to the scientific system of modern times. I am very interested in the study on the modern scientific system. In people's traditional concept; science is only natural science and social science, and someone even regard science only as natural science. It is a one-sided view. I specially admire Mr. Qian Xuesen in his construction of the entire scientific system, as he has juxtaposed system science, human-body science, noetic science, behavioral science and so on with natural science. Only when you have a framework of modern scientific system in mind can you "have a unique view of the general picture" of each aspect of the knowledge, dabble extensively, and take wide eyeshot. On such a basis, you may further study the knowledge most needed for you, and kindle great creative ideas.

I am always satisfied with my own learning efficiency. As a CEO said, *"Nowadays, the most important is to compete; who can learn faster"*. In "learning fast", I feel proud of myself, for I could read very fast from my childhood. In a vacation in junior middle school, I trained my reading speed. At that time, I could read 3 to 4 thousand characters per minute, and my rate of understanding could reach 70%–80%. Later, I found, the goal of those fast reading training classes being puffed to skies turned out to be only 1 thousand characters per minute or so. I surpassed that standard long ago. I read very fast, think very fast, too. Everyone praises me for my agility in thinking, and even my father is amazed! With a fast reading speed and a fast thinking speed, of course, my learning efficiency cannot be slow.

4. Any learner knows, futurology is not easy to access, as it not only covers rich knowledge, but also requires a unique way of thinking and an original theoretical system. Without the great wisdom of a genius and the sharp vision of a hierarch, no excellent works on futurology can come out. We have noticed in *The Second Declaration*, there are so many question marks. Moreover, in your book, you use the analogical method and the inverse deducing method very frequently, which exerts a great shocking impact on the reader. How did you develop your habit of questioning and reflective thinking? How did your home education influence you? Can you give an example from your life?

Answer: Since my childhood, I liked reading books. I particularly like the reading manner of Mr. Li Zongwu reputed as a "wizard", that is, "Take the book as the enemy", try to find out the mistakes in it and refute them. That is a point distinguishing my reading from ordinary people's reading. Hence, I have developed a habit of questioning and reflective thinking gradually. An expert criticized me for my citing examples of Zeng Guofan and Du Yuesheng in my book. Zeng Guofan was a slaughterer stained with the blood of the insurrectionists, but his self-criticizing spirit is greater even than communists, and his strict self-discipline was stricter even than Lu Xun. Du Yuesheng was the chieftain of a sinister gang, but his viewpoints on cooperation and wisdom are very reasonable. Therefore, we should not only consider the positive side of the book. I prefer seeing the other side of the book. Besides, I am very fond of debating. I have invented a "transposition debate" method: after being the supporting party, then be the opposing party. This kind of debate is very helpful for training a person's ability of reflective thinking, for breaking the thinking set, and for developing the habit of questioning and reflective thinking.

My home education has given me a lot of good things, which are also invisible things: good concepts, good habits and so on. The most valuable things are invisible. It is just these concepts and habits handed down from my family that act as an invisible hand helping me taking this creative road. For instance: after I learned to read and write at the age of one, my father and mother taught me to read some simple picture books. After reading them, they asked me to tell the content, and praised me while I was telling the story. In this way, I became more and more interested in book reading. My parents not only encouraged me to tell stories to them, but also encouraged me to tell stories to my little friends. When I was in the first year of primary school, I took the picture book of *Snow White* to the school, told the story to my classmates, and organized some little friends to perform it. I became more and more interested in books, and always felt proud of my wider scope of reading than others. I developed my reading habit in this way. From my childhood, after I told a story to my parents, they would ask me questions. After watching TV, they also asked me questions. And they often discussed all kinds of questions with me. In this way, I developed a meditative habit gradually.

5. For what kind of people is this book written? What help can it give them?

Answer: This book is written for all people who care about the future. It is worth reading for those who hope to grasp their individual's fate, and their enterprise' fate. It is worth reading for those who wish to change their own thinking mode and backward ideas. It is worth reading for those who wish to cultivate a creative spirit and an innovative personality.

This book may help people learn about a lot of front-edge information and latest knowledge on high technology and the future world. It may help people get some correct and advanced concepts and ideas conforming to the times. It may help people widen their field of view, have great foresight for the future, and clearly understand the development trends of future mankind, future society and future economy. It may help people better design their own future, design their enterprise's future, and design the future of the whole world.

6. Since you study futurology, what plans do you have for your own future? You stress planning and designing the future. How do you plan your future and design your future? Do you plan to be a freelance writer for all your life, or find a job? Or design a job for yourself? Was your quitting school a result of your designing and planning for the future? What is your plan for the coming 3 years or 10 years? After writing *The Second Declaration,* what else do you plan to write?

Answer: My present state is just the result of my planning for the future, and also the result of the philosophy "The future decides the present" I have adhered to. Especially my quitting from school, it can be said that the future of that time decided the present of that time. Today's reality is the future of that time, and it was just under the guidance of "the future of that time" that I made the decision to quit school.

I never want to be "a person employed by others", but want to be "a self employed person", that why, I prefer working completely at my own will. I won't be a freelance writer for all my life, as writing is only a way for me to convey a sound, to express my thought to this world. From now on, I think I will find other ways.

I think what I wish to do will not only include academic research, but also include a greater culture undertaking. In the future , I want to get *The Second Declaration* translated into foreign languages so as to let it reach the whole world. Run a culture company, and organize a team, so as to do many more things. Set up a worldwide association for study on mankind's future, together with some people of great insight. I also want to run a web site of my own, to spread my ideas and culture to the whole world. In the near future, I plan to make *The Second Declaration* into a film, using the means of movie and TV to make these thoughts become more powerful and influential. In short, things I plan to do are too many.

In planning and designing my own future, I lay special stress on positioning of the individual's life and positioning of the individual's undertaking, then, according to the positioning, do systematic thinking, systematic planning and systematic designing. In positioning, I pay special attention to originality, differentiation and chance seizing.

I also plan to do in-depth study on the development trends and development strategies of future economy and future culture. If possible, I will do in-depth exchange with 100 scientists, 100 entrepreneurs and 100 scholars, and write something.

7. Everyone has his (her) own future. What good suggestion and advice do you have for people to design their own future?

Answer: My suggestion to those who wish to design their own future is like this: Please think well, and answer five questions for yourself: "Who am I?" "What do I want to do?" "What can I do?" "What should I do?" "How will I do it?"

Moreover, you may design your own future according to a saying by Premier Zhou Enlai, *"The principal point is to do your best to give play to your own strong points"* or with reference to my saying in *Ability Panic*, *"The road suitable for myself is the best road".*

8. Why do you say this book may greatly help to realize world peace?

Answer: The root cause of the mutual killing of the present time is just that we only see our own narrow national interest or interest of our ethnic group, but this book may help people see the common interest of all people on the earth, that is, the interest of mankind. So long as everyone sees the common interest of mankind, we may ensure world peace. The reason is like this: *"When the interest differs, even brothers may fight against each other. When the interest is the same, even enemies may share the same boat."*

On the other hand, the book has designed and developed the new culture of mankind that is most helpful for world peace. Just as I say in the book: When the people are upright, the world will be upright. When the heart is upright, the person will be upright automatically. To set the heart right, we must set the culture right. Similar culture, similar people, and similar world. All the forms of new culture of mankind mentioned in my book, such as the human-centered culture, life culture, ecological culture and peace and harmony culture, mutual benefit culture, great achievement culture and so on, are beneficial to world peace, and may lead to lasting and eternal world peace. Some present prevailing cultures, such as violence culture, killing and conquering culture, are just like viruses in the brain, encroaching on people's thinking method and pattern of behavior. With the new culture of mankind, we may clear away the viruses for the human brain very effectively.

9. So far, there is no Chinese winner of the Nobel Prize yet. Do you strive for the Nobel Prize?

Answer: The Nobel Prize can be said to be a complex of the Chinese. Indeed, as a wise nation with 1.3 billion people, we have to face this sharp problem. Why hasn't China contributed a Nobel winner to the world? However, I think, what is more important for us is not to try every means to get the Nobel Prize. Rather, we should do some practical things, do some creative things to solve the hard problems in world peace and mankind's development, especially we should dare to say what others have not said, study problems others have not studied, and solve problems others have not been able to solve.

10. Your book says, "The future decides the present". Why can the future decide the present?

Answer: That means, our forecast and judgment on the future may determine our present decision making and behavior. In the 1960's, Mao Zedong's forecast of the future was there would be a world war, and according to such a judgment on the future, he mobilized all the people of the country to prepare for the war, and the popular slogan of that time was *"Dig tunnels deeply and store grain greatly"*. Everyday, people lived in an atmosphere that the war might break out at any moment. They spent a great amount of manpower, financial resources and material resources in preparation for the war. All the people were busy digging bomb shelters without production, and many important factories in large cities were moved to remote mountain areas. Such a wrong judgment on the future led to a great waste of money and labor, and our national economy slid nearly to the verge of collapse. But later, Deng Xiaoping's future forecast and judgment was *"The world war cannot break out"*. Thus, he took a wise move of "great disarmament of one million army men", and let all the people concentrate on economic construction, so, a wonder appeared: our national economy took off at a surprising speed! So, it can be seen that whether the judgment and forecast on the future are correct or not will determine directly whether the present decision and action are correct or not.

All our present plans and decisions are for the future, and are based on our understanding of the future. Our understanding of the future determines what we do or not do, and how to do at present, determines whether our present choice and decision are correct or not, wise or not. Such a "future decides present" feed-forward thinking mode is an advanced thinking mode that we need most today.

11. In your book, you say there was a survey on whether "to wish to be a beautiful woman" or "wish to be a talented woman". Then, if you ask yourself: "Would you like to be a beautiful woman or a talented woman?"

Answer: I want to be a beautiful woman! Don't be surprised. Let me tell you a story. An old rich man asked Afanti a question: "Afanti, do you want to be a wise man or a rich man?" Afanti said: "Of course, I want to be a rich man." The old rich man had a good laugh, saying in despise: "It turns out that you are so vulgar too, loving money so much." Stroking his beard, Afanti said: "I am already wise enough, so, I want to be rich man." I am called a "talented woman" now wherever I go, so, I want to be a beautiful woman very much now.

AN OPEN LETTER TO PLUTOCRATS OF THE WORLD

夢星

All Plutocrats in the World Who Have a Kind Heart and a High Sense of Responsibility,

Please accept my greeting.

I am a girl who cares about the fate of mankind, also the author of *The Second Declaration*. For the sake of our mankind's most splendid and magnificent joint undertaking—mankind's Dacheng (great accomplishment) undertaking, I am writing this letter to you specially. What I am eager to say is:

You are the people who are most capable of making money in the world, and I hope you will also become the people who are most capable of spending money in the world.

You are the richest people in the world, and I hope you will also become people who shoulder the greatest responsibility for mankind's fate.

American steel king Carnegie once said: *"Making money requires a great ability, and so does spending money. Only by being so the society can be benefited."* To my great delight, among you, more and more people have begun to take a increasing notice of how to spend the money so as to benefit the society better, while mastering the super ability to make money.

It goes without saying that wealth is a great energy. When it is put into good use, it may help people and benefit people. But when it is not put into good use, it may harm people and ruin people. Bill Gates declared many times that he will only leave ten million US dollars to each of his children, and use all the rest in charities. Among you, more and more people of vision have realized: *"Leaving the omnipotent dollars to the son is nothing else than leaving a curse to him"*. On the contrary, using wealth properly, that is, putting it into the most important undertaking to human fate, will be a great service to mankind, being a boundless beneficence.

Mankind's great accomplishment undertaking, i.e. letting everyone become immortal and letting all mankind evolve into gods, is the greatest progressive undertaking of mankind. There is no greater or nobler undertaking than this in the world, and this undertaking requires great money input. Gates intends to use his wealth to "the study on the vaccines for curing AIDS, scarlatina and so on", as "wide spread of the vaccine will help us whole world realize the dream to get rid of AIDS". Mankind's great accomplishment undertaking will enable all people in the world to live forever and keep youth forever, becoming super humans—gods, so we can realize the greatest dream: to leap into being gods, which is much greater and nobler than the curing of diseases such as AIDS, scarlatina and so on. This noble undertaking most beneficial to all mankind will be more worthy of your donation of enormous wealth. With your huge sums of donation, we shall be able to speed up the progress of mankind's great accomplishment undertaking greatly and realize the great ideal sooner.

To you, world super-capitalists, such a boundless beneficence is not something *"as hard as carrying Mount Tai to subdue the North Sea"*, but something as easy as *"breaking a tree branch to help an old man"* as Mencius said. According to a report by the United Nations published in February 2000, three top plutocrats among you have a wealth that is even more than the total national income of 35 poor countries, and the top 200 plutocrats among you have a wealth that is even more than the total income of 41% of the world population! In your hands, there is nearly one half of the wealth of the world. If you put such a huge wealth into the greatest undertaking for human progress, mankind's Dacheng undertaking, what a great contribution it will be to mankind! Mankind's future and mankind's hope are calling for your "chivalrous mind" and your "great virtues and morality".

For the interest of all mankind, today, we are in dire need to establish a Mankind Dacheng Charitable Foundation. Mankind Dacheng Charitable Foundation will take in donation and support from all people of the world, especially plutocrats, and the donation will be used for promotion of Dacheng undertaking, as the expenditure for the research and application of mankind's Dacheng science and technology (especially the longevity science and technology), for the study and spreading of mankind's Dacheng culture and so on, and also as the common outlay for those who cannot afford the high and new technology products and service of Dacheng undertaking. Turner says: *The reason for his donation of 1 billion US dollars to the United Nations was just "to put money into the future of mankind".* Then, contribution to Mankind Dacheng Charitable Foundation will be considered as a great alms deed "in putting money into mankind's future". As a member of mankind, any person caring about mankind's fate should contribute his (her) strength, wisdom and wealth as much as possible to the greatest progressive undertaking of mankind. Only by gathering the wealth, wisdom and strength of all mankind for mankind's Dacheng undertaking can we realize the greatest common ideal of mankind, the Dacheng ideal. So to speak, you, people possessing great wealth, will naturally try your best to contribute your wealth. "It is a shame for one to die in riches" (Carnegie) whereas using his wealth for mankind's loftiest undertaking will be the greatest glory and happiness.

Mankind Dacheng Charitable Foundation will be in the charge of people of insight chosen by citizens of the whole world, who are the most prestigious, most responsible, and most trustworthy. The Foundation will accept open supervision from the people of the world in all aspects such as fund acceptance, fund management and fund use, in light of the principle of "Fairness, Justice and Openness". Impartiality will lead to prosperity in everything, but partiality will result in obstruction of everything. The basis of "impartiality" lies in openness. All the regulations, procedures and conducts of the Foundation must be transparent and open. Everything will be put in the sunlight. Today's network technology and information technology have created very good conditions for full openness. All the documents and rules of the Foundation, and everything in the process of taking-in and using of the fund will be displayed in a file on the web site, for people to browse, download and inquire freely. We shall use full openness to ensure complete fairness and justice.

Meanwhile, we shall set up "Mankind Dacheng Investment Foundation". "Mankind Dacheng Charitable Foundation" will take in charity contribution from all people to the Dacheng undertaking, and "Mankind Dacheng Investment Foundation" will be engaged in the investment in mankind Dacheng industry.

The principle followed by Mankind Dacheng Investment Foundation is the "sharing" principle, namely "sharing the interest, sharing the work and sharing the fruit". Unlike the "interest" in the usual sense, the "interest" to be shared here covers not only material interest and spiritual interest, but also the life interest that is even more essential than the ordinary material interest and spiritual interest. When someone asked the famous investor Buffett, "Once you have become the richest person in America, what other goal will you have?" He answered without hesitation: *"To become the most long-lived person in America!"* Liu Yonghao, the "Richest Man in Chinese Mainland" rated by Forbes in 2001, said: *"If possible, I am willing to give up all my wealth to change age with a youngster"*.

Mankind Dacheng Investment Foundation takes "mutual benefit" as its basic rule, and any investor may get corresponding material interest, spiritual interest and life interest from it. Such an investment is not only a great benevolence to mankind, but also a saving of one's own life, an investment in one's own youth and a saving of one's own youth.

In recent years, "Responsibility Investment" has already become a new investment philosophy in the world. It values "social responsibility", namely not only paying attention to the earning capacity, but also paying attention to environmental protection and the bottom line of social justice. If the undertaking goes against environmental protection, human rights and public interest, we shall abandon it without hesitation. Mankind Dacheng investment fund is just a "responsibility investment" and "moral investment" in a higher sense, as neither any of the fund will be put into profiteering industries such as tobacco, ammunition, etc, and nor will it be put into restaurants or groceries, but will be completely used in the development of mankind Dacheng undertaking, including longevity economy and longevity technology. By starting up the Mankind Dacheng Investment Foundation, we may utilize the mode of industrialized operation to speed up the progress of mankind Dacheng undertaking effectively. Mankind Dacheng investment fund will be operating together with Mankind Dacheng Charitable Foundation, to boost the fast development of the Dacheng undertaking.

Mankind Dacheng undertaking is an unprecedented, most groundbreaking and pioneering undertaking, and is also the grandest enterprise for changing the fate and future of mankind, so, it needs the concerted efforts of all mankind.

You, the richest people in the world, are shouldering more responsibilities. Please join this undertaking that is most valuable to mankind, to the society, to the populace and also to yourselves! In Aristotle's eyes, *"A great person is someone who contributes wealth"*. To the best of my belief, the great kind mind and strong sense of duty will make you extremely generous, so, you will contribute your wealth generously and contribute your power generously to the fine future of mankind.

Sincerely yours

Wang Xiaoping

AN OPEN LETTER TO ANNAM, SECRETARY-GENERAL OF THE UNITED NATIONS

夢星

Dear Mr. Annam, United Nations' Secretary-general,

Dear Leaders of the United Nations,

Please accept my greeting!

I, Wang Xiaoping, a Chinese girl fond of thinking, am caring about mankind's future, thus caring about the future of the most authoritative international organization, the United Nations. After finishing the book *The Second Declaration*, when I thought of the United Nations, all sorts of feelings welled up in my mind, thus, an idea occurred to me: write a letter to all of you working in the United Nations. As a member of mankind, I deeply believe that the fate of the United Nations is just the fate of mankind.

Mankind's greatest creation in international relations is the establishment of the United Nations. In the half century, the United Nations has played a great role in maintenance of world peace and promotion of cooperation and development of all countries. However, we have also seen, the United Nations is so weak in many ways, and lacks real power.

Only by all-round reforming of the United Nations and enabling it to have the power of leadership and management of the whole world can the United Nations shoulder the important task to govern the whole globe and the mission to integrate the whole world. Therefore, I wish to raise some suggestions to Mr. Annam and all officials of the United Nations on strengthening and reforming the United Nations.

My dear friends, I do not know how you think about the fact that the United Nations has been in a powerless position in many ways for a long time. Most people only see the economic cause, but I think the more important cause is the cultural one. Today's United Nations is not only a dwarf in economy, but also a dwarf in culture.

Dear friends, I think you must also be worrying about the fact that the present-day mainstream culture of the world is still a nationalism culture, a "Darwinism" culture and a cold war culture. It is just these cultures that are inducing the whole world to spend the money, resources and power on mutual fighting and mutual killing. The United Nations should endeavor to develop and spread a new culture that may have the greatest power to promote mankind's union and world's unification, such as the human-centered culture (including individual-centered culture, the-people-centered culture and mankind-centered culture), the life culture, the ecological culture, the great harmony culture, the mutual benefit culture, the Dacheng (great accomplishment) culture and so on. Only by doing so can the United Nations unite all the countries in the real sense, and realize its original intention and mission.

Dear Secretary-general Annam, you have cudgeled your brains and taken great pains in mediating for stopping war-flames and conflicts over the world. This work is very important, but it is not a fundamental remedy. The fundamental way is to develop the culture of the United Nations greatly, so as to make it into a great cultural power. I think, what you should do is to try your best to let the United Nations work out an advanced mankind culture and concept of value, build up the common spiritual tie and ideal tie of mankind, provide the common ideal and common goal for all mankind, and let all people be aware of the common ideal and common goal of mankind. Only mankind's Dacheng culture can provide the common ideal and common goal for all mankind. Only by wide spreading of mankind's Dacheng culture and letting it go deep into the hearts of the people can the peoples of the world be united together for the common ideal and common goal. Hence, only greatly developing and spreading such a culture is the fundamental way to realize world peace and achieve harmony all over the world.

The *Declaration for the New Millennium* published at the summit conference of the United Nations celebrating the new millennium said: *"The major challenge facing us today is to ensure globalization will become an active power beneficial to all people of the world."*

How should the United Nations respond to such a challenge?

Globalization means global integration. Today's globalization is only a single-sided globalization, not an all-round globalization, that is, there is only economic global integration, but there is no cultural global integration and political global integration. Such a single-sided integration will certainly be like a wild horse freed from the bridle, dashing around madly, and it might lead to a catastrophe. The "negative issues caused by globalization" asserted by the opponents of globalization, such as the ever-aggravating environmental pollution, the great poverty-stricken class, the continual expansion of brink-oriented movement, the unordered capital flows and so on, can only be solved by an all-round globalization. If we only have a single-sided economic globalization, the problems facing mankind will become even more serious. The United Nations should carry forward the process of all-round globalization, shoulder the mission to govern the whole globe, make decisions from the height of the whole world and all mankind, break national restrictions, guarantee the rich and care about the poor from a global eyeshot, and make good use of its power to turn the globalization of a few powerful nations and multinational corporations into "globalization of the whole mankind"... All these must rely on the cultural integration of the globe. In the all-round globalization, only by global cultural integration (letting the world culture that helps to unite and integrate the world take the leading position) can we have a real global political integration, and can we make global economic integration advance in the right direction.

Moreover, the United Nations should also make efforts in the following aspects.

I. Design and Set Up the "Cosmopolite" Course System and Education System.

Only by "producing" a new mankind can we create a new world. If the United Nations wishes to create a beautiful new world, it must consider how to "produce" humans and how to produce a new mankind that can create a new world. Such a new mankind is free from the narrowness and trivialness of nationalism, but has a cosmopolitan mind and sentiment. From now on, the United Nations should start to study and design the new course for the cosmopolite or world citizen, carry out the "world citizen education" widely, strengthen the "world citizen consciousness" and the "world citizen ideology" of the people all over the world. It should work out a complete course system for the new mankind characterized by the ideals of the new mankind (such as the ideal to strive for the greatest happiness for all mankind, the ideal to become gods through science, and ideal to realize all-round globalization, and so on), the morality of the new mankind (such as the universal moral of all mankind, including self-love, love of others, universal love, benevolence, justice, good faith and so on), the culture of the new mankind (such the six major cultures of mankind's Dacheng culture, namely; human-centered culture, ecological culture, life culture, harmony and cooperation culture, mutual benefit culture, and great accomplishment culture), and the abilities of the new mankind (ten major abilities, such as super learning ability, thinking ability, creative ability and so on), and then, on such a basis, it should build up the "world citizen" education system.

II. Try to Become the Pacemaker of World Science and Technology.

Science and technology development is the commanding point for future development. The United Nations should have a great foresight for the future, organize and attract top-leveled science and technology personnel, master and develop the most advanced science and technology in the world, especially such high and new technologies as gene technology, nanometer technology, robotics, outer space technology and so on. The power of science and technology will increasingly become the greatest power. If the United Nations becomes the pacemaker in science and technology, it will master the greatest power, and will be able to stand at the commanding point for development. Moreover, if the most advanced science and technology are mastered by the United Nations that cares about all mankind, we shall be able to guarantee the greatest happiness of mankind, and prevent disasters caused by wrong use of high and new technology. Meanwhile, when the United Nations has mastered the high and new technology, it can also get hum sums of fund through technology transfer...

III. Try to Have Its Own Money Pouch.

To have a real power, it will never do without money. If the United Nations wishes to become a world government leading and governing the world, it must have its own money pouch. Mr. Annam, a great headache for you now is that the United Nations has no reliable and ideal financial resource of its own, as it chiefly relies on the contribution and donation from member states for survival. Such a survival relying on the "almsgiving" from great powers has resulted in numerous crises in the United Nations. You see, now many countries especially the USA have been behind in payment of huge sums of fees, which has put the United Nations in a situation verging on "bankruptcy".

In the future, the United Nations should utilize its own advantages to run business, establish enterprises and issue world stock. The United Nations should, according to the future quintet industry theory, choose to develop the human-centered industry and the science and technology industry representing the future. So long as the United Nations make great efforts to develop the human-centered industry and the science and technology industry, it will have its rich money pouch, will not go bankrupt, nor will it become weak, and it will have the real power over the world gradually.

Meanwhile, the United Nations may also collect taxes from its member countries and winners of the world. Along with the advent of the globalized era, from now on, all transnational corporations should carry out the policy of "registration with the United Nations", and the policy of "registration with the United Nations" should also be applicable to intellectual property rights. Now, many transnational corporations are becoming more and more domineering along with their increase of wealth. In future, they should pay taxes to the United Nations in a proper way. It may make it more convenient for them to carry out economic activities worldwide, and at the same time, it may also help to consolidate the dominant position of the United Nations in global economy.

IV. Guide the Consumption and Production of the World, Taking High-Humanity Demands as the Center.

Today, the consumption pattern of "steel - automobile - petroleum" in traditional economy and the traditional development mode of the whole world characterized by high living cost, high flow rate, high pollution and "pollution occurring before pollution control", are just speeding up the mad suicidal steps of mankind! American population is less than 5% of the world population, but its energy source consumption accounts for 34% of the world energy resource! The population of the developed countries only account for 1/5 of the world population, but is consumes 3/4 of the world resources! If people of the whole world consume in this mode and on this scale, mankind will need 20 earths! It can only lead to perdition of mankind! Some scholars assert: It is hard for mankind to escape from this perdition! Mr. Annam, in such a grim situation, the United Nations should guide the global economy from high material consumption to high-humanity consumption, lead enterprises in their turning from high material production to high-humanity production. Therefore, the United Nations should go all out to advocate life economy, eternal life economy, human-centered economy and so on, as such economies consume the least resource and energy, but give the greatest benefit to mankind, and are also most environment-friendly. Hence, the United Nations should turn all consumption and production into the direction that takes high-humanity-centered demands as the center, which will not only release the alarm on mankind's self-destruction, but also turn a brand-new leaf for global economy, offering a new vitality to it!

V. Let the United Nations Become the United Nations of All People in the World.

In the face of the globalized times, the United Nations should not be an empty talk club for governments of all nations, but should cause all people on the earth to regard it as their own organization, as their own "first country". The members of the United Nations are both representatives of the governments of all nations and representative of the peoples of all nations. It should not only take in previous polities, but also take in non-polities, namely non-governmental organizations or transnational corporations having a sense of justice and responsibility. It should also reach to every member of mankind, and all the citizens of the world may become participants of the policy making and decision making of the United Nations through fair competition. Only in this way can every member of mankind develop another kind of "patriotic" feeling, that is, loving of the United Nations, and loving this great family of mankind!

VI. Set Up "An Internet-based United Nations", Using the Global Network to Enhance Its Influence Power, Attractive Power and Cohesive Power to People All Over the World.

The United Nations should take full advantage of the Internet to develop and expand itself. With the advent of Global Internet, "cosmopolitanism" will become true. Nowadays, there is no event that only belongs to only one place, one nation or one nationality without belonging to the world and mankind.

The United Nations may use the Internet to develop the UN culture, the UN education, the UN commerce, and the UN government affairs. In the present Internet times, cultural information is shared so expediently and it circulates to every nook and corner of the world so swimmingly that citizens of different regions and different nations may love the same song, be fond of the same kind of amusement, like the same star. Everyone knows who is Britney Jean Spears, who is Elvis Presley (Cat King), who is Zhang Ziyi, who are Rolling Stones, who is David Beckham... In that case, the United Nations should deliberately use network technology to spread the culture helpful to international cooperation, world peace and mankind's progress, and introduce such a culture into the education of our offspring and the education of the future leaders. Moreover, the United Nations should utilize the Internet to open the United Nations' commerce and government affairs to every citizen of the world, so that every person in the world may be willing to participate in the development of the United Nations, so that everyone in the world will feel the United Nations is a nation of mankind that serves the interest of all mankind, instead of the interest of some countries only.

The above are some suggestions I have made for the development of the United Nations.

May the courage and wisdom of every one of you help the United Nations to improve in all the above aspects!

May the future United Nations avoid mankind's "going down direct to Hell", and lead mankind "direct up to Heaven"!

May the future United Nations become the world state and world government bringing about the greatest happiness to mankind!

Sincerely yours,

Wang Xiaoping

AN OPEN LETTER TO US PRESIDENT BUSH

夢星

January 9, 2003

Dear President Bush,

Please accept my greeting!

I am an ordinary Chinese girl fond of thinking and caring about the future of mankind and the future of the world. Today, I have just got my newly published book *The Second Declaration*, and what occurred to my mind first was to send a copy to you. I take you as a respectful elder. Would you like to listen to the thoughts of a Chinese girl just like your daughter? Recently, I often heard such a message that you think it is "necessary" to start a war on Iraq, and I could not help worrying about the people of Iraq, about the people of America, also about you. So, I cannot help writing this letter to you.

You have two lovely daughters. However, when you are considering the decision to start a war on Iraq, do you ever think that thousands of Iraqi girls just like your daughters will lose their fathers or fathers will lose their beloved daughters?

I remember, in your speech at Tsinghua University last year, you stressed repeatedly: "I feel one important point is that we realize we are flesh and blood after all, we are humans after all". When I heard such words, I could not help feeling a great respect for you. Indeed! So long as we remember we are human beings after all, we shall not start a war, killing our own species! The photo and story of each victim of the 9/11 event has been carried on the *New York Times*, and in the face of the eternal disappearance of these fresh lives, we cannot help shedding tears. But today, because of your decision, thousands upon thousands of Iraqi common people will lose their lives, thousands upon thousands of Iraqi families will disappear. Are their deaths only dry numerals? They are humans too, fresh and animated human beings!

The book I am sending to you is called *"The Second Declaration"*. Why? I think, "The First Declaration" was made by Darwin, declaring that mankind evolved from apes; "The Second Declaration" is the viewpoint in my book, as I think mankind that has shaken off barbarism will evolve into gods, which is called "The Second Declaration". American people's great respect of life has left a very deep impression on us. In America, there are many great futurists, and they have made many wise judgments on the forthcoming development of mankind. I think, in the last analysis, mankind's future is held by mankind itself. What is needed the most for the future is not money, not resource, even not technology, but a common ideal that can join together all mankind! When the people are upright, the world will be upright. When the heart is upright, the person will be upright automatically. To set the heart right, we must set the culture right. In the book, I have uttered our orientals' thinking on the future, and also expressed a Chinese girl's ardent expectation for you.

Throwing thousands upon thousands ton bombs onto the land of a small country and killing uncountable innocent civilians, is unjust in moral, in any case. Even judging from the American interest, it is very inadvisable. I quite understand your great worry in the face of all pervasive terrorism. Yet, if you only regard the lives of your own country as gold but treat the lives of other countries as grass, only take your own absolute safety as No. 1 but disregard others' safety, you cannot possibly have real safety, even if you sweep all before one and assume hegemony over the world. It may just trigger the fire of hatred to burn by ten times or hundred times. Today you kill one hundred of our people, tomorrow we may kill one thousand of your people... It will be a vicious cycle without end. Your anti-terrorism purpose cannot be realized this way, and on the contrary, terrorism will grow more and more serious, causing endless trouble for the future. Even if you have won on the surface, you have lost in fact! Only the water of morality and justice can put out the fire of animosity, uproot the source of terrorism and ensure peace. Using violence to get rid of violence can not get rid of violence at all, but can only lead to an even worse outcome!

In my new book *The Second Declaration*, I have said mankind is faced with a choice to go up direct to Heaven to do down direct to Hell. Mr. Bush, I am sending this book to you, and hope you will read it over, in a position not as a president, but as a father. Your choice now may have also determined America's future to go up direct to Heaven or go down direct to Hell!

Please do think twice before taking action!

Sincerely yours,

A Chinese girl, Wang Xiaoping